Coastal Planning and Management

Coastal Planning and Management

Robert Kay and
Jackie Alder

E & FN SPON
An imprint of Routledge
London and New York

First published 1999 by E & FN Spon, an imprint of Routledge
11 New Fetter Lane, London EC4P 4EE

Simultaneously published in the USA and Canada
by Routledge
29 West 35th Street, New York, NY 10001

Typeset in Palatino by Keystroke, Jacaranda Lodge, Wolverhampton
Printed and bound in Great Britain by Biddles, Guildford, Surrey

British Library Cataloguing in Publication Data
A catalogue record for this book is available from the British Library

Library of Congress Cataloguing in Publication Data
Kay, Robert (Robert C.)
    Coastal planning and management / Robert Kay and Jacqueline Alder.
       p.   cm.
    Includes bibliographical references and index.
    ISBN 0–419–24340–2 (hardbound). — ISBN 0–419–24350–X (pbk.)
    1. Coastal zone management. I. Alder, Jacqueline, 1954– .
    II. Title.
    HT391.K36   1999
    333.91'7—dc21   98–28804
                              CIP

ISBN 0–419–24340–2 (hbk)
ISBN 0–419–24350–x (pbk)

Contents

Foreword

The rapid increase in population across the globe, large-scale exploitation of coastal resources and rapid development of infrastructure has often resulted in severe degradation and decline in the quality of the coastal environment. These pressures on the coastal zone are certain to intensify in the future.

Mangroves, coral reefs, cliffs, beaches, tidal flats and estuaries are just some of the coastlines on which pressure is exerted. The complexity of these ecosystems and their variety across the Asia-Pacific region makes simple management solutions difficult to find. I am pleased this book not only recognizes these problems but also offers a range of planning and management approaches and tools of immense practical value to those responsible for our valuable coastal resources.

The book has a global perspective and will play a significant part in assisting the sustainable development of all coastal nations around the world. Moreover, the coast plays an important part in the lives of many people living in the Asia-Pacific Region, with over two-thirds of its 3.2 billion people living within the coastal zone. As a result, I believe the book has particular value to this region.

The links between global and local activities related to coastal management and planning are important, and are well illustrated in the book. I am pleased the roles of the various UNEP programmes in facilitating these linkages are well recognized. The book will be a valuable resource in assisting the Coastal Zone Management theme of UNEP's Network for Environmental Training at the Tertiary Level in the Asia-Pacific.

The mixture of theory and practice of coastal planning and management demonstrates the importance of combining abstract and technical elements to achieve the best outcome for the coastal zone. The use of case studies shows examples of sound practice and differences in approaches around the world. The case studies also demonstrate the linkage between scales of coastal planning. Many of these case studies are from developing and developed countries in the Asia-Pacific region, including Australia, Bangladesh, Indonesia, New Zealand, the Philippines, Vietnam and Western Samoa.

The great strength of this book is its emphasis on coastal planning at different scales. This approach will appeal to all those involved in coastal zone management, from fisherfolk to government ministers.

Dr Suvit Yodmani
Regional Director and Representative for Asia and the Pacific
United Nations Environment Programme
Regional Office for Asia and the Pacific
Bangkok, Thailand

Preface

An estimated 50 to 70% of the estimated 5.3 billion people alive today live in coastal zones.

(Edgren, 1993)

Today, the world's population in coastal areas is equal to the entire global population in the 1950s.

(Beukenkamp, Gunther *et al.*, 1993)

In 30 years more people will live in the world's coastal zones than are alive today.

(NOAA, 1994a)

Up to 75% of the world population could be living within 60 km of the shoreline by 2020.

(Edgren, 1993)

Coastlines are the world's most important and intensely used of all areas settled by humans. It is this simple fact that directs special attention to the planning and management of coastlines. Coastal resources have been, and will continue to be, placed under multiple, intense and often competing pressures. The use of techniques which attempt to assist in managing the resulting conflicts in a sustainable way will therefore become increasingly important in both developed and developing countries.

Translating sustainable development principles into tangible actions aimed at improving the long-term management of coastal areas is the main purpose of this book. We do this by providing practical guidance through the dual use of theoretical analysis and numerous examples of best practice from around the world. We draw on our personal experience and the contributions of practising coastal planners, managers and academics from three continents.

We have chosen to focus the book on coastal planning, management and the nexus between them. We believe that achieving genuine sustainable development in coastal areas will be extremely difficult, but without proper planning it will be impossible. Planning helps governments to reconcile the apparently conflicting aims of sustainable development: to promote the economic development of coastal resources while attempting to preserve their ecological, cultural and social uses. We believe a key component of coastal planning efforts is to harness the energy of coastal residents and industrial and recreational users in the day-to-day management of coastal areas. We show practical examples of stakeholder participation in coastal planning, including collaborative management and co-management approaches.

One of the biggest challenges faced by governments is to direct financial and human resources effectively to the management of coastal areas through administrative systems established on sectoral lines. Sectoral-based systems of government focus on each part of a government's operations, such as transport, employment, health and environment. These systems do not explicitly focus on the planning and management of discrete geographic areas, such as coastal areas. Governments have chosen to face this challenge through various mechanisms to coordinate and/or integrate fuctions within coastal areas. These mechanisms are critically analysed throughout the book.

Case studies from around the world are used to illustrate sound coastal planning practices and to show differences in approach. Four groups of case studies have been selected to provide constant themes at different planning scales and to provide links between these planning scales, listed in the table.

Planning Scale	Main Case Study	Secondary Case Study
Whole of jurisdiction	• Indonesia • Sri Lanka • Western Australia/ Australia	• United States • New Zealand
Regional (subnational)	• South Sulawesi, Indonesia • Thames Estuary, United Kingdom	• Central Coast region, Western Australia • New Zealand Regional Councils
Local	• Hikkaduwa, Sri Lanka	• Christmas Island, Australia
Site	• Port of Victoria, Seychelles • Apo Island, Philippines • Warnbro dunes, Western Australia • Others site level plans are used to illustrate particular techniques	

The structure of the book, outlined below, reflects our aim of emphasizing the current state of best practice coastal area planning and management.

Chapter 1

Coastal areas are introduced, how they are defined, and a brief history of coastal management is presented; the terminology used throughout the book is discussed.

Chapter 2

The major issues facing coastal managers today are discussed, together with the emerging issues likely to be of importance in the future.

Chapter 3

Principles of coastal planning and management are analysed. The chapter emphasizes sustainable development principles, and how governments are currently attempting to work towards the implementation of sustainable coastal policies and practices.

Chapter 4

In this chapter the overall theory of coastal planning and management is translated into on-the-ground actions. These actions are through a range of techniques, each of which is described with reference to real-world examples.

Chapter 5

Coastal planning aids and coastal management processes are described. The mechanisms and contents of plans and strategies at a range of scales are critically examined.

Chapter 6

This chapter draws together the major findings of the book and outlines possible future directions for the management and planning of the coast.

Robert Kay
Jackie Alder
Perth, Western Australia
3 September 1998

Credits

The authors and publishers would like to thank the following for permission to reproduce material:

Addison Wesley Longman Ltd (Smith and Mitchell, 1993, *Impact Assessment and Sustainable Resource Management*, fig. 4.2); Allen & Unwin (J.M. Owen, 1993, *Program Evaluation: Forms and Approaches*, table 2.1); John De Campo (R. Zigterman and J. De Campo, 1993, *Green Island and Reef Management Plan*, Queensland Department of Environment and Heritage, Great Barrier Reef Marine Park Authority, Cairns City Council, Cairns Port Authority and Department of Lands, Cairns, Australia); Canadian Association of Geographers (R.W. Butler, 1980, The concept of a tourist area cycle of evolution, *The Canadian Geographer* 24(1), fig. 1); Herman Cesar (1996, *Economic Analysis of Indonesian Coral Reef*, Environment Department, World Bank, Washington DC, tables E–1, E–4); Dunwich Museum; Ian Dutton (K. Hotta and I. Dutton, 1994, *Coastal Management in the Asia-Pacific Region: Issues and Approaches*, Japan International Marine Science and Technology Federation, fig. 1.2); Elsevier Science Ltd (B. Cicin-Sain, 1993, Sustainable development and integrated coastal zone management, *Ocean and Coastal Management* 21(1–3), fig. 2, table 2); Great Barrier Reef Marine Park Authority; HMSO (1988, *The Tolerability of Risk at Nuclear Power*, p. 9); Kent County Council (Clive Gilbert, 1996, Local government responds to coastal decline, *Coastlines* 1996, 2, UK issues diagram); Kluwer Academic Publishing (S. Gubbay, 1995, *Marine Protected Areas*, fig. 6.1); Ministry for Planning, Western Australia (*Central Coast Regional Strategy*, fig. 14); *New Scentist* (H. Gavaghan, 1990, The dangers faced by ships in port, *New Scientist* 128(1744)); S. Olsen, K. Lowry and M. Kerr (1997, *Survey of Current Proposals and Methods for Evaluating Coastal Managemetn Projects and Programs funded by International Donors*, Report #2200, The Coastal Resources Center, University of Rhode Island, Narrangasett); Risk Unit, University of East Anglia (B.A. Soby, A.C.D. Simpson and D.P. Ives, 1993, *Integrating Public and Scientific Judgements into a Tool Kit for managing*

Food-related Risks. Stage 1: Literature review and feasibility study, Centre for Environmental Risk Research Report No. 16); School for Resource and Environmental Studies, Dalhousie University, Nova Scotia (H.J. Ruitenbeek, 1991, *Mangrove Management: An Economic Analysis of Management Options with a focus on Bintuni Bay, Irian Jaya*, figs A6.1, A6.3, 4.10, 4.12, tables 2.1, 3.1); Taylor & Francis (R.C. Kay, I Eliot, B. Caton, G. Morvell and P. Waterman, 1996, A review of the Intergovernmental Panel on Climate Change's Common Methodology for assessing the vulnerability of coastal areas to sea-level rise, *Coastal Management* 24(2), fig. 4); John Wiley & Sons, Inc. (Harvey M. Rubinstein, 1987, *A Guide to Site and Environmental Planning*, fig. 1–2); The World Bank (J.C. Post and C.G. Lundin (eds), *Guidelines for Integrated Coastal Zone Management*, pp. 5–6); World Wide Fund for Nature (S. Gubbay, 1994, Local Authorities and Integrated Directions for Integrated Coastal Zone Management, *Marine Update*, Newsletter of the World Wide Fund for Nature).

While every effort has been made to contact and acknowledge copyright holders, any omissions should be reported to the publisher.

Acknowledgements

We would have been unable to produce this book without the help of a large number of people. Their dedication and love of the coast has provided us with the momentum when ours ran low. Our partners – Caro Kay and Reg Watson – deserve special thanks. This book would simply not exist without them. The contributors listed below also deserve special thanks. Their enthusiasm and support lifted us up to carry on with the next task.

In particular we would like to thank for their assistance: Ian Alexander, Ian Briggs, Herman Cesar, Terry Christiansen, Alan Carman Brown, Stephanie Clegg, Richard Congor, John DeCampo, Ian Dight, Rili Djohani, Bill Gladstone, Garry Middle, Mahesh Pradhan, Melanie Price, John Quilty, Hamish Rennie, Gary Russ, Rob Tucker, Mike Williams and Simon Woodley. We are also sorry if we have unintentionally overlooked anyone who helped us.

The editorial team at E&FN Spon, especially Camilla Myers who encouraged us to start writing this book. Gerry McGill's editorial patience and humour is acknowledged.

Robert Kay would also like to acknowledge the support and endorsement of Mike Paul and Reece Waldock from the Maritime Division of the Department of Transport. Robert also thanks Ian Eliot and John Dodson, Department of Geography, University of Western Australia, for supporting his Adjunct Research Fellowship.

Jackie Alder acknowledges the financial support of the Centre for Ecosystem Management, Edith Cowan University, in writing this book.

CENTRE FOR
ECOSYSTEM
MANAGEMENT

Hilary Watson was so patient with her mother. Thank you.

Disclaimer

Contributors

Case study/technical contributors and general editorial assistance

John E. Hay
Woodward-Clyde Professor of Environmental Science
School of Environmental and Marine Sciences
University of Auckland
Private Bag 92019
Auckland
New Zealand

Mick Kelly
Climatic Research Unit and Centre for Social and Economic Research on the Global Environment
School of Environmental and Marine Sciences
University of East Anglia
Norwich NR4 7TJ
United Kingdom

Richard Kenchington
Chief Executive Officer
Great Barrier Reef Marine Park Authority
GPO Box 791
Canberra 2601
Australia

Graham King
Chairman, United Kingdom National Coasts and Estuaries Advisory Group
Director, Coastal Zone Management Associates
2 Newton Villas, Newtown
Swansea SA3 4SS
United Kingdom

Alan White
Coastal Resources Management Project
5th Floor
Cebu International Finance
Corporation Towers
North Area, Cebu City
Philippines

Case study/technical contributors

John Cleary (contributor of Section 4.3.3. Landscape and visual resource
analysis)
Visual Resource Officer
Department of Conservation and Land Management
PO Box 104
Como WA 6012
Australia

Ian Dutton
Coastal Resources Center, University of Rhode Island
Coastal Resources Management Project
Jl. Madiun No. 3 Menteng 10320PO
Jakarta Pusat, Indonesia

Greg Fisk
Coastal Planning Section
Queensland Department of the Environment
PO Box 155
Albert Street
Brisbane QLD 4002
Australia
(formerly: Centre for the Study of Marine Policy, University of
Delaware, USA)

Simon Gerrard
University of East Anglia, Centre for Environmental Risk
Management
Norwich NR4 7TJ
United Kingdom

Sapta Putra Ginting
Director General (Bangda)
Ministry of Home Affairs, Directorate General for Regional
Development
Jl. TMP. Kalibata No. 20
Jakarta 12740, Indonesia

Kathy Kennedy
Coastal Management Consultant
Flat 3, 104 Fentiman Road
London SW8 1QA
United Kingdom

Wayne Schmit
Recreational Planning Officer
Department of Conservation and Land Mangement
PO Box 104
Como WA 6012
Australia

Chapter 1

Introduction

This chapter introduces the importance and uniqueness of the world's coastal areas, with a view to outlining the coastal issues and planning and management tools described in later chapters. Several important terms, including 'coastal area', 'planning' and 'management' are defined, and the use of the terms 'coastal area' and 'coastal zone' is discussed. The fundamentals of the approach taken in the book are described.

1.1 Coastal areas or coastal zones?

The boundary between the land and ocean is generally not a clearly defined line on a map, but occurs through a gradual transitional region. The name given to this transitional region is usually 'coastal zone' or 'coastal area'. In common English there is little distinction between zone or area, but in coastal management there has been some debate as to the implied meanings associated with zone, as used in 'coastal zone management'. The debate has focused on the implication that zone may imply that geographically defined planning zones will be established and become the dominant part of the coastal management process. This implication is not important in many developed countries, where 'coastal zone management' is a phrase commonly used to describe a variety of coastal programmes (OECD, 1992), such as the US Coastal Zone Management Act (1972). But developing countries often equate coastal zone with land-use or marine-park zoning (Chapter 4). Although 'coastal zone' and 'zoning within the coastal zone' are clearly different, to avoid confusion many coastal management initiatives use the description 'coastal area' (e.g. UNEP OCA/PAC, 1982; Chua and Pauly, 1989).

Kaluwin (1996) describes the notion of delineating a zone or area as an essentially western concept which places artificial boundaries on the geographical extent of this transition. He considers it culturally inappropriate for Pacific islands, where the coast has traditionally been viewed as a transitional region between land and ocean; however, few coastal nations, especially in developed countries, take this enlightened traditional Pacific view of the coast. In this book we concur with Kaluwin (1996) that zone could be implied to mean a planning zone, and to ensure consistency we use coastal area or simply 'at the coast' or 'on the coast', except when quoting from original sources which use the term coastal zone.

1.2 Defining the coastal area

Defining the boundaries of a coastal area is of more than academic interest to coastal planners and managers. Governments often create adminis-trative systems, or set out policies to guide decision-making, that operate within a defined coastal policy area. The variety of ways in which such areas may be delineated in order to serve the purposes of particular policies are outlined in this section.

1.2.1 Scientific definitions of a coastal area

The coast is where land and ocean meet. If this line of meeting did not move, defining the coast would be easy – it would simply be a line on the map – but the natural processes that shape the coast are highly dynamic, varying in both space and time. Thus the line that joins land and ocean is constantly moving, with the rise and fall of tides and the passing of storms, creating a region of interaction between land and sea.

There are parts of the coastal environment that clearly have strong interactions between land and ocean, including beaches, coastal marshes, mangroves and fringing coral reefs; other parts may be more distant from the immediate coast (inland or out to sea) but they nevertheless play an important role in shaping it. One of the most important of these is the rivers that bring freshwater and sediment to the coastal environment. In this case, the inland limit to the coast is catchment boundaries that can be thousands of kilometres inland at the head of catchments. For example, the Ganges–Brahmaputra river system whose sediments form much of Bangladesh rises far inland in the Himalayas.

Therefore, the coast may be thought of as the area that shows a connection between land and ocean, and a coastal area defined (Ketchum, 1972) as:

> the band of dry land and adjacent ocean space (water and submerged land) in which terrestrial processes and land uses directly affect oceanic processes and uses, and vice versa.

The key element of Ketchum's definition is the interaction between oceanic and terrestrial processes and uses: coastal areas contain land which interacts with the ocean in some way, and ocean space which interacts with the land. Thus coastal areas:

- contain both land and ocean components;
- have land and ocean boundaries that are determined by the degree of influence of the land on the ocean and the ocean on the land; and
- are not of uniform width, depth, or height.

The three above elements are depicted in Figure 1.1, which shows, for a sandy beach coast, the strength of interaction between coastal and ocean processes and uses, termed here the 'degree of coastalness', against the distance away from the immediate coast. Figure 1.1 could be repeated for other coastal environments, such as delta coasts, beach/barrier systems and estuarine coasts, where the various physical and biological processes of these environments will determine the 'degrees of coastalness'. On deltaic coasts, for example, important determining factors would be the degree of salt water penetration in to fresh surface- and groundwater systems, and the seaward distance to which sediments of terrestrial origin are moved.

As Figure 1.1 shows, the transition between land and ocean is often gradual, depending on local biophysical conditions. The issue here is not

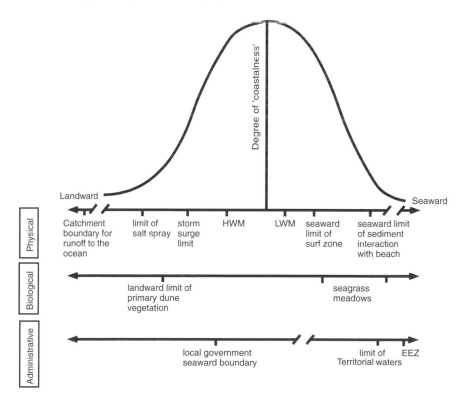

Figure 1.1 Example of 'degrees of coastalness' for a sandy beach coast.

the nature of the actual transition, but what its implications are for defining a coastal area. Choosing the thresholds which define the landward and seaward limits of a coastal area depends to a large extent on why the definition is needed. This 'need-driven' approach to coastal area definition is discussed further in the next section.

1.2.2 Policy oriented definitions of a coastal area

In practice, the [coastal] zone [area] may include a narrowly defined area about the land–sea interface of the order of a few hundreds of metres to a few kilometres, or extend from the inland reaches of coastal watersheds to the limits of national jurisdiction in the offshore. Its definition will depend on the particular set of issues and geographic factors which are relevant to each stretch of coast.

(Hildebrand and Norrena, 1992)

Coastal zone [area] management involves the continuous management of the use of coastal lands and waters and their resources within some designated area, the boundaries of which are usually politically determined by legislation or by executive order.

(Jones and Westmacott, 1993)

At a policy level the limits of coastal areas have been defined in four possible ways:

- fixed distance definitions;
- variable distance definitions;
- definition according to use; or
- hybrid definitions.

Current or proposed examples of each of the above definitions are given in Appendix A.

Fixed distance definitions, as the name implies, specify a fixed distance away from the coast which is considered 'coastal'. Usually this distance is calculated from some measure of the boundary between land and water at the coast, usually the high water mark. Fixed distances defined for the ocean component of a coastal area usually apply to the limit of governmental jurisdiction, for example the limits of Territorial Seas. An example of a fixed definition coastal area as used by the government of Sri Lanka is shown in Figure 1.2.

As for fixed distance definitions of coastal areas, the boundaries of variable distance definitions are set from some measure of the coast, usually the high water mark. However, their boundaries are not fixed, but vary along the coast according to a range of variables such as:

- physical features – e.g. the landward limit of Holocene dunes, or the seaward limit of submarine platforms;
- biological features – e.g. the landward limit of a coastal vegetation complex, or the seaward limit of a fringing reef; and
- administrative boundaries – e.g. the landward limit of local municipalities which front the ocean.

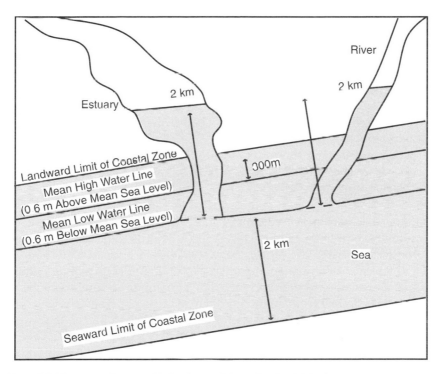

Figure 1.2 The coastal zone of Sri Lanka, as defined by the Sri Lankan Coast Conservation Act (Coast Conservation Department, 1996).

International organizations and large coastal nations often define the limits of a coastal area according to the particular coastal management issue being addressed; that is, the coastal area is defined according to the use to which that definition will be put, and the form of definition is termed 'definition according to use'. For example, tackling the issue of non-point sources of marine pollution would require the definition of an area of attention that included inland catchments and groundwater outflow regions. A coastal area defined for this purpose would be much larger than one defined to manage four-wheel-drive vehicle damage of beaches and dunes. As recognized by the Coastal Committee of New South Wales (1994, p.22):

To a large extent, the definition of the coastal zone depends upon the purpose for which the definition is intended. From both management and scientific viewpoints, the extent of the coastal zone will vary according to the nature of the management issue.

Within the context of defining a coastal area according to what the purpose is, the concept of 'areal foci' used by Jones and Westmacott (1993) is useful. Areal foci include:

- an administratively designated area, in the sense that the political process or the administration will designate the responsibility to manage;
- an ecosystem area;
- a resource base area, e.g. a mineral body, oil fields, fisheries, habitats, etc.; and
- a demand area, i.e. the wider area from which demands are exerted on the designated coastal area, such as for use for recreation, marine transport or waste disposal.

Defining a coastal area according to use has the advantage of focusing attention on particular issues. However, care needs to be taken to avoid multiple coastal area definitions being established in one region to address different coastal management issues, leading to confusion. Defining the coast according to one use only may perpetuate sectoral managerial systems and detract from an integrated management perspective.

Hybrid definitions mix one type of coastal definition for the landward limit of the coastal area and another for the seaward limit. This is relatively common practice by governments that have a fixed limit of jurisdiction over nearshore waters. Australian States, for example, have management responsibilities for coastal waters 3 nautical miles from the coastline. Some Australian State governments use this to define the seaward limit of their coastal areas, while choosing other means to define the landward boundary (see Appendix A). For example, the recent definitions of coastal areas adopted by the Queensland State Government are shown in Box 1.1.

The vertical dimension of any coastal area definition can also be included; that is, the depth below the surface and height above a coastal area considered to be covered by a coastal policy. Usually the vertical dimension is part of the overall legislative framework of governments, and is not explicitly covered by coast-specific policies. Examples include all mineral rights below coastal lands and waters and the atmosphere above it, which are generally covered by laws and regulations that cover all other parts of a government's jurisdiction.

Box 1.1

**Coastal definitions used in the Queensland
Coastal Protection and Management Act (1995)**

- Foreshore means the land lying between high water mark and low water mark as is ordinarily covered and uncovered by the flow and ebb of the tide at spring tides.
- The coast is all areas within or neighbouring the foreshore.
- Coastal management includes the protection, conservation, rehabilitation, management and ecologically sustainable development of the coastal zone.
- Coastal resources means the natural and cultural resources of the coastal zone.
- Coastal waters are Queensland waters to the limit of the highest astronomical tide.
- Coastal wetlands include tidal wetlands, estuaries, salt marshes, melaleuca swamps (and any other coastal swamps), mangrove areas, marshes, lakes or minor coastal streams regardless of whether they are of a saline, freshwater or brackish nature.
- The coastal zone is:
  (a) coastal waters; and
  (b) all areas to the landward side of the coastal waters in which there are physical features, ecological or natural processes or human activities that affect, or potentially affect, the coast or coastal resources.

In summary, a generic definition of coastal areas is not proposed here. Rather, a pragmatic view of defining a coastal area is taken, where the definition reflects the use or uses to which it will be put. If the purpose is to control certain types of development, then fixed, variable or hybrid definitions may be used. If reducing pollution of marine waters is the purpose, then variable definitions including catchment or groundwater boundaries may be more appropriate. By focusing on coastal management issues, and not on problems of definition, simple and workable definitions of coastal areas usually follow.

1.3 The unique characteristics of coastal areas

Stating that the coast is unique because it is where land and oceans meet may appear rather obvious, but it is a fact of great significance. The contrast between land and ocean may be dramatic where ocean swells crash against rock cliffs, or more gradual where tides ebb and flow over marshes. It is this interaction between marine and terrestrial environments that makes the coast unique – and uniquely challenging to manage.

The transition between land and ocean at the coast produces diverse and productive ecosystems which have historically been of great value to human populations. Use of the coast for its resources has long been combined with its value as a base for trading between countries, both across oceans and by the rivers which flow out to sea. Coastal lands and nearshore marine waters have consequently long been at a premium. As populations grow and increase their level of socio-economic development, this premium also grows. The consequence of this intense and long-standing pressure on coastal resources is that problems with the way in which competing uses are managed within a country as a whole tend to become manifest first on the coast.

To make management even more difficult, major administrative boundaries commonly follow high or low water lines, bisecting coastal areas and dividing the management of the land from that of the ocean. Coastal land is usually owned and/or managed by a multiplicity of private, communal, corporate and government bodies, whereas coastal waters are usually owned and/or managed solely by governments. Furthermore, administrative boundaries can follow the centres of rivers and estuaries, dividing their management between two neighbouring authorities.

The uniqueness of the coast is further enhanced by the value of its resources such as fish and offshore mineral reserves, which are considered by the populace to be common property, and in high demand by coastal dwellers for subsistence use, recreation and economic development (Berkes, 1989; Feeny *et al.*, 1990). Exploitation of such resources raises their value, with a consequential demand for equitable resource allocation. Therefore, resource planning often forms an integral part of coastal management programmes.

1.4 A brief history of coastal management and planning

A brief history of the development of coastal area management and planning is presented for two main reasons. First, history provides a framework for understanding how current approaches to the planning and management of coastal resources have evolved, and the constraints these approaches are operating within. Second, by looking back at how coastal planning and management have developed, trends become evident. Projecting such trends provides an insight into the possible future development of coastal management and planning.

Humans have deliberately modified the coastal environment and exploited its resources for thousands of years. Ancient civilizations throughout the world built ports and seawalls, or diverted river water flowing into the sea; they also evolved various management systems for their fisheries, use of rich coastal soils for agriculture, trading through ports, and other coastal resources. Examples include: ancient Greek and Roman port cities throughout the Mediterranean; the diversion of the Yangtze (Yellow) River, China in AD1128 (Ren, 1992); and the reclamation of mangrove areas over 1000 years ago on Pohnpei, Federated States of Micronesia (Sherwood and Howarth, 1996).

Ancient interventions such as these in the coastal environment were all works of civil engineering. That is, structures were built to modify the flow of water and/or sediment. Given that such structures were all essentially hand built, the scale and intensity of their impacts on the coastal environment were limited, but over the centuries the ability of humans to influence coastal processes increased as construction techniques improved. Perhaps the most famous example of diversion of water courses and construction on the coast was the building and maintenance of the current urban form of Venice, Italy, from the seventh century AD (Frassetto, 1989).

For these civilizations an informal form of resource planning was undertaken either by community consensus or by a leader who decided when, where, how and how much resources would be exploited. Resources were abundant but sparsely exploited because of limited technology. Hence resources were generally allocated on a social rather than on an economic basis.

Technological limitations were dramatically reduced as a result of the industrial revolution, which started in Europe in the mid-nineteenth century. The industrial revolution brought machines that could be used to construct grander civil engineering works. Major modifications of the coastal environment were now possible: large rivers could be dammed or diverted and vast areas of coastal wetlands could be converted to urban or agricultural land.

The industrial revolution also altered the community's view of its resources. Viewing them as tangible elements or objects of nature led to the use of the term 'natural resources', and management, including planning, now focused on supply and demand, and the options for managing these factors. This was linked to the pervasive western cultural attitude at the time of human dominance over other animals and natural systems.

Concentrating on economic factors, very little attention was given to the ecology (including habitats), social demands or public perceptions (O'Riordan and Vellinga, 1993). The underlying objective was to maximize profits, which usually translated into maximizing production. The weakness of this approach was the assumption that resources are easily valued, single purpose and static in value over time, which we now know is not valid (Chapter 4).

During the industrial age the market place began to dominate resource allocation, while social norms no longer guided resource use. Resources were perceived as limitless and there to be consumed for profit (Goldin and Winters, 1995; Grigalunas and Congar, 1995). It was not until late last century that this view began to change. Resources came to be considered finite, a change in attitude attributable to:

- advances in economic theories on supply and demand;
- the developing realization that society had the ability to destroy the environment, ultimately affecting its survival;
- social reforms; and
- studied attempts to plan for resource management.

In contrast, deliberate human intervention in the coastal environment to preserve components of its natural character or ecological integrity is a much more recent activity. Coastal ecological management grew from the national park movement of the late nineteenth century. During this era, protected areas or parks were perceived as places of significant scenic or natural value set aside for the enjoyment of visitors or for scientific pursuits (MacEwen and MacEwen, 1982). The first such parks in coastal marine areas were established in the 1930s. Since then, protected areas with significant coastal components have been established throughout the world, with most being terrestrial. Currently there are approximately 4500 recognized protected areas (as defined by the IUCN) around the world, of which only about 850 include a coastal or marine component (Elder, 1993).

Expansion of land use planning in the late nineteenth and early twentieth centuries also influenced coastal area management in developed and colonial 'new world' countries (Platt, 1991). Important influences included the notion of separating conflicting land uses through zoning, planning open space areas for the public good and health, and sanitation problems which affected waste disposal into coastal waters. While the main way to effect such interventions was through the use of the engineering works described above, it is the role of land use planners in directing the expansion of urban environments into coastal areas, and their enthusiasm for embracing engineering interventions, that is important here. Urban expansion brought with it the need to develop the coast for new residential areas and industries, as well as a need to cater for increased recreational use of the coast.

Different streams of human endeavours in coastal areas, such as ecological management, resource management, engineering intervention and urban/industrial development, operated relatively independently for many years. The coastlines of developed nations had been planned and managed using land use planning and environmental management techniques which had evolved within their various governmental and cultural settings. Each can be considered as a form of coastal area management, and their proponents as coastal managers. However, it was not until

Table 1.1 Phases in the development of coastal management (adapted from O'Riordan and Vellinga, 1993)

Phase	Period	Key Features
I	1950–1970	• Sectoral approach • Man-against-nature ethos • Public participation low • Limited ecological considerations • Reactive focus
II	1970–1990	• Increase in environmental assessment • Greater integration and coordination between sectors • Increased public participation • Heightened ecological awareness • Maintenance of engineering dominance • Combined proactive and reactive focus
III	1990–present	• Focus on sustainable development • Increased focus on comprehensive environmental management • Environmental restoration • Emphasis on public participation
IV	Future	• Establish coastal area management based on ecological empathy, precautionary management and shared governance

the 1960s and 1970s that these, and other disciplines, were brought together under the banner of 'coastal zone management', a phrase credited to those involved in the development of the US Coastal Zone Management Act in the late 1960s and early 1970s (Godschalk, 1992; Sorensen, 1997).

Realization around the world that environments were being continually degraded by a rapidly expanding human population led to the concept of sustainable development in the late 1980s and early 1990s. The basis of sustainable development is 'development that meets the needs of the present without compromising the ability of future generations to meet their own needs' (World Commission on Environment and Development, 1987), a concept now central to most coastal management efforts worldwide, as will be shown in Chapter 3. Given the importance of sustainability principles in coastal management, the topic is discussed further in the next section.

Today, it is generally accepted that coastal resources can only be effectively evaluated and managed in the total context of the social and cultural environment (e.g. Ehler, 1995). Hence, effective resource planning provides for decision making which allocates resources over space and time according to the needs, aspirations and desires of society, taking into account society's ability to exploit resources, its social and political institutions, and its legal and administrative arrangements.

O'Riordan and Vellinga (1993), in reviewing the history of coastal area management, as outlined above, summarized into four phases its development over the past 40 years as a professional activity (Table 1.1).

1.4.1 Sustainability – the dominant paridigm in coastal planning and management

> Sustainable development requires a broader view of both economics and ecology than most practitioners in either discipline are prepared to admit, together with a political commitment to ensure that development is 'sustainable'.
>
> (Redclift, 1987, p. 33)

Sustainability has emerged as the dominant paradigm of the world's coastal management programmes in the late twentieth century. The historical context of this emergence is described in the previous section; here we describe the concept of sustainability and discuss its influence on coastal programmes, from broad scale strategic planning to day-to-day management regimes. This discussion forms the basis for the more detailed treatment in Chapters 3, 4 and 5 of tools and techniques to help to achieve the sustainable development of coastal areas.

The concept of sustainability came into prominence with the publication of the World Commission on Environment and Development (WCED) report called *Our Common Future* (World Commission on Environment and Development, 1987). The WCED group was chaired by Gro Harlem Brundtland, hence the report came to be known as the Bruntland Report. The message of the Brundtland Report (WCED, 1987, p. 8) was that:

> it is possible to achieve a path of economic development for the global economy 'which meets the needs of the present generation without compromising the chances of future generations to meet their own needs'.

A central precept of sustainability, to quote Pearce *et al.* (1989, p. xiv), is that sustainable development leaves 'future generations a wealth inheritance – a stock of knowledge and understanding, a stock of technology, a stock of man-made capital, and a stock of environmental assets – no less than that inherited by the current generation'. Young (1992) recognizes a number of themes underlying the sustainability concept, summarized by his 'three Es':

- environmental integrity;
- economic efficiency; and
- equity, defined to include present and future generations and recognize cultural as well as economic considerations.

Though precise definitions of sustainability may be rather elusive, it is clearly not a set of prescriptive actions; rather it is the basis for a fundamental reassessment of the way in which resource, environment, social and equity issues are considered in decision making. The profoundness of its implications has caused sustainability to be compared with such basic societal values as freedom, justice and democracy (Buckingham-Hatfield and Evans, 1996a). Seen in this light, sustainability becomes a 'way of thinking', helping to modify the context to which it is applied (Turner, 1991). Thus, sustainability principles can 'highlight unsustainable systems and resource management practices' (Turner, 1991, p. 209). The tests of sustainability having been applied and unsustainable practices revealed, the way opens for new, sustainable management approaches to coastal area management to be devised and adopted (Figure 1.3).

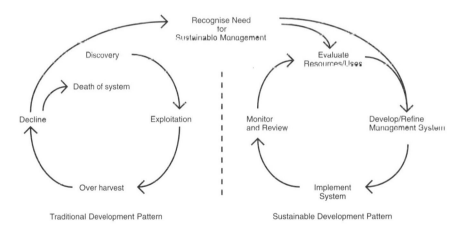

Figure 1.3 Sustainable and unsustainable approaches to coastal resource use (Dutton and Hotta, 1994).

As a 'way of thinking', sustainability has not only become part of the mainstream of decision-making processes; it has also in many nations become a political reality (Buckingham-Hatfield and Evans, 1996a) – though remaining elusive in many others (Kirkby *et al.*, 1991). However, the idea that the present generation can through the application of sustainability principles act as stewards of the earth for future generations is as much an act of faith as it is one based on technical or scientific evidence (Buckingham-Hatfield and Evans, 1996a). This raises two important issues: the weight to be given to technical information, and the time-dependence of decision making.

Sustainability has acted as the catalyst for a new mix in the information sources on which decisions are based. It has seen the 'hard science' emphasis of the 1970s and 1980s evolve into a more balanced appreciation of scientific and non-scientific inputs into decisions. This balancing has manifested itself in various ways – for example, the Best Practicable Environmental Option system in the United Kingdom (Gerrard, 1995) – but its most pervasive expression is the 'precautionary principle' (Cameron, 1991), commonly defined in the language of Principle 15 of the Rio Declaration (UNCED, 1992):

> In order to protect the environment, the precautionary approach shall be widely accepted by the States according to their capabilities. Where there are threats of serious or irreversible damage, lack of full scientific certainty shall not be used as a reason for postponing cost effective measures to prevent environmental degradation.

This principle is now incorporated into the London and Hague Declarations dealing with marine pollution.

Precautionary action has three central components. First, there is an economic dimension of cost-effectiveness; second, decisions which may have irreversible impacts, so providing a legacy for future generations, gain heightened importance in the decision-making process; and third, the lack of a requirement for complete scientific information in the face of economically inefficient and/or irreversible impacts – a substantial shift from a rational-comprehensive view of decision making, as will be shown in Chapter 3. It is important to note that a precautionary approach to guiding decision making is a very recent phenomenon and its use is not uniform around the world (O'Riordan and Cameron, 1994). However, its current use in some coastal nations, and probable spread to many more, is likely to see precaution entering the lexicon of most coastal managers in the next few years.

A central part of the 'way of thinking' introduced in this section is the consideration of time dependence in decision making; that is, consideration of the effects of present-day activities on future generations (Young, 1992). Sustainability thinking requires that future effects and impacts of decisions, and not simply those in the present day, be considered. Relating to this concept, many planners have seized upon sustainability with the notion that planning and sustainability principles are similar, and that a convergence of planning and sustainable development is emerging under the banner of environmental planning (Blowers, 1993; van Lier et al., 1994; Buckingham-Hatfield and Evans, 1996b).

Having looked at the general principles of sustainablity, three specific effects of sustainable thinking on coastal management can now be briefly considered. The first is the effect on the use of economics and economic instruments in decision making. Environmental economics, as we shall demonstrate in Chapter 4, is rapidly becoming one of the mainstays of the practical use of sustainable development in decision making. Sustainability has provided many economists with a basis for implicitly including equity, environmental considerations and a long-term view into the cost–benefit and other economic analyses (Jacobs, 1991). Likewise, sustainability has also allowed environmental issues, such as conservation of biodiversity, to become a central part of decision making, especially in those areas previously the exclusive domain of economists – most notably economic development. Finally, sustainable development explicitly recognizes the quality of human life of both current and future generations. Thus, social and cultural equity is recognized as an equal partner with economic and environmental considerations.

In summary, sustainable development principles have had four main effects on the way the coast is managed: one general and three specific. The general effect is the influence 'sustainability thinking' has on the overall decision-making context. The mixture of equity, environmental and economic concepts moves the decision-making paradigm away from considering economic and environmental decisions in isolation from each other. The three specific impacts are in the fields of economics, environmental resource management and social and cultural development, summarized by Reid (1995) as requiring the following characteristics:

- integration of conservation and development;
- satisfaction of basic human needs;
- opportunities to fulfil other non-material human needs;
- progess towards equity and social justice;
- respect and support for cultural diversity;
- provision for social self-determination and the nurturing of self-reliance; and
- maintenance of ecological integrity.

Clearly, these are major issues which go to the heart of the human cultural, spiritual and developmental aspirations as well as fundamental issues of governance, democracy and the relationship of humans and the environment. They are weighty issues but, nevertheless, ones which must be confonted to ensure a viable future for the world's coastal regions. Sustainablity, then, is 'not just about managing and allocating natural capital. It is also about deciding who has the power both to do this and to institute whatever social, economic and political reforms are considered necessary' (Reid, 1995, p. 231). Any discussion of approaches to the sustainable development of coastal areas must, as a result, analyse techniques for environmental management, systems of governance and the role of individuals in decision making and planning processes. It is the interplay between these factors, which is explored at length in the coming chapters.

1.5 Summary

Coastal management programmes have generally developed in response to problems experienced in the use and allocation of coastal resources. Development of a coastal programme usually follows a period of mounting public, political and scientific pressure on governments to tackle problems, usually resulting in a time lag between the identification of problems and the development of responses. Development of the US Coastal Zone Management Act in 1972, for example, followed a period of intense pressure for improvement in coastal land and water management which started more than 12 years earlier (Godschalk, 1992). A similar pattern was followed in the United Kingdom during the late 1980s to the early 1990s (King and Bridge, 1994). Much of the history of coastal management and planning illustrates similar reactions to problems experienced in coastal regions. Many other national and international initiatives can be traced to the time when the problems could no longer be ignored.

A further stimulus to the development of coastal programmes was the realization that coastal area management programmes could be used to avoid future problems. However, unlike the development of programmes which respond to existing problems, it is unclear when this proactive approach became important. It may be inferred that there were some important forward-looking parts of the US Coastal Zone Management Act; this is not formally reflected in its aims (Godschalk, 1992). Proactive management, through the use of various coastal planning approaches, is now one of the most important components of coastal area management around the world, with modern programmes blending proactive and reactive elements to address current problems, such as ecosystem degradation, and to avoid future problems.

It is worth re-emphasizing at this point that the deliberate actions of humans to influence the natural processes of the coast have been occurring for thousands of years. Coastal management choices made during this time reflected the cultural and spiritual relationship between people and the coastal environment. Historically, it is the perception of how the coast should be managed, and for what purpose coastal resources will be used, that has shaped management of coastal areas. These perceptions are culturally and politically influenced; they have clearly changed over time, will continue to change, and are demonstrably different around the world. The diversity of coastal area management approaches reflects these differences.

The documented development of coastal planning and management described above is largely a history of western nations, or those countries colonized by western nations. In this group of nations, the evolution of coastal programmes, as they reflect changes in cultural values, have been well described (Table 1.1). In contrast, the traditional coastal management systems of indigenous cultures in other parts of the world are relatively poorly documented. Although culturally appropriate coastal management programmes are making something of a resurgence in many developing countries (Chapter 3), there remains much to be done in understanding how these traditional practices evolved and how they have been extended into the modern age. This is especially so since their integration with western management practices is becoming increasingly important with the pervasive spread of western technologies and management approaches. These are recurring themes of the following chapters.

In summary, the early development of coastal area planning and management programmes in the early 1960s and 1970s was generally in response to urgent problems on the coast. As these reactive coastal programmes became more established, they gradually evolved into a combination of reactive and proactive programmes during the 1980s and 1990s. This evolution may reflect the heightened influence of planning on the management process, or it may reflect the need to manage existing problems by addressing possible future pressures. However, perhaps the key lesson to be drawn from this brief history is the need to combine present and future perspectives; that is, attempting to address present day problems whilst preventing new ones, an aim which fits well within the techniques described in Chapters 3, 4 and 5.

Chapter 2

Coastal management issues

> Man has only recently come to realize the finite limitations of the coastal zone as a place to live, work, and play and as a source of valuable resources. This realization has come along with overcrowding, overdevelopment in some areas, and destruction of valuable resources by his mis-use of this unique environment.
>
> (Ketchum, 1972, p. 10)

This chapter provides an overview of the major coastal management issues, problems and opportunities in coastal management. Consistent with the general focus of this book, particular emphasis is placed on describing and analysing management tools and planning techniques to assist in dealing with the issues.

The chapter does not attempt to analyse and describe every issue at length. For a more detailed treatment of coastal issues, refer to texts specifically devoted to this subject. The most recent and comprehensive is the 694-page text of Clark (1996), which lists the many complex and interrelated problems that face coastal managers, and updates his earlier work (Clark, 1977). Ketchum (1972), Ditton *et al.* (1978) and Beukenkamp *et al.* (1993) also provide useful treatments of the issues. In addition, there are numerous conference and workshop proceedings which contain specific examples of coastal problems from around the world (Appendix B). Further information on the range and depth of coastal issues can be obtained through reference to the sources of the many case studies listed throughout the book.

Coastal management initiatives are usually a response to a demand to resolve problems such as conflicting uses of coastal resources, urbanization, access, pollution and environmental degradation. Problems may also be related to poor liaison or inefficient coordination between those responsible for making decisions on the allocation of coastal resources; or they may even be a perception among decision makers that a problem does not exist. A sound understanding of such issues is integral to planning an effective approach to coastal management.

The issues described in this chapter are those common to many coastal areas around the world. Inevitably, they are more critical in some places than in others, and hence will be of differing levels of interest to managers in different places. Nevertheless, they are all relevant to the development of an understanding of coastal problems and the approaches to avoiding or mitigating their impacts.

Issues are discussed under the broad groupings of population growth, coastal use, the impacts of coastal use and impacts on coastal uses, and administrative issues. The groupings are not mutually exclusive, but are designed to give a general feel for the major challenges facing coastal managers today.

A useful introduction to the range of typical issues for coastal nations is provided from the United Kingdom (Figure 2.1) (Local Government

EXAMPLES OF LANDWARD ISSUES

- port and harbour works
- land take
- marinas and moorings for leisure craft
- power generation (c.g. wind)
- major developments (e.g. refineries, container terminals)
- coastal defences (e.g. groynes)

EXAMPLES OF SEAWARD ISSUES

- waste disposal
- increased leisure sailing
- sea fishing
- water sports and bathing
- marine aggregate extraction
- oil and gas production
- tidal and wave power generation marine fish farming

EXAMPLES OF IMPACTS ON COASTAL SYSTEMS

PHYSICAL CHARACTER

- loss or decline of landscape value
- disruption of sediment transport
- decline in amenity resources (beaches, dunes, etc.)
- impacts on character of coastal towns

NATURAL HERITAGE

- loss or decline of habitat
- disturbance of coastal ecosystems
- decline in fish/shellfish resources
- loss of treasured landscapes

COASTAL USE

- conflicts with rights of sea users
- incompatible uses need other locations
- pressure for services and facilities (e.g. car parks, moorings etc.)
- impacts on existing businesses and employment

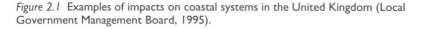

Figure 2.1 Examples of impacts on coastal systems in the United Kingdom (Local Government Management Board, 1995).

Box 2.1

Issues and topics addressed by the Thames Estuary Management Plan

The Thames, one of the world's most famous estuaries, has multiple management conflicts. It is the United Kingdom's busiest and most commercially significant tideway; 12 million people live within easy reach of it and the port alone supports 37 000 jobs. Nowhere in the country are environmental pressures and competing demands for space and resources greater than on Thames-side. Despite the enormous pressure, the Thames is also internationally important for wildlife. The estuary supports 114 different species of fish, and its mudflats and marshes are home to an estimated 170 000 birds.

In recognition of the need to plan for the future, many of the users of the Thames have worked together to produce an estuary management plan, described in Chapters 3 and 5 (Boxes 3.5, 3.10 and 5.27). The general issues and specific topics addressed by the Thames Estuary Management Plan are:

- General issues:

  - communication between different sectors is poor;
  - there is little understanding of different organizational cultures;
  - a need exists for shared technical information of agreed standards;
  - there is enormous administrative fragmentation;
  - a shared realization among stakeholders to ensure impending problems do not occur.

- Specific topics:

  - agriculture;
  - coastal processes;
  - commercial use of the estuary;
  - fisheries;
  - flood defence;
  - historical and cultural resources;
  - landscape;
  - nature conservation;
  - recreation;
  - waste transfer and disposal;
  - water management;
  - public awareness;
  - enhancement opportunities;
  - targets and monitoring.

Management Board, 1995). Here, coastal issues have been considered as either essentially landward or seaward in character. Together with the topics addressed by the Thames Estuary Management Plan (Box 2.1, discussed further in Chapters 3 and 5), they provide a concise introduction to the issues outlined in the following sections.

2.1 Population growth

Population growth is the driver behind many, if not most, coastal problems. The scale of this growth in recent years has been staggering (Haub, 1996), with estimates putting the world's present population in coastal areas as equal to that of the entire global population of the 1950s (Edgren, 1993). Growth in coastal populations is not limited to developing countries: an estimated 50% of the population of the industrialized world is now living within 60 km of the coast (Turner *et al.*, 1995). These growth trends are set to continue, with scenarios of future populations estimating that in 30 years more people will live in the world's coastal zones than are alive today (NOAA, 1994a)

Population growth in coastal areas has two main causes. First, it reflects the general trend of population growth in developing countries, linked to rural–urban migration; and second, the migration from inland areas to the coast, which often offers people more economic, social and recreational opportunities than inland areas (Goldberg, 1994). Examples of coastal population growths and their impacts in Florida and California (USA), and in the Indonesian province of Sulawesi Selatan, are shown in Boxes 2.2 and 2.3.

The clearest result of population growth in the coastal zone is the accelerating rate of urbanization: by the year 2025 more people are projected to live in cities than occupied the whole world in 1985, while the physical size of cities in developing countries is expected to be double what it was in 1980 (World Resources Institute, 1992).

Cities on the coast are often associated with major ports which facilitate cheap sea transport of goods, which in turn attracts major industries. Economic growth provides employment and investment opportunities, coastal cities acting as a magnet for people looking to improve their economic status (Ehler, 1995). The coast's attractiveness also draws people for holidays, retirement and those seeking coastal lifestyles. In response, many urban areas are being developed or expanded to meet the needs of new coastal residents for housing, sanitation and transport.

Many specific resource allocation and planning issues are raised by the urbanisation debate: urban residential densities, the development of high rise buildings, and public versus private access to beaches and foreshores are among the more prominent. These in turn impact on the visual landscape , and create increased pressure on coastal resources and the use of facilities such transport, land fill and sewerage.

Box 2.2

Coastal issues in Florida and California (Fisk, 1996b)

California and Florida are among the fastest growing states in the United States. Their warm and sunny climate and resulting outdoor lifestyles have attracted migrants from northern states. Many settle on the coast, creating coastal development and management issues which have required concerted efforts for many years.

California

California has one of the longest coasts in the United States, made up of spectacular sea cliffs, rocky shores and beaches. The coastal area contains abundant living and non-living resources as well as one of the largest bay-estuary systems in the world – San Francisco Bay. The major impacts on the California coast include increased residential and commercial development, the effects of relative sea-level rise on coastal structures, and degraded coastal water quality from urban and industrial runoff.

Florida

Florida's tropical and subtropical coastal area contains the most extensive mangrove and wetland areas in the United States as well as the greatest concentration of coral reefs, found around the Florida Keys. Major impacts to the Florida coastal area include rapidly expanding commercial and residential construction, tropical storms, increased erosion and loss of life and property due to primary sand dune removal, and threats to the preservation of Florida's unique wetland and coral reef areas.

The administrative mechanisms for organizing coastal management programmes to tackle the above problems in California and Florida are described in Box 3.8.

Management of urban areas expanding along the coast can be one of the most difficult tasks of coastal planning. The often enormous values of coastal land which can be developed for residential and tourist developments can see the widespread conversion of agricultural, forestry and other low intensity land uses to urban. A result can be urban 'strip development' as tentacles of urban sprawl spread monotonously up and down the coast from urban centres. Ultimately, cities hundreds of kilometres apart can become joined, effectively becoming one coastal 'megacity' (e.g. Toyko–Osaka in Japan).

Urban and regional planning attempts to resolve these competing demands (Box 2.4). Techniques for consideration of such issues are presented in Chapters 4 and 5.

Box 2.3

Coastal pressures in Sulawesi Selatan province, Indonesia

Indonesia is a rapidly developing country. Like many Asian nations it had until recently a strong economy, experiencing an annual real economic growth rate of 7.4% in 1990 (Department of Information, 1992). Corresponding with this growth has been an expanding urbanization and an annual population growth of 1.8%. However, coastal populations have been growing at twice the national rate (Asian Development Bank, 1987). This rapid economic growth, continuing population growth and urban expansion have strained coastal environments. Eastern Indonesia has been the focus of many economic initiatives and rapid urban development; one area which has experienced rapid growth is the province of South Sulawesi (locally called Sulawesi Selatan or SulSel).

More than 80% of Sulawesi Seletan residents live in coastal settlements, most are located on the fertile coastal plan adjacent to the Makassar Straits (Bangda, 1996). Many of these residents are economically dependent on fisheries resources, especially the Spermonde Archipelago and Taka Bone Rate reef systems (see Box 5.18). These coral reef systems are considered to contain some of world's highest marine biodiversity. The highest number of coral reef species are found here; they also support one of the world's most intensive reef fisheries.

These rich waters have enabled coastal communities in Sulawesi Seletan to develop a strong marine and coastal culture. Many communities rely on coastal and marine resources for subsistence and income generation. These communities, especially the Makassanese and Buginese, have developed innovative approaches to resource use and established pioneering trade routes throughout the Asia-Pacific region (Bangda, 1996). This marine culture continues today with the provincial capital, Ujung Pandang, firmly established as the hub of marine transport in Eastern Indonesia and an emerging economic centre.

As a consequence the demands for access and use of coastal and marine resources has increased with significant costs to the environment. Fifty-one per cent of the Province's mangroves have been destroyed since 1982. Many of the mangroves have been converted to aquaculture ponds which operate with no environmental controls. Other marine environments have been destroyed due to destructive fishing practices such as blasting and cyanide.

Shipping within the Makassar Straits has grown and is expected to continue expanding now that the Straits are an international shipping lane. The demand for access to the coast and islands for tourist developments has increased; many developments will displace local residents and place a burden on existing water supplies. In addition, many developments are not required to provide sewage treatment facilities.

To address these pressing issues, the Indonesian governments are working to develop a coastal planning and management framework, including national guidelines and regional and local plans described in Boxes 3.6, 5.9, 5.13 and 5.18.

Box 2.4

Coastal urban expansion issues north of Perth, Western Australia

The 1.7 million population of Western Australia is concentrated in the State's south-west, with 1.2 million people living in the capital, Perth. With the State's economy rapidly expanding at 5.1% per annum (Western Australian Planning Commission, 1995b), its population is expected to continue to grow. Projections are for a total population of 2.62 million by 2026, of which it is predicted 1.92 million will live in Perth (Western Australian Planning Commission, 1995a).

The Central Coast region, immediately north of Perth, is currently sparsely populated. A risk for this area as Perth expands is an unplanned urban sprawl northwards along the coast. The Central Coast Regional Strategy was developed for this 250 km of coastline with the aim of balancing urban expansion pressures with conservation, recreation and tourism opportunities (Western Australian Planning Commission, 1996a). Four major issues prompted the strategy:

- access, protection and use of the coastline;
- the need for new road connections;
- the future use and management of the large amount of public land; and
- the impact of metropolitan development on the future of the region.

Coastal management issues and values addressed by this study were:

- the scenic attractions and natural recreation opportunities of the coast which are valuable to the region and make it a desirable place to live and visit;
- the illegal squatter developments causing significant land management problems and jeopardizing recreational and conservation opportunities;
- development associated with settlements occurring too close to the coast;
- loss of seagrass possibly affecting marine environments;
- the multipurpose nature of coastal activities, requiring different facilities and access considerations;
- the attractions of the coast for recreation and tourism, necessitating low key, low impact development, taking into account environmental and social considerations; and
- the potential, without adequate rehabilitation and planning, of mining and extraction of basic raw materials to damage the coastal environment.

The outcomes of the strategy are discussed in Chapter 5.

2.2 Coastal use

Coastal uses are considered under four main categories: resource exploitation (including fisheries, forestry, gas and oil, and mining); infrastructure (including transportation, ports, harbours, shoreline protection works and defence); tourism and recreation; and the conservation and protection of biodiversity. Each category is described in turn. The use of land for residential purposes was outlined in the previous section, and is not considered further in this section.

2.2.1 Resource exploitation – fisheries, forestry, gas and oil, and mining

Coastal renewable resources are primarily exploited in the fisheries sector by commercial, subsistence and recreational fishers and the aquaculture industry. Worldwide attention has been focused on the sustainability of today's fisheries. Industry, resource managers and conservation groups are concerned with overfishing of most stocks, especially inshore fisheries, and the long-term sustainability of these fish stocks. Indeed, an estimated 70% of the world's commercially important marine fish stocks are either fully fished, overexploited, depleted or slowly recovering (Mace, 1996; World Wide Fund for Nature, 1996).

Current trends in the development of new fisheries such as the live fish trade, which has been responsible for the collapse of a number of reef fisheries throughout Asia and the South Pacific, are also of concern (Johannes, 1995). Coastal management has a critical role to play in managing fisheries since many coastal habitats such as mangroves and seagrass beds are part of the life cycles of many commercially important species.

Aquaculture, pond and cage culturing have been practised in Asia for centuries. The last 50 years has seen an exponential expansion of this industry, not just for fisheries, but for other emerging marine resources such as seaweed, prawns and sea cucumbers. Sea cage culturing has also developed in a number of areas. There are a number of issues associated with both forms of culturing. The conversion of land to ponds and the consequential loss of productive agricultural land is a major concern amongst coastal managers (Figure 2.2), especially as in some areas pond production is sustainable for less than 20 years; and the conversion of coastal habitats such as mangroves leads to a loss of fish habitats (Hay et al., 1994). Pond systems produce high nutrient levels which ultimately enter coastal waters, a problem which is compounded when antibiotics, algicides and other chemicals are used. Cage culturing in marine areas causes local pollution and can introduce diseases into wild populations. The introduction of exotic species and the consequential displacement of native species is a potential problem with all forms of culturing.

Figure 2.2 Aquaculture ponds, South Sulawesi, Indonesia (credit: Reg Watson).

Coastal forestry focuses on the commercial and subsistence exploitation of mangrove stands. Historically, exploitation of mangroves for charcoal, furniture and other uses was sustainable, but current demand for fuel far exceeds supply in many parts of the developing world. The result is that mangrove stands are commonly no longer a sustainable supply of cooking fuel. These issues are evident in Indonesia, as shown in Box 2.5. Clearly the loss of mangrove forests is a loss in biodiversity and habitat with potential impacts on adjacent commercial fisheries. When mangroves are cut, sediments from upland areas entering coastal areas are no longer trapped, and shoreline stability can be adversely affected.

Inland forestry practices in many developing countries can have indirect impacts such as increased sedimentation due to soil loss, especially in poorly managed rainforest extractions. Agricultural land-uses in both the developing and developed world can have similar effects, as well as the potential impacts of herbicides and pesticides.

Oil and gas are the major non-renewable resources exploited in many coastal areas, and are a major source of revenue for many coastal nations. Ancient coastal deposits and sedimentary basins adjoining continents commonly favour oil and gas accumulation. Examples include deposits found under or adjacent to modern deltas, such as the Mississippi, Niger and Nile.

The siting of oil and gas facilities on the coast requires careful planning and management. The facilities themselves can conflict with commercial

---

Box 2.5

Mangrove conversion to prawn aquaculture issues – South Sulawesi, Indonesia

Mangroves are an important coastal resource and serve a number of functions. They are critical to maintaining foreshore stability and trapping sediments from river runoff. Many commercially important fish spend a part of their early life cycle in mangrove areas. Mangroves are also important habitats or sources for other marine products. For many people, mangroves are a source of cooking fuel, subsistence and income generation (Table 4.13).

In Indonesia, as in many areas of the world, the maintenance of mangroves is threatened, mostly by competing resource uses. The harvesting of mangroves for charcoal as a cooking fuel, their conversion to ponds for aquaculture production, or their infilling for development, industrial or urban, are just a few examples of the competing uses facing coastal managers.

Many competing uses limit the production of mangroves to a single activity; the harvesting of mangroves for charcoal cannot be maintained if the forest is converted to a port. Uses which convert mangroves to other forms of land use such as pond aquaculture, urban expansion or industrial estate development are permanent. There are no options to rehabilitate the area back to a mangrove, with the result that biodiversity is lost, a source of food production and cooking fuel is reduced, shifting and exacerbating the problem in another area, and the elimination of a source of income generation for a group who are already considered the worst off socially and economically in Indonesia.

In the past, decisions to convert mangroves were made without due consideration of the long-term impacts. In the Province of South Sulawesi the area of mangroves has been reduced by 51%, with conversion to pond aquaculture systems the primary reason.

Measures such as maintaining a buffer zone of mangroves between the converted land and open water, selective cutting and encouraging re-planting have been promoted to address the loss of mangroves throughout the country. The implementation of these measures, however, has been variable (Box 4.25) (Ruitenbeek, 1991).

---

and recreational fishing areas, and can affect visual amenity and reduce recreational potential. Access roads and shipping channels to facilities dug through deltas and other sensitive coastal environments can significantly alter ecosystems and sediment balances. The risk of blow-outs and oil spills is a major environmental issue associated with this industry. There are, unfortunately, numerous examples of spills associated with both the production and transportation of oil and gas products. Other issues include the impacts of seismic surveys on marine communities. A longer-term problem of oil and gas production can be the subsidence of land due to the collapse of sub-surface reservoirs (Dolan and Goodell, 1986). In response to these concerns, the oil and gas industry has been active in monitoring various marine and coastal parameters, providing much needed information for managing the coast.

An emerging issue with the oil and gas industry is the decommissioning of offshore facilities as fields reach the end of their production lives. This is becoming a major issue in the North Sea, as shown in Box 2.6.

Box 2.6

Oil rig decommissioning in the North Sea (Gerrard, 1997)

Following the controversy surrounding the disposal of the Brent Spar Oil Platform there has been considerable debate about the future options for managing the oil and gas platforms in the North Sea (International Offshore Oil and Natural Gas Exploration and Production Industry, 1996).

In terms of the characteristics of these platforms the North Sea can be divided into two main areas. In the Northern area the sea is relatively deep and so the platforms are larger. Conversely, in the Southern North Sea the depth is only 30–40 m, hence the platforms are much smaller, pylon-like structures.

Decommissioning or recommissioning?

In the Southern North Sea there are about 150 existing structures. Due to the shallow water and the high frequency of shipping there is no question that these platforms could be disposed of by toppling. The entire structures have to be brought onto land and managed either by decommissioning through recycling into component materials (largely metals and concrete) or, more innovatively, through recommissioning whereby platforms are renovated and reused. There are about 50 sites already licensed for new exploitation and these sites will require new platforms. With new approaches to lifting and towing these smaller platforms, the opportunity has arisen for platforms to be brought onto land in one piece rather than being dismantled offshore and transported in pieces. This, in turn, allows for much greater levels of renovation and reuse and it seems likely that oil companies will take seriously the option of recommissioning. Recommissioning minimizes the waste stream and closes the loop much more effectively than conventional recycling options, which have the potential to generate significant impacts associated with land-based traffic.

continued . . .

The larger platforms in deeper water are a different matter. Here very careful trade-offs have to be made between removing the entire structures and leaving behind a proportion of the platforms below 50 m depth. The arguments for and against are complex and have yet to be played out sufficiently for final decisions to be made. The oil companies prefer to see each platform as an individual entity and thus decide the fate on a case-by-case basis. Others see the not insignificant moral and ethical issues associated with extracting oil and gas from the environment as being of primary concern. From this perspective there is no question that any remnants of the extraction process should be left and the oil companies, who have made significant profits from the exploitation of oil and gas, have a moral obligation to pay for total clean up with the aim of returning the environment to its former state. Several arguments against this stance have been posited, not least the relatively high risks for divers having to work at great depths to remove the structures below 50 m. Experience has shown that decommissioning these larger platforms is not as simple as reverse-commissioning. Removing the structures is an engineering feat in its own right. In addition to the technical complexities of removing the structures, there are environmental impacts of transporting vast quantities of concrete onshore.

The basis for making these decisions, identifying the Best Practicable Environmental Option (BPEO), is a relatively new technique and will take some time to become established. What is clear from the Brent Spar example is that the BPEO is not solely a technical and economic exercise but must account for much wider views from a broader range of stakeholders. Much of the acceptance of the final decisions about exactly how best to manage the platforms will come from the procedural rather than the substantive aspects of the decision-making process. The implications of these findings for risk assessment and risk management are discussed in Chapter 4.

Other resources such as mineral sands, coral and salt are exploited at the coast and can result in major environmental impacts when improperly managed. Again there are conflicting uses when land is used in conjunction with these activities. Waste products from mining operations can enter the system either through runoff or leakage from settling ponds or tailings sites.

In many tropical nations, coral is a cheap source of building and road-making material. Many of the coastal erosion problems in the developing world are due to unmanaged mining of fringing reefs. Mined coral reefs lose their ability to stabilize the coast, since wave energy is no longer dissipated by the reefs but acts directly at the beach edge, causing the re-distribution of vast amounts of sand from reef flat areas to deeper waters (Box 2.7).

Box 2.7

Critical coastal management issues in Sri Lanka

Sri Lanka, like many developing countries, has a range of coastal management issues centered around the mix of subsistence uses of the coast combined with increased industrial and tourist developments (Kahawita, 1993). An estimated 80% of the country's tourist infrastructure is sited on the coast to capitalize on its beaches, marine waters and coral reefs. Poorly planned tourist developments have aggravated pre-existing natural coastal erosion problems, especially on the south coast, which faces the Indian Ocean. The erosion has been found to be very sensitive to sand and coral mining, improperly sited coastal protection structures and loss of coastal vegetation. Other critical coastal management issues in Sri Lanka include (Kahawita, 1993):

- degradation and depletion of natural habitats caused by physical impacts of fishing and tourism on coral reefs, over-exploitation of resources, some land reclamation, pollution, dredging and other causes;
- loss and degradation of historic, cultural and archaeological sites and monuments due to building construction; and
- loss of physical and visual access to the ocean caused by siting of hotels and other facilities impeding access.

These issues prompted the development of a Sri Lankan coastal management initiative described in Boxes 3.7, 5.10 and 5.16 and in Table 4.1.

Conflicting uses can be effectively managed within a planning framework. Planning can be at the strategic level if conflicts apply on a wide geographic scale, or at the site level if issues are local in nature. Chapters 3 and 5 describe these planning approaches.

2.2.2 Infrastructure – transportation, ports, harbours, shoreline protection works and defence

Major infrastructure developments on the coast include:

- ports and harbours;
- support facilities for and operation of various transport systems;
- roads, bridges and causeways; and
- defence installations.

Ports have historically been the link between inland and marine transport. As transportation technology has evolved with larger ships and advanced cargo transfer capabilities (e.g. containers, bulk handling), ports have expanded from the natural sheltered waters of estuaries and inlets to the open ocean, and in some cases new artifical offshore islands (Couper, 1983).

The thousands of ports around the world can be multi-functional or used for a single commodity such as mineral exports or containers. Irrespective of the type, port development results in a number of environmental and social impacts. Generally, port developments involve the manipulation of coastal areas by dredging, land reclamation and clearing of coastal forests. Socially, port development can displace pre-existing coastal inhabitants, limiting areas for subsistence and recreation, and creating increased local traffic (road and rail). In worst case scenarios, port development constrains dwellers from using the area for subsistence and income generation.

Port development can act as a driver for regional economic growth and employment opportunity, mainly for skilled workers (Figure 2.3). Once a port and associated infrastructure is established, port-related industries develop, which in turn enhances trade through the port, fuelling more industrial development and job growth. This feedback mechanism has been one of the most important drivers of coastal urban grown for thousands of years. The benefits of ports must be balanced with natural habitat loss, pollution, changes to visual amenity, increased road and rail traffic, and loss of recreation sites.

Figure 2.3 Container port, Yokohama, Japan.

Maintenance dredging and channelling of ports and harbours, and the dumping of the dredged material which affects water quality, can raise environmental issues and associated pollution concerns such as oil spills, hazardous cargo, and dumping of ballast water. Port traffic can also conflict with recreational boating. An example of environmental issues in the planning and mangement of ports is shown by the Port of Victoria in the Seychelles (Box 2.8).

Transportation within coastal areas consists of domestic and international shipping, and passenger ferry services. Efficient and safe ships combined with state of the art navigation systems have the potential to ensure the industry has minimal environmental impact. Unfortunately collisions and sinking of ships do occur, especially those under 'flags of convenience'. In 1991, 258 ships with greater than 100 gross registered tonnage were lost in a total fleet of 80 030 ships (Jones *et al.*, 1995).

Management of oil spills and other pollution problems associated with transportation is addressed in the MARPOL Convention. For example, in the Great Barrier Reef, which is an Environmentally Sensitive Area in MARPOL, pilotage of international ships (cargo and passenger) is compulsory. Other cases where pilotage is compulsory is in the approaches to ports which are inherently dangerous, or where shipping lanes conflict with other users in the area. Pollution associated with transportation is discussed in section 2.3.1.

The operation of seaplanes, helicopters, hydrofoils, jet foils and other ferry services within coastal areas can be a source of conflicts between users. Some services are visually disruptive and noisy, while others can be hazardous. Environmental concerns regarding the operation of vessels, especially hydrofoils and jet foils, may include disruption to whale and dugong populations, both of which can be a focus for marine tourism. The operation of the foils may also damage fragile benthic communities and sensitive areas such as those used for recreation.

The location of scenic drives, bridges and causeways at the coast can also raise environmental and amenity concerns. Development of road works provides easier access to the coastal area and consequently the natural features or wilderness setting of the area may be diminished. Improved access may also result in demands for amenities in the area and a consequential loss in the area's scenic value. Increased access to coastal areas, especially those in sensitive areas, raises environmental issues such as dune erosion.

Finally, the infrastructure associated with military and defence uses of the coast can be significant. Defence infrastructure on the coast includes ports and harbours, repair yards, surveillance and communications facilities, and training grounds.

Box 2.8

Issues in the Port of Victoria, Seychelles

Victoria, the capital of the Seychelles, is on the island of Mahe and contains the country's two international ports – one for fishing and the other for commercial services. Both ports are well sheltered by reefs and some inner islands, share a common channel and an outer anchorage, and are divided by yacht basins. The physical layout of the port reduces conflicts between the various users.

The fishing port services the industrial tuna vessels, artisanal fishing boats and most inter-island vessels; the commercial port services container vessels, other cargo boats and cruise liners, and a few larger inter-island vessels (Shah, 1995). Shipping traffic and freight handled through the two ports has increased steadily since the 1970s. In line with the Seychelle government plan to establish the country as an international business centre it is anticipated that the ports may need to expand and make provisions for a container terminal.

Like most small island states, the Seychelles is constrained by the availability of land for development, including transport infrastructure. The same applies to Victoria, the country's major port facility, which is partially built on land (approximately 200 ha) reclaimed from nearby reef flats over a 20-year period.

Dredged reef and limestone material were used throughout the reclamation project. In the initial reclamation works, areas of the harbour were dredged and corals and other obstructions removed. Reefs siltation and live coral community dredging resulted in degradation or loss of coral communities. Land runoff may also have contributed to siltation (deGorges, 1990). However, silt screens and filter cloth were used toward the end to reduce siltation and trap suspended solids (Porcher and Millon, 1991).

In addition to using reef flats for landfill for the fishing port, other issues are associated with the two ports. A tuna canning factory is located at the fishing port, where 3000 tonnes of tuna are processed annually for export (Shah, 1995). Wastes from the tuna vessels and the canning factory are disposed directly into port waters. Until recently ship-generated garbage was dumped offshore as well. Garbage is now separated for composting, landfill, recycling and incineration. Small oil spills occur during bunkering operations, but these are dealt with using oil spill management equipment.

The approach taken to address many of these issues is focused on applying the MARPOL 73/78 Convention. How this convention is applied is discussed in Chapter 4 and its use in planning at the site level is discussed in Chapter 5.

2.2.3 Tourism and recreation

International and domestic tourism is recognized as a growth industry, and much of it is focused in the coastal zone. World tourism grew by 260% between 1970 and 1990, with annual growth projections of 2% to 4.5% (Brandon, 1996). Estimates for 1995 indicate that travel and tourism will generate 10.9% of world GDP and employ an estimated 10.6% of the global workforce (World Travel and Tourism Council, in Brandon, 1996).

Many developing nations see tourism as a potential source of foreign revenue, but lack the expertise to plan for a sustainable and well managed industry. Many have embraced tourism, especially on the coast to meet the Northern Hemisphere's demand for tropical destinations close to the coast. The tourism industry in the Red Sea region has, for example, expanded rapidly as European holidaymakers seek an alternative destination to the Mediterranean (see Box 4.13).

Tourism can be an environmentally appropriate industry if managed correctly. There are many examples of where tourism has not been well managed, and not only have the natural resources of the area diminished, but local communities and economies have suffered (Chapter 4). But there are successes in developing sustainable tourism which also benefits local communities. An example of planning for sustainable coastal tourism in Sri Lanka is given in Chapter 5.

Most of the issues associated with tourism development fall into two categories: environmental and social. Environmental issues include the

Figure 2.4 Floating hotel, John Brewer Reef, Great Barrier Reef (credit: GBRMPA).

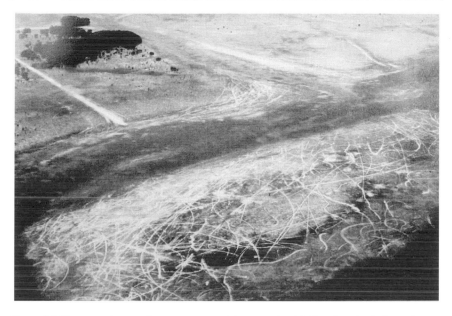

Figure 2.5 Seagrass damage from recreational boating, Florida (credit: Curtis Kruer).

Figure 2.6 Anchor damage, Great Barrier Reef Marine Park (credit: Geoff Kelly).

impacts of developing tourist facilities such as resorts, caravan parks, golf courses, marinas and offshore structures (Figure 2.4). Tourist facilities alter the natural landscape, disturb natural areas and, if they are not properly managed become a source of pollution (Figure 2.5). Throughout the developing world, coastal resorts are often established with little consideration of environmental issues such as sewage disposal. In areas where there are a number of resorts without some form of treatment and poor flushing of systems, sewage can be a public health hazard.

Other environmental impacts of increased use of coastal and marine resources by recreationalists include anchor and mooring damage to benthic communities, overfishing and littering (Figure 2.6).

Social issues related to coastal tourism development and recreational activities include: the displacement of indigenous residents, restricted access to coastal resources for income generation and subsistence, loss of wilderness opportunities, conflicts between users, changes to the area's amenity and possible life style changes.

2.2.4 Conservation reserves and protection of biodiversity

Only a small proportion of the biodiversity of coastal areas is held in parks and reserves which aim to protect flora and fauna. Despite these small percentages current and proposed future parks and reserves have the potential to meet the conservation objectives set out in Agenda 21 (UNCED, 1992). How to capitalize on such reserves is the subject of current research efforts, especially how protected areas can be linked to the conservation values of coastal areas without specific habitat protection. The level of protection of natural coastal systems versus the level of human development and use of such systems is an ongoing debate with any coastal project. Often a coastal development will be required to include a foreshore reserve/buffer zone, the purpose of which is to act as a buffer for physical processes, provide recreation for local residents and meet conservation requirements.

The ability of reserves to meet the multiple-use demands of coastal users and provide for conservation is questionable. Multiple-use plans have been effective for broadly managing large marine areas, but transferring these plans into coastal systems which need more detailed planning has not been well tested (Chapter 5).

2.3 Impacts of human use

As shown in Figure 2.1, a number of problems can result from the coastal uses listed in the previous section. In this section these problems are considered under the headings of pollution (including industrial, sewage and runoff) and coastal hazards (climate change and liability).

2.3.1 Pollution – industrial, sewage and runoff

Major coastal pollution issues are:

- diminished water quality from urban and industrial sources;
- oil pollution, including the risk of oil spills;
- transport of hazardous goods and wastes;
- dumping at sea; and
- ballast water and hull fouling.

Monitoring in coastal areas throughout the world has detected declining water quality, especially in proximity to urban areas. In Indonesia, Tomasick *et al.* (1993) have demonstrated a decline in water quality and consequential loss in reef habitats offshore of the nation's capital, Jakarta. Australia, which is noted for its clean marine and coastal environments, also concedes that water quality around major urban centres has declined over recent years (Zann, 1995).

Changes in water quality can be attributed to several sources: sewage outfall from primary and secondary treatment directly into the oceans or via river systems, storm water drainage, industrial wastes, runoff from pastoral lands and groundwater inputs (Box 2.9). Diminished water quality can lead to a loss of important coastal habitats, such as seagrasses, or an increase in unwanted species such as toxic algae, with a corresponding decrease in fish populations and resultant loss of coastal values for human recreational and amenity values. Concern has also been raised regarding the pumping of sewage from vessels, especially in sheltered embayments and estuaries. The disposal of garbage from ships, cargoes and ferries is a major source of litter washing up on beaches. Siting of landfill sites in close proximity to coastal areas, where leachates can be a source of pollution, exacerbates this problem.

The potential impact of oil spills is a major pollution issue in coastal areas. Emergency oil spill response plans are in place in several countries, and when implemented they can reduce the impact of most spills. Oil pollution also occurs from other sources – shipwrecked vessels, oil exploration, bilge pumping and recreational craft.

In nations where landfill sites are limited or the community is opposed to disposal of particular wastes (e.g. radioactive), the sea is often viewed as an easy and cheap dumping ground. Clearly this is not an acceptable practice except under very strict controls. International agreements such as MARPOL prohibit dumping at sea, and many nations have also enacted national legislation banning disposal at sea. Disposal of toxic substances such as radioactive wastes carries considerable risks since our knowledge of the long-term storage of such materials in marine environments is very limited. Disposal of landfill waste or dangerous

Box 2.9

Pollution of urban coastal waters – the case of Jakarta, Indonesia

Pollution of nearshore waters adjacent to coastal cities has long been a problem. Since the 1960s, when critical pollution levels were reached in the developed world, a number of concerted efforts have been made to improve urban coastal water quality and to remediate polluted bottom sediments.

Like many capital cities in developing countries, Jakarta has experienced rapid population and industrial growth over the last 50 years. This growth, however, has been at a cost to the coastal environment of Jakarta Bay and adjacent coral reefs (Kepulauan Seribu), primarily from pollution.

A number of studies have demonstrated that human impacts have severely degraded coral reefs in the Jakarta Bay area (Tomasick *et al.*, 1993; Harger, 1986; Moll and Suharsono, 1986). The studies have shown:

- water transparency increases with increasing distance from the Jakarta Bay, which also corresponds with the maximum depths where corals are found;
- low water transparency reduces the maximum depth at which coral communities can survive;
- algal blooms are spreading further offshore (in 1986 blooms were only reported within 2 km of Jakarta's port, whereas in 1991 blooms were reported 12 km offshore); and
- a decline in fish landings from the muro-ami reef fishery.

Causes of these impacts include the lack of sewage facilities throughout most of Jakarta and surrounding urban areas, where a series of canals above and below the ground carry raw sewage to the Bay. A city of at least 9.5 million without a sewage treatment system is clearly a significant source of nutrient input into the Bay. Existing waste disposal facilities, where much of the waste ends up as coastal landfill or in the rivers emptying into the Bay, are inadequate for the city. Port activities including dredging and dry-docking have also contributed to the decline in water quality.

The impact of adjacent land use has been analysed by Tomasick *et al.* (1993). They found that nutrient runoff from land contributed to coral growth, but that wastes from industrial, agricultural and urban land uses impacted detrimentally on corals. Until recently, coral reefs were a source of building material and road construction in Jakarta. Coral extraction from shallow reef flats in 1982 totalled 840 000 m^3 (Tomasick *et al.*, 1993) and continues today.

The price paid by the environment for the rapid development of Jakarta

continued . . .

Bay and Kepulauan Seribu is typical of many coastal areas throughout the world. Considerable resources will be needed to reduce these impacts, let alone rehabilitate areas. Impact mitigation measures required include Environmental Impact Assessment and Strategic Environmental Assessment, which are discussed in Chapter 4, while Chapter 5 highlights how integrated coastal planning at the local and regional level can also contribute.

wastes is prohibited in many nations; however, enforcement of regulations is difficult. The difficulty of waste disposal in coastal environments is illustrated by the example of the coral atoll of Cocos Island (Box 2.10, Figure 2.7).

Box 2.10

Waste disposal on a coral atoll – Cocos Island

The disposal of waste generated by residents of coral atolls is often a major problem. Disposal options are limited, and with an increased use of consumer goods and packaged foods in island nations the problem is likely to remain in the foreseeable future. Disposal options, and their problems include:

Waste disposal options and problems on coral atolls

Disposal option	Problem
Deep landfill	Shallow freshwater lenses would be readily polluted
Shallow landfill	Potential to contaminate groundwater; unsightly; odour problems
Dumping at sea	Pollution risks; ecosystem damage
Removal to mainlands	Expensive; can just shift the problem elsewhere
Recycling	Not practical for all wastes

The waste management system adopted on Cocos Island (Australia) is to use a combined strategy of waste minimization, recycling and dumping at sea. Larger items such as cars and stoves are cleaned of toxic substances, which are transported to the Australian mainland, and then the remainder dumped at sea with the authority of a permit granted by the Australian Federal Government.

Figure 2.7 Coastal landfill, Cocos Island.

The introduction of exotic species through the pumping of ballast water in port is a major environmental issue since in many places these exotic species have virtually destroyed the native fauna, reduced the bio-diversity and altered the port's ecosystems (and subsequently adjacent marine ecosystems). Eradication of introduced pests is impossible and in many cases it is difficult even to control populations. Various countries have (mainly voluntary) guidelines which require mid-ocean exchange of water, where there is a greater chance that conditions will not favour survival of the exotic species, and taking relatively clean water on board for disposal close to port. Similar voluntary guidelines exist for the management of the impacts of toxic anti-fouling paints used on the underside of vessels.

2.3.2 Coastal hazards and climate change

The coast is highly dynamic and subject to natural forces which have the potential to damage property and threaten public safety. For those living on the coast, cyclones, storm surges and tsunami hazards are inherent and damaging natural events (Box 2.11). Hazards like these are difficult to manage and pose liability problems to managing agencies. The question

---

Box 2.11

Bangladesh cyclone hazards

The Bangladesh coastal zone could be termed a geographical 'death trap' due to its extreme vulnerability to cyclones and storm surges. The massive loss of life from cyclones is due to the large number of coastal people living in poverty within poorly constructed houses, the inadequate number of cyclone shelters, the poor cyclone forecasting and warning systems, and the extremely low-lying land of the coastal zone. Approximately 5.2 million people live within coastal areas of high risk from cyclone and storm flooding within an area of 9,000 km^2.

(Kausher *et al.*, 1996)

Nearly one million people have been killed in Bangladesh by cyclones since 1820 (Talukder *et al.*, 1992) due to there being an estimated 10% of the world's cyclones developing in the Indian Ocean (Gray, 1968) – an average of just under two (1.77) cyclones occurring each year (Talukder *et al.*, 1992). Once the cyclones have been formed they generally move in a direction between north-west to north-east and can cross the coast in either Burma, Bangladesh or India.

The last devastating cyclone to hit Bangladesh occurred on 29 April 1991. An estimated 131 000–139 000 people died, with the majority of those dying being below the age of 10, and a third of them below the age of five; also more women than men died (Talukder and Ahmad, 1992). An estimated 1 million homes were completely destroyed, and a further 1 million damaged. Up to 60% of cattle and 80% of poultry stocks were destroyed and up to 280 000 acres of standing crops destroyed; 470 km of flood embankments were destroyed or badly damaged, exposing 72 000 ha of rice paddy to salt-water intrusion. Coastal industries and salt and shrimp fields were also badly damaged. The flood waters brought disease and hunger to the survivors. The total economic impact of the cyclone was US$2.4–4.0 billion (Kausher *et al.*, 1996).

How the Government of Bangladesh is attempting to plan for the impacts of future cyclones within their coastal management programme is described in Chapter 4.

---

managers need to discuss with the community is: who pays to manage these natural events? In developing nations, often there is no compensation for coastal dwellers who lose their property from cyclones and similar events. Depending on the nature of the event, in some developed countries compensation is provided. In these cases the question arises as to whether the wider community should fund those who choose to live close to the coast and therefore risk damage. The answer to this question is not easy to formulate since it will depend on the social, economic and political culture of each country.

Other coastal hazards can either be permanent, such as cliffs and headlands (Figure 2.8), or intermittent, such as rip currents on sandy ocean-beaches. In either case, they pose serious risks to public safety. Public liability needs careful consideration when access to hazardous areas is provided by managing agencies, and when rescue aids are provided.

This raises the question of liability, indemnity and compensation. Liability in the event of accidents or damage to property is a complex question. In many countries the agency which has vested control over the area is responsible for public safety and protection of property. Similarly, for major developments in the coastal zone, it is still unclear who is responsible in the event of a natural disaster or climate change.

Governments throughout the world are dealing with an increase in the number of litigation cases in the courts. General answers to questions of liability for specific aspects of coastal developments, planning and management are not possible. Often advice from the legal profession is sought for situation specific problems.

As shown above, planners and decision makers face many hazards in coastal areas in the here and now. On the horizon, though, is the serious possibility that, over coming decades, the scale of the threats faced on the interface between sea and land may escalate as a result of global environmental change.

Over recent decades, a firm scientific consensus has emerged that pollution of the atmosphere by greenhouse gases such as carbon dioxide and methane may bring about a significant change in the earth's climate – global warming – which could have widespread consequences (Houghton, et al., 1996). Coastal areas may face primary impacts as a result of, for example, a change in the risk of storm impacts, changes in ocean temperatures or rising sea level alongside secondary effects as regional changes in climate influence economic performance and other aspects of human well-being (Watson et al., 1996).

The problem of global warming highlights the difficulties that coastal planners face in getting to grips with the broader issue of sustainability. How can we 'climate-proof' coastal management? One of the major difficulties lies in the uncertainty of the climate predictions. Although the threat posed by the changing composition of the atmosphere is clear, understanding of the problem is not sufficient to provide the kind of definite forecasts that are needed if effective adaptive strategies are to be developed.

To give one example, there is concern that tropical cyclone frequencies may rise as the oceans warm and the sea surface temperature conditions that favour storm development occur over a larger area. But ocean

Figure 2.8 Erosion of Dunwich, United Kingdom 1886–1919 (credit: Dunwich Museum).

(a)

(b)

(c)

temperature is just one factor affecting cyclone occurrence. How will the winds which steer the storm towards a particular stretch of coastline alter as climate changes? This question cannot be answered with any certainty due to limited understanding of the mechanics of climate change, leaving those responsible for management at the level of an individual stretch of coast at a loss as to whether to plan for an increase in cyclone risk, no change, or even a decrease.

It is for this reason that the climate community has recommended that a 'precautionary' approach be taken to the global warming problem at this time (Chapter 1). This means that measures should be adopted which cost little or nothing or which result in immediate benefits above and beyond minimizing the impact of long-term climate change.

To pursue the tropical cyclone example, Tri *et al.* (1996) have demonstrated through benefit–cost analysis that rehabilitating mangrove in northern Vietnam represents a sensible precautionary response to the threat of global warming as it is 'win–win' strategy, providing additional storm protection, reducing dyke maintenance costs over time and, managed sustainably, providing an immediate boost to local incomes through the provision of extractable resources such as fish, crabs, fuelwood, honey and so on (Chapter 4).

The broader lesson here is that many measures which might be taken to protect the coastline against long-term climate impacts are precisely those which should be adopted on the basis of more immediate priorities. At this precautionary stage, there need be no contradiction between present-day goals and the longer-term aim of 'climate-proofing' management plans (Kelly *et al.*, 1994).

If the climate threat does prove as serious as some projections suggest then there will be a point at which coastal management has to move from precautionary action on this issue to a more concerted response. If this transition is to be handled effectively, it is important that management plans made today contain the degree of flexibility necessary if they are to be modified at a later date (Chapter 5). Options should be kept open where possible. More generally, it must be recognized that, alongside the dynamic of demographic change, social evolution and economic development, the dynamic of long-term environmental change must always be borne in mind.

As far as planning for sea level rise is concerned, the Intergovernmental Panel on Climate Change – the scientific body charged with advising the international community on the climate issue – has identifed a set of response options (IPCC, 1990, 1992):

- protection;
- accommodation; and
- retreat.

It is at this stage, beyond the point of precautionary action, that the most difficult challenges will be faced by coastal planners and that compromise and trade-off between short-term goals and long-term objectives may come to the fore. This issue is explored throughout the following chapters.

2.4 Administrative issues

As this chapter has shown, there are many complex and overlapping problems along the world's coastlines. This complexity, linked with government administrative systems that are designed to address issues on a subject-by-subject basis, can create problems in the effective management of the coast. The implications of these administrative issues for the design of coastal planning and management programmes are described in Chapter 3.

2.5 Summary – coastal conflict

Today's coastal managers face a plethora of problems, challenges and demands, many of which were unheard of only a few decades ago. As coastal populations grow in both developed and developing countries, the scale and intensity of coastal issues is also likely to increase.

Two important conclusions can be drawn from this chapter. The first is that most, if not all, coastal management problems centre on the issue of conflict. Obvious examples include the conflict between the conservation of mangrove areas and their conversion to shrimp ponds. A less obvious example is where land-based activities bring about a decline in water quality, creating a conflict with natural ecosystem values. Viewing the majority of coastal issues as conflicts is useful in that mechanisms for their management become in effect strategies for conflict resolution. This conclusion forms a useful basis for describing and analysing the development of coastal planning and management practices, the subject of the next chapter.

The second, and perhaps the key conclusion, is simply that coastal issues are now recognized as problems for which solutions must be sought. Having crossed this threshold, the principal issue is now not what the problems are, but how they should be tackled. This orientation towards management action requires clear guidance, a well organized government structure, and – most importantly – a well defined set of objectives and actions; all of which are introduced and analysed in the following chapter.

Chapter 3

Concepts of coastal planning and management

Development of specific coastal planning and management initiatives is a common response by government to the many issues discussed in the previous chapter. These issues will only be effectively resolved if managers are guided in their decision making and can plan to avoid future problems by taking a proactive approach. This chapter provides a conceptual framework for decision making and a common understanding of terms and definitions. Tools for tackling individual problems are discussed in Chapter 4 and coastal planning approaches are analysed in Chapter 5.

The chapter has five main sections. First, the most important terms and guiding statements for coastal management and planning are outlined. Second, the development and application of overreaching concepts are discussed, with examples of how they have been interpreted and implemented by governments. Third, coastal planning concepts are described and analysed. Fourth, choices in the design of administrative arrangements to implement coastal management and planning programmes are discussed. Finally, the monitoring and evaluation of coastal programmes are described and analysed.

3.1 Terminology

> One of the difficulties of writing about a process of management is that many of the words which form the vocabulary of management are hopelessly overworked. Words of common usage have been taken and given a specific meaning by different authors: unfortunately they have not all been given the same interpretation. The result is a problem of semantics, which can act as a barrier to a common understanding.
>
> (Hussey, 1991, p. 38)

A review of the words used by coastal managers and planners reveals that the same terms are frequently given different meanings. In most cases it is clear what is intended by their use, but it nevertheless makes comparison of coastal programmes from different parts of the world difficult. Three areas of terminology used in coastal management and planning are discussed in turn below, and standardized terminologies are developed for use in later sections. These three groups of terms focus on the difference between coastal planning and coastal management; the meaning of integration; and statements which provide guidance to coastal programmes.

3.1.1 What is coastal planning, what is coastal management and what is the difference?

As with many widely used words, 'planning' and 'management' can have various meanings depending on the context in which they are used. Here we briefly discuss their various interpretations and subsequently define the terms 'coastal planning' and 'coastal management' as they will be used in this book.

Everyone, every day, undertakes some form of planning. Deciding what to eat for lunch, or what time to go fishing, requires planning. So 'planning' is usually taken in everyday language to mean the process of charting future activities. To 'have a plan' is to be in possession of a way of proceeding. In this context planning has two components: first, the determination of aims for what is to be achieved in the future; and second, clarifying the steps required to achieve these aims. These two components may be viewed as common to all plans and planning exercises. However, different types of plans and planning initiatives may interpret these two components in contrasting ways.

There are perhaps as many types of plans as there are planners attempting to classify them. Businesses produce business plans, management plans, corporate strategies and so on. Some governments have a Department of Planning which, as the name suggests, has as one of its core activities the production and administration of formalised systems of planning – usually land-use planning and/or economic planning. However, despite the large number of plans and different approaches to planning, the vast majority of plans and planning initiatives can be characterized as either strategic or operational. Those that do not readily fall into either of these categories generally combine both strategic and operational components (Hussey, 1991).

Strategic planning is the highest order of planning; it attempts to provide a context within which more detailed plans are designed to set and achieve specific objectives. Strategic planning sets broad objectives and outlines the approaches required to achieve them; it does not attempt to give detailed objectives, or to give a step-by-step description of all actions required to achieve the objectives.

There are two main types of strategic planning initiatives relevant to the management of the coast: geographic focused (integrated area plans); and sector-based strategies (focusing on one subject area or the activities of one government agency). Each of these types of strategic planning is described in Chapter 5.

In contrast to strategic planning, operational planning sets the directions and steps to achieve on-ground management actions. As the name suggests, operational planning dictates localized operations – such as the rehabilitation of a mangrove area, or the building of walkways through dunes. They have to detail exactly where, and how, operations will be carried out. Contents of typical operational plans include details such as site designs, costings and schedules of works.

'Manage', like planning, also has a number of meanings. It can mean the ability to handle a situation (as in 'yes, I can manage'), or it can indicate control or the wielding of power. Managers in business circles are people who are in control of the organization.

Thus 'coastal management' could be interpreted to mean directing the day-to-day activities occurring on coastal lands and waters, or it could be used to mean the overall control of the government agencies (organizations) that oversee these day-to-day activities. Both of these interpretations appear to be valid. As is the case with planning, management can be divided into strategic and operational management, the former being the processes of being in control of an organization's affairs with respect to the coast, the latter being the activities of controlling on-the-ground actions.

In this chapter the terms coastal planning and coastal management are taken to be inclusive of both strategic and operational components. This is partly for ease of use, and partly because the overall concepts of coastal planning and management described later in the chapter apply to both strategic and operational processes. Also, most of the literature describing the conceptual framework for coastal management and planning does not distinguish between operational and strategic planning or management, from which we may infer that the authors included both in their analyses. Where either operational or strategic planning and/or management is being explicitly described, the relevant prefix is used; the implications of the use of the terms are explained more fully in Chapter 5, where the division of both planning and management into strategic and operational components provides a very useful framework for the analysis of different styles of coastal management plans.

3.1.2 Placing an emphasis on 'integration'

Many governments and international organizations choose to include the word 'integrated' as a prefix to describe their efforts in bringing together the various parts of their coastal planning and management initiatives into a single unified system. Others choose to use 'coordinated' or similar words, while yet others opt for no specific word to describe such efforts. Hence the description of many of the world's coastal management initiatives as 'integrated coastal management'. Use of 'integrated' in this way has been popular for many years, but has expanded greatly since its adoption in Agenda 21, where the introduction to the chapter on ocean and coastal management describes the need for new approaches to marine and coastal area management and development which 'are integrated in content' (UNCED, 1992)

Interpretation of the word 'integrated' (Box 3.1) can have a bearing on whether governments choose to attach it to their programme descriptions. For example, in much of the Pacific and south-east Asia the use of 'integrated' has become widespread because many have found that it conveys an appropriate policy goal, is culturally and administratively appropriate and is widely understood. In contrast, Australian governments have chosen not to use it because of the inference that it could be interpreted to mean the amalgamation of different levels of government – an extremely sensitive political issue in that country. This sensitivity is reflected in the difference between integration and coordination as defined by Kenchington and Crawford (1993, p. 112):

> an integrated system is complete or unified although it will generally have subordinate components. A coordinated system involves independent, generally equivalent components working to a common purpose.

Another way at looking at the use of integrated, coordinated and other descriptors of coastal management programmes is outlined by Cicin-Sain (1993) who has set up a continuum of terminology describing the degree to which coastal programmes bring together disparate elements (Box 3.2).

There are clear similarities between the various approaches adopted by Cicin-Sain (1993), Kenchington and Crawford (1993) and Scura (1994) to the use of integration and other words implying bringing together. All approaches stress the amalgamation of disparate elements into a single coastal management system. The various words to describe this amalgamation concentrate on its degree and to a certain extent the mechanisms by which it is achieved. Finding ways to achieve this amalgamation is a key theme of this book, and hence will be visited many times in the following Chapters. However, the above discussion shows that the term integration has been used in such a variety of contexts that its strict meaning has become confused. So, to avert confusion, we deliberately avoid attaching any prefixes to the term coastal management unless

---

Box 3.1

The meaning of 'integration' in coastal management

An interesting discussion and definition of 'integrated management' is provided by Scura (1994) in her work for the United Nations Development Programme on integrated fisheries management. Her discussion has wide application to overall coastal management.

The term integration is used differently by various disciplines. For example, at the micro production level, integration can focus on production technologies such as byproduct recycling and improved space utilisation. Integrated farming also uses the term in a predominantly technical sense, where the focus is on the use of an output or byproduct from one process as an input into another process. In a more macro sense, an integrated economy is one which is organised or structured so that constituent units function cooperatively. In a sociological or cultural sense, integration pertains to a group or society whose members interact on the basis of commonly held norms or values.

A broad interdisciplinary definition of integration is adopted here, which incorporates several disciplinary and sectoral concepts. Integrated management refers to management of sectoral components as parts of a functional whole with explicit recognition that human behaviour, not physical stocks of natural resources such as fish, land or water, is typically the focus of management. The purpose of integrated management is to allow multisectoral development to progress with the least unintended setbacks.

---

quoting original sources. The terms 'coordinated coastal management' or 'integrated coastal management' will therefore only be used when referring to its use by other authors, or in Chapter 5 to described the integrated style of coastal management plans.

3.1.3 Guiding statements for coastal management and planning

Fundamental to the success of coastal programmes is the use of statements which clearly enunciate the purpose, directions and expected outcomes of the programme. Well planned coastal programmes therefore carefully consider such guiding statements so that stakeholders know exactly what ends they are working towards. Various terms are used to describe these direction setting statements – such as mission, vision, goals, principles, objectives, targets, expected outcomes and actions.

Box 3.2

A continuum of integration in coastal management (from Cicin-Sain, 1993)

less integrated ———— more integrated ————→

fragmented   communication   coordination   harmonization   integration
approach

Continuum of policy integration in coastal management (from Cicin-Sain, 1993).

The five locations on the continuum shown in the figure are described by Cicin-Sain as:

1. fragmented approach – presence of independent units with little communication between them;
2. communication – there is a forum for periodic communication/meeting among the independent units;
3. coordination – independent units take some actions to synchronize their work;
4. harmonization – independent units take some actions to synchronize their work, guided by a set of explicit policy goals and directions, generally set at a higher level; and
5. integration – there are more formal mechanisms to synchronize the work of various units which lose at least part of their independence as they must respond to explicit policy goals and directions (this often involves institutional reorganization).

The choice of guiding statements depends on the particular coastal issues being considered, political imperatives and management scale. The choice will also be influenced by local languages and cultural settings: some English words are more readily translated or locally understood. However, being clear about the purpose to which these phrases are to be put is more important than what they are to be called. Whether the overall direction of a coastal programme is articulated by a mission statement, vision statement or goal will matter little as long the purpose of using such a statement is clear. As will be shown in Chapter 5, the processes by which these statements are derived is also important. A major exception to this is if guiding statements are to be used in legislation or other formal documents, where there may be tight legal requirements for the use of particular words to describe direction-setting statements, and reasons why others should not be used.

Despite differences around the world in the use of particular terms, there is general agreement that planning and management should use a

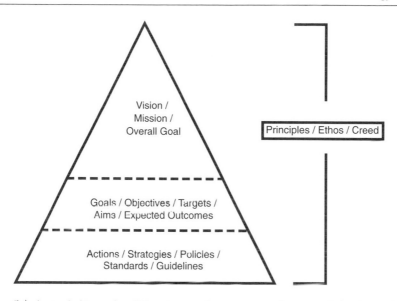

Figure 3.1 A simple hierarchy of direction-setting statements for coastal planning and management.

hierarchy of direction-setting statements. A simplified version of such a hierarchy is shown in Figure 3.1.

Overarching a hierarchy of direction setting statements are general expressions which describe the philosophy behind the direction of the coastal programme. These are expressions of the philosophical background which provides the basis to the implementation of a coastal programme (Figure 3.1). In some cases these are statements of moral or ethical issues, which in the business planning world are often called statements of ethos or creed. However, for coastal programmes they are most often called statements of principle. While statements of principle often provide the philosophical climate for the development of a well defined hierarchy of guiding statements, they are generally not strictly part of that hierarchy. Nevertheless, statements of principle are often a critical part of the family of guiding statements.

At the top of the hierarchy is a statement which describes the overall direction, or purpose, and which will guide all subsequent actions. Such a statement can be given various names, including vision, mission, or overall goal.

The choice of words will depend on the particular interpretations attached to them by the programme initiators. For example, the word vision implies deliberate foresight, and some element of inspiration. A government may deliberately use 'vision' to imply that they have such attributes. The use of 'overall goal' suggests that there is some overall target which can be met. Likewise, a 'mission' suggests that there is a well defined campaign ahead in order to develop and implement a coastal programme.

The next, and probably the most important, set of guiding statements are those which describe exactly what a coastal programme is trying to achieve. Such statements are most commonly referred to as goals, objectives, targets, or expected outcomes. The critical issue in formulating these statements is the degree to which they are measurable, or specific as to time. For example, there is a distinct difference between describing an objective for the improvement of coastal marine water quality as 'safe for swimming', and defining specific targets such as 'ensuring the level arsenic in sea water is less than 50 ($\mu g \, l^{-1}$)' (see Box 3.14). The latter objective is clearly something that can be measured, while the former would require additional performance standards to determine whether it has been met. The advantages and disadvantages of different types of goals, objectives, targets or expected outcomes are discussed further in section 3.4.3c.

At the lowest level of the hierarchy of coastal programme statements are Action Statements. These translate the overall directions set higher in the planning hierarchy into tangible on-the-ground or on-the-water activities, and are designed to meet the goals, objectives, targets or expected outcomes that achieve the mission, vision or overall goal. Where possible, action statements should be designed to meet specific goals, objectives, targets or expected outcomes. This has the major advantage of clearly showing how the threads of a coastal programme will be pulled together by following, for example, the mission statement through to an objective and then through to a set of actions designed to meet both the objective, and subsequently the mission. Examples of how these linkages are achieved in coastal programmes are discussed in section 3.4.2.

The above description of the hierarchy of guiding concepts for coastal management and planning assumes a single organizational tier: a single organizational unit which can develop and implement a set of guiding statements for a coastal programme. A single organizational tier is analogous to a self-contained business developing a business plan in which it can write various statements of mission, objectives, etc. and then implement these through its own business practices. However, this self-contained business environment is not usually the case for governments managing the coast, where a single tier of government solely responsible for coast management is unusual. There may be constraints placed on, for example, a local level of government by higher government levels.

Coastal management goals and objectives may be written into national legislation, in which case local government has a limited ability to develop its own guiding statements. A national hierarchy of guiding statements may therefore include an interaction of guiding statements of different levels of government. Three such 'sub-hierarchies' may be required within a federal system of government (with national, state and local governments) in order to develop truly national guiding statements.

The concept of sub-hierarchies can also be applied within a single level of government, where the various agencies may have their own guiding statements, such as performance criteria for the discharge of their specific coastal management and planning responsibilities.

Coastal programmes around the world use different combinations of the guiding statements in each level of the hierarchy illustrated in Figure 3.1. There is no universal set of guiding statements; however, to simplify the use of language throughout this book the following standard set of terms will be used: overall goal, objectives; and actions, guided by statements of principle.

How the above terminology is applied to actual coastal programmes is described in section 3.4.2.

3.1.4 Summary of terminology

The previous sections have shown that different terminology is used in the day-to-day practice of coastal planning and management around the world. While this is to be expected as the coastal initiatives of different cultures and language groups are translated into English, decisions have to be made about whether to standardize the use of language for the purposes of analysis in this book. For simplicity, our decision is to use the shortest and most flexible terminology – and use 'coastal planning' and 'coastal management'. We do not use the prefix 'integrated' to describe the bringing together of participants, initiatives and government sectors. Nor do we insert 'zone' or 'area' to define that a broad geographic area is the focus of attention in coastal planning and management, and not the immediate boundary between land and sea. We take the pragmatic view that the use of area', 'integrated', 'coordinated', 'zone', etc. will be made when it is useful to do so within the social, cultural and political circumstances of a coastal nation. In other words, we strongly advocate using terminology as a means to an end – a particular set of words should be used if this is the optimum means of ensuring the sustainable development of a particular section of coast.

3.2 Concepts of coastal management

While coastal management practitioners have fashioned a set of concepts which guide their actions, this cannot be construed to be a rigorous theoretical framework in the sense that, for example, a pure scientific discipline is governed by physical laws. The broadly accepted concepts of coastal management described below are a combination of the general theory and practice of resource management as applied to the coast, mixed with pragmatism. This mix provides a set of coastal management concepts which describe a set of practices which help achieve desired management outcomes.

The broad concept of coastal management, as distinct from simply managing activities at the coast, encompasses the management of everything and everyone on the coast within some form of unified system or approach. So what makes the practice of coastal management distinct from other forms of resource management or planning?

First, and perhaps most importantly, coastal management focuses on the management of a distinct geographic area – the coast. As described in Chapter 1 this focus led many to define a 'coastal zone' or 'coastal area' within which specific coastal policies or procedures apply. These coastal areas can be defined through legislation, policy and planning documents, as shown in Appendix A, and usually contain both areas of nearshore waters and land close to the immediate land/ocean boundary. The issue is not the extent of the coastal area involved, but that specific management initiatives are undertaken which focus on a defined region – the coast. This distinguishes coastal management initiatives from other government programmes, such as forestry and fisheries management, the provision of education and health care, for example, which are not targeted to the coast.

As previous chapters have shown, the coast has many unique attributes, the most important (and obvious) of these being the dynamic interaction of land and ocean. However, in terms of the overall concepts of coastal management, defining a geographic area – the coast – and then applying special coastal management tools is analogous to the management of other parts of the world which can also be separated geographically from one another. Examples include the management of mountain ranges, or areas of significant groundwater resources, both of which can be mapped and which require sensitive and distinctive management arrangements. Perhaps the closest analogy to coastal management is river catchment management: catchment and coastal management are both concerned with the integrated management of land and water resources.

The point we want to emphasize here is that coastal management *per se* is not unique. There are management approaches and techniques for other environmental systems which bear close resemblance to the coastal planning and management tools and approaches described in this book. Hence, coastal management is concerned with the application of techniques which attempt to clearly focus the efforts of governments, private industry and the broader community onto coastal areas. These techniques centre around ways to bring together disparate planning and management techniques on the coast, to form holistic and flexible coastal management systems.

Thus, it is the combination of developing adaptive, integrated, environmental, economic and social management systems which focus on coastal areas which are the core coastal management concepts.

In recent years a number of governments and international organizations have developed guidelines on their perceptions of what are appropriate concepts of coastal management. These include guidelines produced by different parts of the United Nations (UN Department of International Economic and Social Affairs, 1982; UNEP, 1995; IWICM, 1996), the Organization for Economic Co-operation and Development (OECD) (OECD, 1992, 1993), the International Union for the Conservation of Nature (IUCN) (Pernetta and Elder, 1993) and the United States Agency for International Development (1996). These documents are generally structured to begin with the philosophy underlying the coastal programme, followed by a list of guiding statements, issues to be addressed, and steps to be taken to tackle these issues. Recent examples of such documents are those of the World Bank (World Bank, 1993; Post and Lundin, 1996) and the United States Agency for International Development (1996). Together they provide a good summary of the present thinking on the concepts guiding coastal management (Box 3.3).

There is a range of techniques used by coastal nations to assist with incorporating the various coastal management concepts listed in Box 3.3 within their decision-making systems. These are described in section 3.4.1, in which the importance of coastal planning as a mechanism to achieve flexibility is analysed, as is the importance of 'learning' approaches to make certain that coastal programmes are dynamic and evolutionary, ensuring that complex and/or emerging coastal issues are addressed.

Integration here is used as outlined in section 3.1.2 – that is, the bringing together of different, often disparate elements into some overall unified coastal management system (Box 3.2).

Cicin-Sain (1993), building on the work of Underdahl (1980), has undertaken a useful analysis of the meaning of integration as it applies to coastal management. Underdahl's work concentrates on 'integrated policy' in the sense that 'constituent elements are brought together and made subject to a single unifying conception' (Cicin-Sain, 1993, p. 23).

According to Underdahl and Cicin-Sain, a coastal management approach 'qualifies' as integrated when it satisfies three criteria: the attainment of comprehensiveness, aggregation and consistency (Table 3.1). If these criteria are satisfied, 'integrated policy' (Underdahl, 1980) must:

1. recognize its consequences as decision premises;
2. aggregate them into an overall evaluation; and
3. penetrate all policy levels and all government agencies involved in its execution.

Box 3.3

Two recent examples of generalized concepts of coastal management (adapted from World Bank, 1993; Post and Lundin, 1996; United States Agency for International Development, 1996)

World Bank

Currently accepted principles and characteristics associated with the Integrated Coastal Zone Management (ICZM) Concept are that ICZM:

- focuses on three operational objectives:

    - strengthening sectoral management, for instance through training, legislation, staffing;
    - preserving and protecting the productivity and biological diversity of coastal ecosystems, mainly through prevention of habitat destruction, pollution and overexploitation; and
    - promoting rational development and sustainable utilisation of coastal resources.

- moves beyond traditional approaches which tend to be sectorally oriented (each dealing with a single factor) and fragmented in character and seeks to manage the coastal zone as a whole using an ecosystem approach where possible;
- is an analytical process which advises governments on priorities, trade-offs, problems and solutions;
- is a dynamic and continuous process of administering the use, development and protection of the coastal zone and its resources towards democratically agreed objectives;
- employs a holistic, systems perspective which recognizes the inter-connections between coastal systems and uses;
- maintains a balance between protection of valuable ecosystems and development of coast-dependent economies (it sets priorities for uses, taking account of the need to minimize the impact on the environment, to mitigate and restore if necessary, and to seek the most appropriate citing of facilities; these are the activities contained in Environmental Impact Assessment);
- operates within established geographic limits, as defined by governing bodies, that usually include all coastal resources (it seeks the input of all important stakeholders to establish policies for the equitable allocation of space and resources in the coastal zone; an appropriate governance structure is essential for such decision-making and oversight);
- is an evolutionary process, often requiring iterative solutions to complex economic, social, environmental, legal and regulatory issues (the main function is integration of sectoral and environmental needs; it should be

continued . . .

---

implemented through specific legal and institutional arrangements at appropriate levels of the government and the community);

- provides a mechanism to reduce or resolve conflicts which may occur at various levels of the government, involving resource allocation or use of specific sites, and in the approval of permits and licenses;
- promotes awareness at all levels of government and community about the concepts of sustainable development and the significance of environmental protection; is proactive (incorporating a development planning element) rather than reactive (i.e. waiting for development proposals before taking action);
- also embraces certain general principles in the course of developing the programme by a given nation. Note that most of the principles listed here are among the recommendations contained in UNCED's Agenda 21 action program. These include:
  - the precautionary principle;
  - the polluter pays principle;
  - use of proper resource accounting;
  - the principle of trans-boundary responsibility; and
  - the principle of intergenerational equity.

United States Agency for International Development (1996)

USAID has identified Integrated Coastal Management strategies which have proven to be successful and can be adapted to the unique qualities of different nations and sites.

1. Recognize that coastal management is essentially an effort in governance. Coastal programmes follow a policy process where the challenge lies in developing, implementing and adapting sustainable solutions to resource use problems and conflicts.
2. Work at both the national and local levels, with strong linkages between levels.
3. Build programmes around issues that have been identified through a participatory process.
4. Build constituencies that support effective coastal management through public information/awareness programmes.
5. Develop an open, participatory and democratic process, involving all stakeholders in planning and implementation.
6. Utilize the best available information for planning and decision making. Good Integrated Coastal Management programmes understand and address the management implications of scientific knowledge.
7. Commit to building national capacity through short- and long-term training, learning-by-doing and cultivating host country colleagues who can forge long-term partnerships based on shared values.

continued . . .

8. Complete the loop between planning and implementation as quickly and frequently as possible, using small projects that demonstrate the effectiveness of innovative policies.
9. Recognize that programmes undergo cycles of development, implementation and refinement, building on prior successes and adapting and expanding to address new or more complex issues.
10. Set specific targets, and monitor and self-evaluate performance.

These three criteria are discussed later in this chapter, especially as they relate to the organization of governments to assist in integrated decision making at the coast.

In the context of coastal management, Cicin-Sain (1993) interpreted Underdahl's dimensions of policy integration (Table 3.1), stressing that several groups of issues were important (Cicin-Sain, 1993, p.25):

1. Integration among sectors

   - among coastal/marine sectors (e.g. oil and gas development, fisheries, coastal tourism, marine mammal protection, port development);
   - between coastal/marine sectors and other land-based sectors such as agriculture.

2. Integration between the land and the water sides of the coastal zone.
3. Integration among levels of government (national, subnational, local).
4. Integration between nations.
5. Integration among disciplines (such as the natural sciences, social sciences, and engineering).

A further concept in coastal management is the clear articulation of the overall philosophy of a coastal programme. This philosophy, often called guiding principles, ethos or creed, underpins the entire basis of coastal programmes. In the 1970s and 1980s the concept of 'balance' was the dominant philosophy underpinning coastal management programmes. Balance in coastal management programmes attempts to weigh up, and reconcile, opposing or conflicting forces. Most often these opposing forces are those of conservation and development (Figure 3.2). For example, although the US Coastal Zone Management Act (1972–1990) does not make specific reference to the concept of balance, this is widely seen as the CZM Act's intention (Keeley, 1994). Indeed, the CZMA was seen as striking the middle ground between earlier proposals for coastal management legislation in the United States which emphasized either conservation or development (Beatley et al., 1993).

Table 3.1 Dimensions of policy integration (from Cicin-Sain, 1993, following Underdahl, 1980)

Stages in the policy process		
Inputs	*Processing inputs*	*Consistency of outputs*
COMPREHENSIVENESS	AGGREGATION	CONSISTENCY
• **Over time** – Long-range perspective • **Space** – Extent of geographic area for which consequences of policy are recognized as relevant • **Actors** – Relevant interests incorporated • **Issues** – Interconnected issues incorporated	Extent to which policy alternatives are evaluated from an overall perspective rather than from the perspective of each actor, sector, etc., i.e. basing decisions on some aggregate evaluation of policy	Consistent policy = different components accord with each other • **Vertical dimension** – consistency among policy levels; specific implementary measures conform to more general guidelines and to policy goals • **Horizontal dimension** – for any given issue and policy level, only one policy is being pursued at a time by all executive agencies involved

In the late 1980s and early 1990s balancing the opposing conservation and development forces in coastal management (Figure 3.2) became viewed as being essentially fixed in time. The danger with this view was that each balancing decision was not seen in a long-term context of overall changes to the coast caused by incremental tipping of the balance in one direction. This was one of the many reasons why sustainable development has become the principle underpinning most coastal management programmes today. Sustainability is effectively the concept of balance extended to also include the notion of time dependency and combine elements of social justice.

Since the Brundtland Report (World Commission on Environment and Development, 1987) and the Rio Earth Summit (UNCED, 1992) sustainable development has been a central theme of numerous policy and planning initiatives at all levels of government throughout the world (section 1.4). The challenge facing those involved in planning for the coast is defining what the term sustainable development actually means in a planning context and what are the practical steps required to 'achieve' sustainable development (Buckingham-Hatfield and Evans, 1996b).

In summary, the various conceptual elements of coastal management have been described. The four key concepts for the effective management of coastal areas can be summarized as follows (Fisk, 1996a):

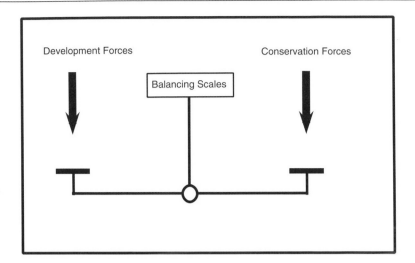

Figure 3.2 Simplified concept of balance in coastal management.

1. An adaptive decision-making process.
2. A recognition of the special nature and value of coastal areas.
3. A comprehensive strategy for integration of sectoral activities.
4. Emphasis on sustainable development.

3.3 Concepts of coastal planning

Planning was described earlier (section 3.1.1) as a process for determining what is aimed to be achieved in the future, and clarifying the steps required to achieve the aims. Thus, planning examines a range of possible directions and explores the nature of uncertainties that inhibit our ability to choose a particular course of action with confidence.

Similarly coastal planning provides for strategies and policies based on the inherent character of the coast, its resources and use demands; it also provides a consistent framework for decision making which considers these factors. Therefore, a well designed coastal planning process should allow managers to decide on a desired direction, while maintaining a range of options for the future.

Coastal planning concepts are much less well developed than those for coastal management (section 3.2). This reflects both its relative newness as a distinct area of activity, and its nature as a hybrid of planning approaches. Contemporary coastal planning is made up of elements from urban/town planning and regional development, protected area (conservation) planning, strategic environmental planning, resource planning and marine planning. The background to the development of these planning approaches was described in Chapter 1. The following analysis of the concepts of coastal planning is somewhat preliminary in that a clearly defined theoretical framework does not yet exist. Nevertheless, a broad description of the major influences on coastal planning and how these affect current coastal planning practice can be given.

3.3.1 The theoretical basis of planning

Much has been written on the theoretical basis of planning, mostly as it relates to planning for the development of urban centres. This literature is based on trying to explain, and in some cases influence, the form of cities around the world which have developed since the Industrial Revolution. These studies first concentrated on Europe, then North America as that continent's population expanded, and now encompass urban centres in the developing world. There is a wealth of specialist literature in planning theory, and it is well summarised in the texts of Faludi (1973), Paris (1982) and Campbell and Fainstein (1996) and the textbook of Alexander (1986). In addition to these is Platt's (1991) lucid historical background to the development of land use planning and its theories. A useful marine-oriented balance to the above land-use planning literature is supplied by Gubbay (1989) and Miles (1989).

Despite the considerable amount of literature on the subject there is still no clearly defined or widely accepted set of planning theories. The reasons for this are clearly articulated by Campbell and Fainstein (1996, p. 2), reproduced in Box 3.4.

Campbell and Fainstein (1996) add to their description of the difficulties, and maybe even impossibilities, of delineating meaningful planning theory by describing planning theory as 'the assimilation of professional knowledge' (p. 2). In this sense modern planning theory effectively represents a mirror held up to current planning practice, with planning practice itself being formed by historical, social and political circumstances which can themselves be subject to theoretical analysis.

What, then, does this mean for coastal planning theory? Principally it must be recognized that there is no single unifying theory which guides coastal planning practice. Instead, there is a range of planning theories which have shaped coastal planning, and provide a 'menu' of theoretical approaches to choose from. These approaches can then be fashioned by coastal managers into coastal planning approaches appropriate for particular cultural, economic, administrative and political circumstances – and of course, the issues being addressed by a coastal planning initiative.

Consequently, the coastal management planning approaches described in Chapter 5 tend to borrow from, and merge, a number of planning theories to provide the best planning solution for a particular stretch of coast. The most important of these are rational, incremental, adaptive and consensual planning, explained in turn below.

Box 3.4

The problems of defining planning theory

Campbell and Fainstein (1996, p. 2) attribute the difficulty of defining planning theory to four principle reasons:

First, many of the fundamental questions concerning planning belong to a much broader inquiry concerning the role of the state in social and spatial transformations. Consequently, planning theory appears to overlap with theory in all the social science disciplines, and it becomes hard to limit its scope or to stake out a turf specific to planning.

Second, the boundary between planners and related professionals (such as real estate developers, architects, city council members) is not mutually exclusive; planners don't just plan, and non-planners also plan.

Third, the field of planning is divided into those who define it according to its object (land-use patterns of the built and natural environments) and those who do so by its method (the process of decision making).

Finally, many fields are defined by a specific set of methodologies. Yet planning commonly borrows the diverse methodologies from many different fields, and so its theoretical base cannot be easily drawn from its tolls of analysis. Taken together, this considerable disagreement over the scope and function of planning and the problems of defining who is actually a planner obscure the delineation of an appropriate body of theory. Whereas most scholars can agree on what constitutes the economy and the polity – and thus what is economic or political theory – they differ as to the content of planning theory.

(a) Rational planning

Rationality has been the primary way western society has thought since the Renaissance era. This was the era of modern scientists such as Galileo and Copernicus, who promoted a scientific approach to problem solving. In its simplest terms, 'rationality is a way of choosing the best means to attain a given end' (Alexander, 1986).

When problems are relatively simple, one can choose the best means to accomplish a given goal. This simple approach is termed 'instrument rationality'. Problems where this form of rationality is used generally have a determinate solution – a solution which is definite and can be defined or explained in tangible terms. For example, engineering problems often have a determinate solution.

When rationality includes evaluating and choosing between goals as well as relating the goals to individual organizations or society's values, it is termed 'substantive' or 'value' rationality. This form of rationality has a significant influence in planning, especially where there are conflicting and multiple objectives. Rational decision making assists planners to make choices within a framework which is consistent and logical; to validate assumptions about the problem and choices; to collect and analyse information, theories and concepts; and to provide a mechanism to explain the reasons for the choices made.

The rational decision model consists of a number of stages linking ideas to actions (Figure 3.3):

- identification of problems;
- defining goals and objectives;
- identifying opportunities and constraints;
- defining alternatives; and
- making a choice and implementing that choice.

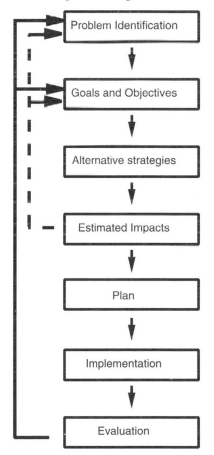

Figure 3.3 Rational (comprehensive) model of planning and decision making (Smith, 1993).

Rational planning theory requires an infinite amount of knowledge in order to make logical decisions when assessing all possible alternatives. Hence, the rational planning model is also called the 'comprehensive' model. Without 'perfect' knowledge there are inevitably value judgements made which reflect the biases and values of the decision maker. Generally, in coastal planning and management there is rarely complete information and knowledge of all possible alternatives. In order to counteract these limitations of rational planning theory, some minor modifications have been proposed, including:

- considering the options one at a time with flexible goals and objectives which can be modified with the options considered – called 'satisficing'; and
- considering a few possible options which are formed and analysed based on their differences and the status quo – called 'disjointed incrementalism' (see below). This avoids information overload and also avoids suggesting radical solutions which may be socially or politically unacceptable.

Currently, the rational planning model generally applies only to the early stages of the coastal planning process – identifying problems, defining goals and objectives, defining opportunities and constraints and sometimes specifying alternatives. But making and implementing choices is often achieved with the assistance of other planning theories which explicitly recognize the influences of value judgements of the participants in the planning process.

(b) Incremental planning theory

Incremental planning is sometimes described as the 'science of muddling through' (Campbell and Fainstein, 1996). It adapts decision-making strategies to the limited cognitive capacities of decision makers and reduces the scope and cost of information collection and analysis. This method looks at alternatives with limited deviation from the status quo. The main components of incremental planning theory are:

- choices are derived from policies or plans which differ incrementally from existing policies (i.e. the status quo);
- only a small number of alternatives are considered;
- only a small number of significant consequences are investigated;
- ends and means are adjusted to make the problem more manageable; and
- decisions are made through an iterative process of analysis and evaluation.

This model is considered by many as a better reflection of how planning decisions are actually made. However, a countering view is that incremental planning is focused on managing present issues and not on the promotion of future goals. As such, it can be considered as pro-inertia and anti-innovation.

(c) Adaptive planning theory

The concept of adaptive planning was first popularized by Holling (1978). It is based on the concept of adaptive control process theory which focuses on decision making founded on experience. As new information is obtained and current management processes are reviewed, new management methods are formulated. Adaptive planning is based on the concept of learning from events of the past, including recognizing society's limited knowledge of ecosystems and the uncertainty in predicting the consequence of using resources within the ecosystem.

Adaptive planning is also an opportunistic form of planning which is responsive to the ongoing management environment in which planning is taking place. It allows planners and managers to anticipate or take advantage of surprise and the results of management activities as learning tools (McLain and Lee, 1996).

However, there are problems in using this approach. These include a reluctance by managing agencies and users of resources to adopt experimental approaches to management. In addition, there may be suspicion of using non-scientific information, such as the perceptions and opinions of coastal users. Finally, adaptive planning requires that shared values amongst diverse interests are formed. This can contrast with the perception of some constituents in the planning process, most often professional planners, that they 'know best'.

(d) The consensual planning approach

> The emergence of consensus building as a method of deliberation has provided the opportunity to reformulate comprehensive planning.
>
> (Innes, 1996, p. 461)

Consensual planning is now used in many coastal planning initiatives in developing and developed countries, including Australia, Indonesia, Sri Lanka, and The Philippines (Chapter 5). Its use has expanded rapidly in Europe since the early 1990s and is now the most widely used coastal planning technique in the United Kingdom (King and Bridge, 1994). Consensus planning uses tools from dispute resolution, pragmatism and education which emphasize the importance of learning communities, empowerment and communicative rationality to effectively involve stake-holders (deHaven-Smith and Wodraska, 1996; Innes, 1996). Communicative rationality focuses on decision making based on reaching a consensus with stakeholders. It assumes stakeholders are fully informed, equally empowered and sincere about the plan. This represents the theoretical ideal for a consensus planning framework; however, rarely does this situation exist in real life. Consensual planning nevertheless draws on this theory's need for deliberation between decision makers.

Consensual planning cannot be viewed as a separate planning theory, unlike those above, but it is perhaps only time until it is provided with a theoretical basis in the same way as other planning approaches. However, its widespread use in coastal planning and management justifies a separate section here.

As the name suggests, consensual planning attempts to develop plans through the building of consensus between the various parties taking part in the planning process. This model is the nearest to a purely pragmatic planning model – that is, it deliberately approaches planning from the view that everyone taking part in the plan has an equally important role to play (Box 3.5). Through consensus building, the planning process strives to reach a win–win situation and to provide mutually beneficial outcomes (Potapchuk, 1995; Williams, 1995). This approach takes a deliberate 'learning' view of the planning process which explicitly realizes that the final form of the plan will be determined by the participants. This way, any number of other planning models can be integrated into the consensual process, including rational, incremental and adaptive planning models.

3.3.2 Summary of the concepts of coastal planning

This section has shown that coastal planning does not have a coherent set of theoretical concepts, but rather has a range of planning theories and practices to choose from. The overriding theme which appears to be emerging amongst planning theorists is that planning theory and pro-cesses are inseparable from the culture, society and politics with which they are so closely tied. As a society changes, so will its approaches to coastal planning.

Indeed, a change over the past 20 years from rational planning theories to more participative approaches, such as adaptive and consensual planning, reflects the overall changes to how societies, especially western societies, relate to the environment (Table 1.1). These changes to planning practice have recently been summarized by King (1996), shown in Table 3.2.

Similar changes in planning practice are reflected in recent trends in the

Box 3.5

The consensus building process used in the Thames Estuary Management Plan (Kennedy, 1996b)

For the Thames Estuary Management Plan, information was gathered via the production of a series of 10 topic papers, each paper drafted by a practitioner from an organization with relevant expertise (e.g. Fisheries paper by the Environment Agency) under the guidance of a topic group. Topic papers were then integrated into a multi-use estuary management plan for the Thames.

One quite widely held concern about this process was that it would be difficult to integrate all of the papers fairly. The non-governmental organizations in particular felt that their views would not be heard when put up against the negotiating ability and financial weight of some of the other stakeholders. In order to allay fears and overcome this problem, the following steps were taken.

1. A small group was established. The group examined in detail a list of 'conflict habits' (see Chapter 5) and between them tried to identify different scenarios under which project participants might adopt each of the different habits. From this exercise we developed a list of Guiding Principles for Achieving Agreement (see Chapter 5), each of which is aimed at counteracting one or more of the more negative conflict habits.
2. The guiding principles were then presented to the project steering group. This generated a discussion on group dynamics (e.g. who is good at negotiating, how is the fact that conflict exists acknowledged, is compromise the best option?).

   The steering group are signed up to respecting the guiding principles. This creates a more level playing field and is also useful for the project manager to refer back to, if any attempt is made to abuse the process.
3. In addition, the programme for integrating topic papers has been carefully thought out with long periods of time set aside for debate, un-oppressive venues selected, a pro forma for rewording policies, etc.

relative power of participants in the United Kingdom's land use planning system (Table 3.3).

3.4 Administrative arrangements for coastal planning and management

Any system of management only survives in the long term when a great deal of attention is paid to its administration. This is especially true of

Table 3.2 Changing coastal planning practices (King, 1996)

Old planning practices	New or emerging planning practices
Mechanistic	Organic/cybernetic
Imposed control	Self-organizing/adaptive
Compartmentalizes	Interdisciplinary/holistic
Reductionist models	Complex/probabilistic
Closed systems	Open systems
Means–ends causality	(Sub) systems functions (multiple causation)
Elimination of uncertainty	Accept and learn from uncertainty
Planning creates order	Order is there already – work with it
Hierarchical order	Market type coordination
Avoid overlap	Semi-autonomous systems need to overlap
Ends given	Goals developed within process
Fixed course	Flexibility and learning
Exploitation of nature	Participation with nature – sustainable use
Programming the future	Flexible frameworks for a changing future
	Subjective judgements required
Consistent goals	Consensus building
Neutral to politics	Planning is politics
Power for others	Power with others
Institutional control	Self-help with government
Government monolithic	Government of many departments, perspectives, agencies
Rational, linear	Intuitive and rational
Entrenched agencies	Experimentation encouraged
Either pragmatic or visionary	Pragmatic and visionary

Table 3.3 Trends in power and land use planning in the UK (Marris *et al.*, 1997)

- Increasing level of conflict/protest about environmental decisions (especially planning).
- Local authority decision-making power is decreasing.
- Central government emphasis on deregulation shifts decision-making power to the market (commerce and industry).
- Public involvement in decision-making attempts to legitimize decisions.
- To be involved requires access to information, resources and time.
- Decision-makers are becoming less accountable and decisions require less justification.
- Decision-making and the process are less satisfactory (for all sides?).

coastal management, where the range and complexity of issues involves many players. These players include those charged with legal responsibilities for managing the coast, such as different levels of government with land under their direct control (such as national parks, or public beaches), and coastal industries which may be required by law to restrict pollution into coastal waters. People who live on the coast or use coastal resources for recreation are also becoming increasingly important in the design of coastal programmes. All participants in coastal management programmes and initiatives are commonly termed 'stakeholders' to stress that they have a stake in the future of the coast, either because they live there, earn a living from the exploitation of coastal resources, or it is their job to administer rules and regulations controlling coastal use. Stakeholders also include vicarious users who may never use or access the coast but still value it, and those who may not reside on the coast but use it for recreation.

This section first analyses the various ways to organize government to deliver coastal programmes, then discusses mechanisms for linking coastal users and residents with government initiatives. However, before doing so it is worth reiterating the factors which are distinctive to coastal management programmes and their administration. These have been summarized by Sorensen and McCreary (1990) as:

1. Initiated by government in response to very evident resource degradation and multiple-use conflicts.
2. Distinct from a one-time project (it has continuity and is usually a response to a legislative or executive mandate).
3. Geographical jurisdiction is specified (it has an inland and an ocean boundary).
4. A set of specified objectives or issues to be addressed or resolved by the programme.
5. Having an institutional identity (it is identifiable as either an independent organization or a coordinated network of organizations linked together by functions and management strategies).
6. Characterized by the integration of two or more sectors, based on the recognition of the natural and public service systems that interconnect coastal uses and environments.

The background to points 1–4 (above) were discussed earlier in this chapter, providing an introduction to points 5 and 6, the focus of attention here. The two key issues drawn from these points are that coastal management programmes should be identifiable within a government's administrative system, and include elements which bring together different sectors of government. These two issues form the basis for discussion in the next section.

Importantly, it is now commonly accepted that 'there is no "best" institutional arrangement for managing coastal resources' (Jones and Westmacott, 1993). Instead, the contemporary focus on institutional arrangements for coastal management is outcome oriented, in that 'the "goodness" of an institutional arrangement can best be judged by the effectiveness and efficiency with which coastal use conflicts are resolved' (Jones and Westmacott, 1993, p. 130), a pragmatic approach to the design of coastal programmes which is adopted in the following sections.

3.4.1 Organizing government

Coastal nations should be in a position to develop Integrated Coastal Zone Management structures uniquely suited to that nation – to the nature of its coastal areas, to its institutional and governmental arrangements, and to its traditions and cultures and economic conditions.

(World Bank, 1993)

Many coastal nations have developed, or are in the process of developing, their own approaches to coastal management. This section describes, analyses and extracts the common threads from these approaches.

The role of government is doubly important because of the dominance of common property in at least the oceanic component of the coast, especially in developed countries (Boelaert-Suominen and Cullinan, 1994). A central question for the administration of coastal management and planning programmes is, then, how government is organized to deliver its programmes and how these programmes interact with private companies and the wider community.

The core issue with organizing government to efficiently and effectively deliver a coastal programme is focusing the activities of many different government sectors in an integrated manner. As alluded to in the introduction to this chapter, this is not an easy task. The great majority of governments are established along sectoral divisions, assigning the responsibility for delivering services and functions to different government agencies. This concept of 'differentiation' is one of the central elements of how most public sectors around the world are organized (Heady, 1996). For example, the Indonesian Government does not have a single agency responsible for coastal management but uses a combination of line agencies, coordinating agencies and non-governmental organizations (NGOs). Line agencies have legislated responsibilities for management of various coastal resources or sectors. Coordinating agencies, despite having no legislated powers, have a government mandate to bring various line agencies together with other relevant parties and formulate coastal management initiatives. NGOs provide cost-effective debate and support to a range of coastal initiatives especially at the local level (Box 3.6).

As the Indonesian example has highlighted, it is important that the design of institutional arrangements within a level of government includes fostering cooperation and/or coordination between government agencies with responsibilities on the coast (Rogers and Whetten, 1982). This can be perceived as a threat to the power and autonomy of individual agencies; however, if well managed and the coordinating agency has limited legislated power, then the power struggle problems can be reduced.

Nevertheless, the relative power of line agencies versus coordinating agencies or coordinating bodies is generally skewed markedly towards the line agencies. The power of horizontally oriented line agencies has been likened (Tasque Consultants, 1994) to 'rods of iron' versus the 'threads of gossamer' which act to pull them together. A number of mechanisms used to balance these relative powers are described later in this chapter.

As of 1993 an estimated 57 sovereign or semi-sovereign states were undertaking a total of 142 coastal zone management programmes (Sorensen, 1993, 1997). These programmes are at various stages of development and implementation, meaning that there is a relatively large pool of information to draw on in order to analyse the performance of the different institutional arrangements used to develop and implement coastal management programmes.

In organizing government to develop and implement coastal management the issue is not how institutions are arranged, but rather what is achieved through those institutional arrangements. This focus on outcomes is the reason for Jones and Westmacott (1993) concluding that there is no 'best' way to organize governments in order to manage the coast. In practice the diversity of cultural, social, political and administrative factors around the world confirms that there is indeed no single best way. Instead, the designers of the administrative arrangements for new coastal management programmes must tailor administrative structures to take advantage of the particular cultural, social and political factors within their jurisdiction as they interact with issues being addressed. For example, what may be the best system of coastal zone management programme governance for a European coastal nation may be disastrous for a country in the Pacific, and vice versa. This section therefore concentrates on addressing the factors which are usually considered in the design of the administrative arrangements for coastal management programmes. How these factors have been applied is shown by referring to case studies drawn from around the world.

Detailed analysis of institutional arrangements for coastal programmes was initiated by Sorensen et al. (1984), and subsequently updated by Sorensen and McCreary (1990). These two texts remain the standard

Box 3.6

**Example government organizations structure –
Indonesia (based on Sloan and Sugandhy, 1994)**

Line Agencies	Responsibilities relevant to Coastal Management
• Department of Agriculture/Directorate General of Fisheries	• Fish and aquaculture management
• Department of Forestry/Directorate General of Forest Protection and Nature Conservation	• Marine conservation and protected areas, mangrove management
• Department of Communications/ Directorate General of Sea Communication	• Ports, shipping, navigation, safety including emergency responses
• Department of Mining and Energy/ Directorate General for Oil and Gas	• Gas and oil exploration and production
• Department of Education and Culture/Universities	• Education and research
• Department of Security and Defence/ Hydrographic and Oceanographic Service	• Territorial water security, hydrography
• Department of Industry	• Development and waste management
• Department of Public Works	• Engineering and erosion control
• Department of Tourism, Post and Telecommunication	• Tourism

Coordinating Agencies	
• Ministry of State for Environment	• National coordination
• Environmental Impact Management	• Environmental Impact Assessment
• National Development Planning Board	• National development plans
• Department of Home Affairs/ Directorate General of Regional Development	• Regional development
• Ministry of State for Science and Technology/Technology Assessment and Application	• Natural resource inventory
• National Coordinating Agency for Surveys and Maps	• Coastal mapping
• Indonesian Institute of Science/Research and Development Centre for Oceanology	• Marine science and research
• Coordinating Committee for National Sea Bed Jurisdiction	• National boundaries and Law of the Sea
• Coordinating Board for Marine Security	• Security in national waters

continued . . .

Non-Governmental Organizations	
• Indonesian Forum for Environment	• National coordination of non-governmental organizations
• World Wide Fund for Nature	• Marine conservation, public education
• Asian Wetland Bureau	• Coastal wetland management, aquaculture development, Environmental Impact Assessment (FIA)
• United Nations, Educational, Scientific and Cultural Organisation	• Marine pollution, education
• International Union for the Conservation of Natural Resources	• Marine protected areas
• United Nations Development Program	• Marine pollution, International Maritime Organization
• Clean Coastlines and Beaches	• Public awareness, coastal litter and pollution
• Mangrove Foundation	• Mangrove conservation and sustainable use
• Indonesia Green Club	• Sea turtle conservation; public awareness

works on institutional arrangements and are drawn on in the following discussion. Sorensen *et al.* (1984, p. 1) describe institutional arrangements as 'the composite of laws, customs, organizations and management strategies established by society to allocate scarce resources and competing values for a social purpose, such as to manage a nation's coastal resources and environments'.

A useful way of broadly describing institutional arrangements for coastal management is to divide a nation's system of government into 'horizontal' and 'vertical' components (Figure 3.4). [This differentiation follows the 'scientific management' school of Taylor (1911) (Kraus and Curtis, 1986).] Levels of government are shown as the vertical component, while the different sectors comprising a single level of government form the horizontal component. In the example shown in Figure 3.4, there are three levels of government, as is common in large and/or populous countries. In many countries the division of power between levels of government is not purely linear, in the sense that higher levels of government exert power over lower levels of government, as inferred in Figure 3.4. Thus, in many federal systems of government the different governments (federal, state/provincial and local) are termed 'spheres' in order to reflect their non-hierarchical nature.

Horizontal components of government are separated according to function, which in turn is reflected in division of government into

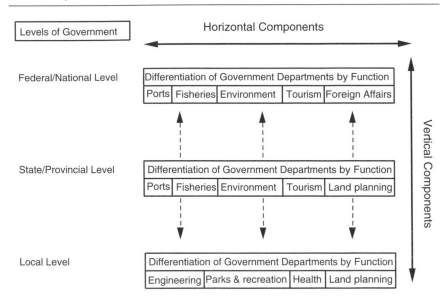

Figure 3.4 Example of national system of coastal management governance divided into vertical and horizontal components.

various agencies and departments. For example, a government may choose to create separate departments, such as environment, transport, energy and primary industry. This horizontal differentiation can lead to gaps and overlaps between the various government departments with responsibilities for coastal management.

Roles and responsibilities for coastal management are usually divided both horizontally and vertically. That is, some activities will be carried out by one level of government, and not another (vertical division), while others are carried out by one particular sector of government (horizontal division). In reality, this horizontal and vertical differentiation is very complex. Indeed, this complexity often provides one of the prime motivations for developing a coastal management system in the first place and the need for cooperation and coordination.

Mitchell (1982) developed a classification method of the governance arrangements for coastal management. Mitchell analysed national coastal management systems according to three criteria:

- Coastal focus: either coastal specific programmes or coastal issues are addressed as part of overall agency responsibilities.
- Strength of national control: strong or weak national government control.
- Policy orientation: programmes oriented towards economic development or environmental/amenity considerations.

Mitchell used these three criteria to develop an eight-fold classification. For example, Sri Lanka's coastal programme (see Box 3.7) was classified as 'coast specific, with a strong national structure and environmental orientation' (Mitchell, 1982).

Sorensen *et al.* (1984) adapted Mitchell's classification system to develop five 'types' of governance arrangements. These effectively combine the degree of integration between government sectors with the degree of the programme's coastal focus. The five types of institutional arrangements are shown in Table 3.4 as Types 1–5 together with countries judged to accord to one of the five programme types. Hay and Kay (1993) added a further two programme classification types to show both the development of new integrating mechanisms, through the use of sustainable resource management legislation with a coastal focus, developed in New Zealand (Box 5.14) and the use by some coastal nations of cross-sectoral units within government with a coastal focus. These additional coastal governance types are shown as Types 6 and 7, respectively, in Table 3.4.

The seven types of institutional arrangements shown in Table 3.4 do not necessarily reflect the complete range of possibilities for broad classification. Indeed, despite recent attempts to collate the current status of institutional design worldwide (Sorensen, 1997), there appears to be little critical examination of the findings of such work.

Another way of describing coastal management governance is to focus on how various coastal management activities are controlled (Born and Miller, 1988). This approach has been used to classify American State coastal programmes developed under the Coastal Zone Management Act (Knecht *et al.*, 1996). Using this method, two main types of governance are produced:

- Networked: existing government sectors and institutions remain. No new specific coastal management legislation is enacted. Sector co-ordination is improved though 'networking' of existing legislation and policies.
- Legislative: new specific coastal management legislation is enacted. This legislation can have a variety of purposes. New institutions or the enabling of existing ones enacted.

Networked coastal management programmes are those which bind together a range of pre-existing approaches to the management of coastal resources into a well defined coastal programme (Taussik and Gubbay, 1997). The networked approach was originally developed in the United States (see Box 3.8) and has been adopted by other coastal nations around the world (Kay *et al.*, 1997). Born and Miller (1988) distinguish four attributes of the networked coastal programmes in the United States:

Table 3.4 Governance arrangements for coastal management (from Hay and Kay, 1993, after Sorensen et al., 1984, and Sorensen and McCreary, 1990)

Type of governance arrangements	Sectoral planning and development	Integrated planning – no coastal focus	Integrated planning – coastal focus	Integrated coastal management	Sustainable resource management – coastal focus	Cross-sectoral units with coastal focus
TYPE 1 Many, if not most, developing countries	X					
TYPE 2 Most developed countries (e.g. Japan, Netherlands, Poland, Sweden, Singapore)	X	X				
TYPE 3 (e.g. Cyprus, France, Norway, Thailand, United Kingdom)	X		X			
TYPE 4 (e.g. United States)	X			X		
TYPE 5 (e.g. Brazil, Costa Rica, Ecuador, Greece, Israel, Sri Lanka)	X	X		X		
TYPE 6 (e.g. New Zealand)	X		X		X	
TYPE 7 (e.g. Bangladesh, Fiji)	X					X

1. The programme emphasizes making pre-existing authorities work better and in a more coordinated manner.
2. The designated 'lead agency' has broad policy formulation and co-ordination responsibilities and a horizontal (cross-cutting) orientation.
3. The lead agency tends to be an Executive Department staff agency and not an operating agency.
4. The lead agency relies significantly on other agencies and/or different levels of government (dispersed programme management), especially regarding regulatory powers.

An important addition to the above four points is the role of coordinating committees, or councils, which help pull together the various threads in the network (see Box 3.8). These coordinating groups play a vital role in the success of networked coastal management systems, especially if the membership or mandate of the groups is powerful enough to ensure the cooperation of its member government agencies, such as land use planning, land management, environmental protection, transport, infra-structure development, primary industry or mining. For example, some Australian State Governments rely on Coastal Councils to coordinate their coastal programmes. Membership of those committees (Table 3.5) reflects the structure of their respective bureaucracies, politics and the relative importance of coastal management issues within their jurisdictions.

Perhaps the central issue in any networked system of coastal manage-ment is the critical role of the people involved in tying the network together. By its very nature a network is a system which requires the commitment of the people staffing the various groups and agencies within the network. Without this commitment, at a personal and agency level, the functioning of the network becomes vulnerable. Conversely, this reliance on goodwill can also be the network's greatest strength, but only if people play their part with the knowledge that without their input there is no safety net of legislation to underpin their actions.

Coastal management systems relying on the networked approach may appear to be less efficient than a legislative-based approach – especially when there is not an apparent force of law to demonstrate a government's commitment. However, comparative studies of the various approaches taken by states in the United States have indicated that networked systems may be more (Born and Miller, 1988) or (at least as) efficient (Knecht et al., 1996) than fixed, legislated programmes. The advantages and disadvantages of networked systems require further evaluation, however, before their efficiency relative to legislative approaches can be fully evaluated.

Separate coastal management legislation, or coast-specific sections of broader legislation, can be enacted to assist programmes in a number of ways. The most common legislative mechanism is to pass enabling

Table 3.5 Current membership of three Australian State Government Coastal Councils (from Donaldson et al., 1995; Kay et al., 1997)

Western Australia Coastal Zone Council	New South Wales Coastal Council	Victorian Coastal and Bay Management Council
Western Australian Planning Commission (Chair)	Independent Chair	Chair (appointed by Minister)
Local Government (Perth Metropolitan Region)	Department of Planning	Department of Planning and Development
Local Government (Country Region)	Public Works Department	Department of Transport
Community/Private	Department of Conservation and Land Management	Department of Conservation and Natural Resources
Department of Transport (Maritime Division)	National Parks and Wildlife Service	Municipal Association
Department of Conservation and Land Management	Environmental Protection Authority	
Department of Environmental Protection	Tourism Commission	Six representatives of the community with experience in conservation, tourism, recreation, commerce, issues relating to indigenous peoples, community affairs or coastal engineering
Department of Resources Development	Department of Minerals Resources	
Fisheries Department of WA	Department of Local Government	
Waters and Rivers Commission	Department of Fisheries	
	Elected State Member	
	Representative of Royal Australian Planning Institute	
	Industry representative	
	Conservation Council	
	Elected local government representative (1)	
	Elected local government representative (2)	
	Elected local government representative (3)	

legislation which defines the shape of a coastal programme by delegating power and/or money to a lead agency or coordinating body. In these cases sufficient regulatory and enforcement mechanisms are already present; for example, planning and environmental impact assessment requirements. Where this is not the case, as in many developing countries, coastal management legislation can be enacted to establish various forms of regulatory instruments, such as (Jones and Westmacott, 1993):

- licenses and permits, e.g. for construction and concessions;
- physical planning regulations for the establishment of developments, water supply of conservation zones and setbacks;
- standards for a range of parameters including water quality (related to environmental and/or human health, construction, the provision of amenities;
- quotas, such as on fish catches.

An example from Sri Lanka of a legislated basis for coastal management is shown in Box 3.7. The use of these various legislative tools, including those used in Sri Lanka, is described in more detail in Chapter 4.

The US coastal management system allows flexibility in whether each State bases its coastal management efforts on legislative or networked approaches (Box 3.8). The two cases of California and Florida show that both state programmes conform to the minimum standards set out in the US federal Coastal Zone Management Act (1972–1990) and have had success, despite major differences in programme structure (legislative versus networked) and level of implementation (local versus state).

A further method for classifying the institutional arrangements of coastal programmes is the primary level of implementation. American state programmes were classified by Knecht et al. (1996) according to whether they are implemented primarily at state level, or at both state and local levels. Initial findings of programmes which were implemented at these two levels suggest that there is no measurable difference in performance between the two. However, these are initial findings only, and as yet there has been no systematic analysis of how the various types of coastal planning initiatives analysed in Chapter 5 influence these outcomes.

The culture and social structure of a coastal nation is often the hidden determinant of its organizational approach to coastal management. This driver is most clearly seen by those examining the organization of coastal management programmes in countries other than their own, particularly when there is a strong cultural contrast – for example, a westerner visiting a Pacific island nation. It would be tempting to interpret the coastal management efforts in the Pacific as 'culturally driven', while classifying those of the western nation according to the various schemes described

Box 3.7

Governance arrangements for the Sri Lankan Coastal Management Programme

Unlike many of its neighbours, Sri Lanka has developed a national coastal zone management programme in response to its numerous coastal management problems (Box 2.7). The Sri Lankan approach to coastal management was developed in partnership with international aid agencies and coastal specialists, most notably those from the United States, Germany, Denmark and Holland (Lowry and Wickramaratne, 1987; Kahawita, 1993; Lowry and Sadacharan, 1993).

Sri Lankan coastal management has been undertaken through various government initiatives since the early 1970s. These initiatives during the 1970s concentrated on the management of critical coastal erosion problems, and hence focused on planning coastal engineering works and attempting to place controls on the construction of new building at the coast. There was a change in emphasis from 'coast protection' to 'coastal zone management' in 1981 with the passing of the Coast Conservation Act (Kahawita, 1993, repeated in Clark, 1996). The Coast Conservation Department was mandated under the Act with responsibilities for the administration of a permit system for development control, the formulation and implementation of works and research within a 'coastal zone', the boundaries of which are shown in Figure 1.2 (Kahawita, 1993).

The Act also enabled a range of coastal planning and management tools to be used, which are listed in Table 4.1. These tools were found to address certain issues well, such as ensuring well planned major coastal development (e.g. tourist hotels). However, a broader, more encompassing approach was needed to ensure that issues external to the permitting system and outside the legal 'coastal zone' could be addressed. The new approach is based on the development of a 'second generation' programme founded on a hierarchy of national, provincial, district and local coastal management plans, linked with enhanced institutional capacity, public education and community involvement (Olsen et al., 1992). This strategy, called Coastal 2000, was endorsed in 1994 and contains a range of strategies described in Box 5.10 (Coast Conservation Department, 1996).

The Coast Conservation Department (CCD) will guide implementation of the policy at the national level and be supported by other national agencies such as Fisheries, the Central Environmental Agency and the National Aquatic Resources Agency. Provincial governments will focus on regional coastal planning with the assistance of the CCD, Urban Development Authority and other national agencies. At the local level, coastal management partnerships will be formed between national and provincial government agencies and user-groups through community level coastal management and planning initiatives.

continued . . .

The development of institutional arrangements in Sri Lanka demonstrates a maturing from a national, agency-driven programme to a multiple-level community focused system. These changes have allowed the development of an integrated coastal management planning programme at a range of planning scales.

Box 3.8

Legislative and networked coastal management programme structure in the United States (Fisk, 1996b)

At the heart of coastal zone management in the United States is an agreement between Federal and State governments. The Federal Government offers financial and technical assistance to states who voluntarily choose to develop and implement coastal management programmes, but they must adhere to a set of minimum federal standards. These standards are made up of four core issues:

• protection of coastal resources;
• ensuring public access to the coast;
• managing development along the coast; and
• managing coastal hazards.

As an added incentive to the states to develop coastal programmes, the Federal Government assumes the legal responsibility that all of its activities including permitting activities are consistent with approved state programmes.

Coastal management in the United States is strongly process-oriented, allowing states to tailor their coastal programmes in an individual and non-regulatory fashion. At the time the federal Coastal Zone Management Act was passed in the early 1970s, several states had already enacted state-wide comprehensive laws and policies for management of their coastal zones that conformed well with national guidelines. However, the majority of states asserted control over coastal activities through sector-specific statutes and laws. For these states, the Federal Government conceived the networking approach. Networking gave states the option of incorporating existing state laws and policies into a network of management controls for the coastal zone, assuming that a lead state agency was named to oversee operation of the program.

The networking concept was very popular; making up the legal basis of over half of the current participating state programmes. Only a handful of the participating states base their programmes on comprehensive pieces of coastal legislation or policy. Other states share elements of both

continued . . .

approaches: specialized coastal laws and policies networked with other state legislation. To illustrate the difference between these two approaches to programme structure in the United States, two states are discussed. The first case, the California Coastal Management Program, is a good example of a programme with a legislative approach. At the opposite end of the spectrum, the Florida Coastal Management Program epitomizes the networked approach with over 26 single-purpose statutes bundled together to establish its legal basis.

Case I: California

The California Coastal Management Program (CMP) is divided into two segments. The first segment, the San Francisco Bay and surrounding area, is managed under the Bay Conservation and Development Commission (BCDC). The remainder of the coastal area (extending seaward from the coast three nautical miles and landward 1000 yards so as to include coastal estuaries and recreation areas) is managed by the second segment, the California Coastal Commission (CCC). The CCC is the lead regulatory body in the state that operates the statewide permit system. A third coastal body, the California Coastal Conservancy, is a non-regulatory body involved with land acquisition, ensuring public access and critical area restoration (NOAA, 1994b).

The CCC administers the California Coastal Act of 1976, which, as amended, requires all coastal cities and counties to develop their own Local Coastal Program (LCP) through a land use plan or zoning ordinance. The CCC then determines if these plans and ordinances conform to state standards set out in the 1976 Act. Once a county or local plan has been approved, municipalities can issue their own permits for coastal building. As a result, the bulk of land use planning and day-to-day coastal management activities occurs at the local level.

Following implementation of the 1976 Act, federal approval of the California Coastal Program followed shortly thereafter in 1978. Since that time the programme has made great strides in ensuring public access to the coast, protecting valuable wetlands, and preserving coastal areas of archaeological significance. By means of the consistency provision, offshore petroleum leasing and development by the federal government have been halted or occur with the necessary provisions to protect the environment. Recently, the CCC has launched several major education and awareness initiatives related to coastal protection, beach clean-up, and marine conservation (NOAA, 1994b).

Case II: Florida

The Florida coastal zone is defined to include the entire state and its coastal waters (three nautical miles seaward on the Atlantic coast; nine nautical

continued . . .

miles seaward on the Gulf of Mexico coast), making it the second largest in the United States.

The Florida Coastal Management Program (FCMP) is based on 26 existing laws and regulations. To obtain federal approval of their coastal programme, these laws were networked together, and a new Florida Coastal Management Act of 1978 was enacted. The 1978 Act established the lead coastal agency, the Department of Environmental Regulation, to coordinate and review current plans and to provide a clearing house and other services requested by the other 11 agencies that administer laws and regulation relevant to the coastal zone. By a joint resolution of the Governor and state Cabinet, an Interagency Management Committee was formed including the heads of all FCMP agencies and a non-government chairperson from the Florida Citizens Advisory Committee on Coastal Resource Management. The Interagency Management Committee works to coordinate policies and resolve disputes among the various agencies and user groups. Based on these programme improvements, the FCMP was approved to enter the national coastal zone management programme in 1981.

Since 1981, significant institutional reform has occurred within the programme. The Department of Community Affairs in charge of local land use planning has become the lead agency, while two former line agencies, the Department of Environmental Regulation (previously the lead agency) and the Department of Natural Resources (in charge of coastal permitting), have been combined to form the Florida Department of Environmental Protection. To formalize a partnership in coastal management, the Departments of Community Affairs and of Environmental Protection have signed a memorandum of understanding to coordinate their respective activities. While local governments are being pushed to develop more comprehensive coastal plans, the bulk of coastal management activities in Florida remain at the state or regional level (NOAA, 1994b).

With the help of federal funds, the FCMP has helped to prepare coastal counties against the threat of hurricanes and other coastal hazards, with hurricane evacuation plans. Critical estuaries and aquatic preserves have been protected, as have wetland resources through the implementation of the comprehensive Henderson Wetlands Protection Act (1988). The FCMP remains a major player in both the National Estuarine Research Reserve Program (operating two sites) and administration of the Florida Keys National Marine Sanctuary.

above. It is intriguing to consider what a Pacific islander visiting a European country, for example, would make of the cultural setting influencing the organization of coastal management there, especially the emphasis on coastal engineering. It could be argued that most, if not all, coastal programmes are influenced by cultural beliefs. Nevertheless, given the importance of customary beliefs in many developing countries, and their inability to be classified as either networked or legislative

Box 3.9

Coastal zone management decision-making framework in Western Samoa (from Kay et al., 1993)

The customary system of decision making in Western Samoa predominates over the majority of coastal land, and is effectively semi-autonomous from national decision making (Cornforth, 1992) (see figure). Only in the capital city, Apia, is the customary decision-making system of reduced importance.

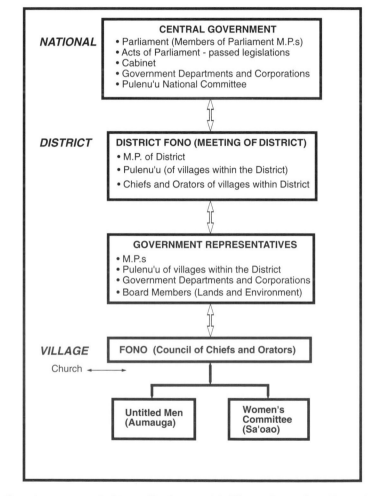

Coastal management decision-making framework in Western Samoa (from Kay et al., 1993).

continued . . .

Coastal resource management decisions are therefore expressly made by villages. National-level decisions are made by the elected members of parliament.

At the village level, decisions made by the Village Council of Chiefs and Orators are usually expressed by the formulation of rules. Rules can be either long-standing, forming an integral part of village culture, or short-term in response to immediate village concerns. Rules are enforced and policed through the village Council of Chiefs and Orators (fono) and heads of families, and non-compliance results in punishment depending on severity. Punishment can be through various forms of shaming, and in extreme cases banishment from the village. Rule breaking is not usually referred to the police, and is instead settled according to custom.

There is no lack of national legislation for sustainable management of coastal and marine areas in Western Samoa, or lack of implementation ability by village fono. However, a recent analysis of environmental management legislation in Western Samoa found that very few 'laws are complied with, and even fewer enforced' (Cornforth, 1992). Nevertheless, there is a widespread belief in Western Samoa that harmonizing the two systems is gradually being achieved and that present government policies and legislation provide a sound basis for successful coastal management in the future.

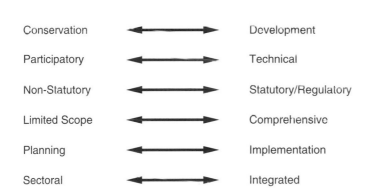

Conservation	Development
Participatory	Technical
Non-Statutory	Statutory/Regulatory
Limited Scope	Comprehensive
Planning	Implementation
Sectoral	Integrated

Figure 3.5 Range of orientation of coastal management programmes (Scura, 1993, adapted by White, 1995).

approaches, coastal management programmes largely controlled by customary beliefs may provide a different category of coastal programme organization. However, this issue has been little studied, and hence no firm conclusions can be drawn at present.

The importance of tailoring a coastal management and planning programme to reflect cultural and social conditions is shown by the development of a coastal programme in Western Samoa, where there are two styles of government: a Westminster style parliamentary system of national government, superimposed upon the traditional village-based decision making structure of the Fa'a Samoa (Box 3.9).

Finally, an alternative method for analysing coastal programme design has recently been developed by considering various programme 'orientations' (Figure 3.5) (Scura, 1993, adapted by White, 1995). Some of these orientations reflect previous work on programme focus, but also inject a component of coastal planning, especially the balance between planning and the implementation of plans. Thus, a useful link between the development of coastal management programmes and their planning components is formed, which is explored in later chapters.

(a) Integration and coordination between levels of government for coastal management

The previous section described and discussed the overall approaches to the administration of coastal management programmes. In this section discussion focuses on the coastal management roles and responsibilities of different levels of government. General principles, and case studies of how different powers for coastal management are shared between levels of government, will both be examined.

Rationally dividing vertical responsibilities for coastal management activities between levels of government is often much more difficult than resolving horizontal differentiation problems. Political, administrative and budgetary clashes between levels of government drive conflict, and often lead to confusion in allocating responsibilities. Such vertical imbalances of power, money, and differences in political affiliation often dictate the overall shape of coastal management governance of a nation, both horizontal and vertical. This is because horizontal differentiation will, to a large extent, be controlled by the relative degree of vertical power held by a particular level of government. Thus a lower, poorer level of government will be unable to create a complex horizontal differentiation of its sectors, in contrast to a larger and richer higher level of government.

A central issue to the vertical distribution of management authority is the degree of centralization in decision making – a fundamental management question not restricted to the coast. The advantages and disadvantages of centralism and localism are summarized in Table 3.6.

One compromise usually struck is to attempt to delegate decision-making powers to the lowest level of decision making consistent with the scope of the problem, but to constrain those decisions within a framework articulated by the next higher level. Some coastal management programmes attempt to achieve a 'controlled devolution' of powers through planning. In

Table 3.6 Advantages and disadvantages of centralism and localism in coastal management (adapted from Ketchum, 1972)

Advantages of centralism	Advantages of localism
Increased general perspective	Intimate knowledge of the problems
More objectivity	More localized outlook
More experts available	Greater likelihood of living with the effects of a decision, creating an incentive to be successful
Increased funds	
Greater political will	

these cases integrated or subject plans are formulated jointly by different levels of government. Once a planning framework is established, ongoing management activities can take place directed by local-level decision makers if they are consistent with the plan. The development of such coastal planning frameworks is described in more detail in Chapter 5.

3.4.2 Linking government with the private sector and community

The vast majority of coastal planning and management programmes in operation today attempt to link the efforts of government with both private industries and the wider community.

By the late 1990s the relative roles of government, industry and the community had changed in most developed countries due to increased environmental awareness of the community and industry, and the desire of both to be more closely involved in decision making. Governments, private industry and the wider community tend now to work more closely together in developing and implementing coastal programmes – albeit with a guiding, and often firm, hand from government. Of course, this does not negate the realities of power, and the influences that powerful government and private sector interests can have in coastal programmes. Nevertheless, the growth of community advocacy groups, and the increasingly rapid access to information by community members, has changed the relative balance of power in coastal programme development and implementation in many coastal nations.

A similar changed relationship between government, industry and the wider community has also occurred in many developing countries as a result of the re-emergence of interest in indigenous cultures. Consequently, the relative degree of power between the formalized systems of government

Table 3.7 The Special Area Management (SAM) framework used in Sri Lanka (from White and Samarakoon, 1994)

Policy Steps*	Outputs	Process/Methods
Consensus on use of SAM by national agencies	Concept Paper approved	Discussion and workshops
Site selection and criteria of choice	Two sites approved based on issues and practicability	Discussion and workshops
Issue identification and analysis for each site	List of issues and causes	Local workshops, interviews, training and education
	Environmental profile on immediate and surrounding area of management	Secondary information, key informants, rapid area assessment, local government and non-governmental participation
	Boundaries for area of work, planning needs and research identified	Planner analysis and inter-agency discussion
Goals and objectives for resource management	Clearly stated objectives and indicators for completion	Techniques to gain consensus through community planning and dialogue with government
Policy selection for resource management	Appropriate policies and their implications	Legal advice and planner analysis, consultation with community and local government
Management strategies and actions	Draft management plan	Workshops, and inter-agency coordination, local participation
Implementation	Field project for education, training, research, people, organisation, small public works, resource management	Public involvement, political support, professional assistance as needed
Evaluation	Monitoring of key indicators and trends, information to revive management plan	Participatory monitoring with professional guidance
Readjustments to plan and implementation	Revised plan and procedures	Planners and local participants.

* Lynne Hale of the Coastal Resources Center of the University of Rhode Island contributed to this framework.

established by colonial powers, and customary land tenure and resource-use practices, has altered. In many cases, this alteration is continuing (e.g. Kay and Lester, 1997). The various tools for eliciting community views for the development of coastal plans will be described in Chapter 5. However, it is worth examining how important community participation has become in the development of such plans through reference to Table 3.7, where community participation is woven through a plan production process.

The degree to which the general public and/or indigenous people take part in the decision making process has been described according to a spectrum of citizen participation by Arnstein (1969). Arnstein drew the analogy of a ladder of citizen participation, as shown in Figure 3.6.

The top two rungs of the ladder are thought of as 'rubber stamp committees'. Here the community's opportunity to participate is only allowed if it agrees with those in power. The degree of interaction between stakeholders and decision makers increases down the next three rungs of the ladder. 'Informing' identifies citizen's rights and options, while 'consultation' allows for citizens to express their concerns. 'Placation' allows for citizens to advise on management decisions, but decision makers do not necessarily act on these concerns. These three levels are characterized by people being tolerated by those in power. At the 'partnership' level, citizens participate actively in decision making through negotiations or 'trade-offs' with managers. On the next rung, 'delegated power', citizens are given management power for selected parts of a programme. In the last rung of the ladder citizens have total control of the decision-making process.

Private industries are also being viewed by governments as playing a key role in supporting the development and implementation of coastal

Figure 3.6 Arnstein's ladder of citizen participation in decision-making (adapted from Arnstein, 1969).

Box 3.10

**Coastal management and the commercial sector –
the case of the Thames Estuary (Kennedy, 1996b)**

An important challenge that British coastal managers have been faced with
is ensuring the private sector becomes involved in projects, alongside public
and voluntary organizations. Usually an element of persuasion is required
to get companies fully on board with the coastal planning process, especially as they are also frequently asked to contribute financially towards
projects. The following list, based upon experience on the Thames,
highlights some of the commercial reasons for becoming involved in coastal
management initiatives:

- greater coordination of planning and management activities will lead to
  a reduction in time and resources required by individual organizations
  when planning their own activities and consulting on their proposals;
- more certainty for private sector interests looking to develop, alter or
  change uses in particular areas of the estuary;
- more efficient and responsive action by management agencies to
  proposed actions;
- greater clarity on the subjects that will need to be explored as part of
  Environmental Assessments on specific proposals;
- once established, coastal plans will provide an information source
  that may be used by the commercial sector, thus cutting consultancy
  costs;
- a strong partnership will allow information to be disseminated between
  participating organizations that could otherwise need purchasing and
  collating, again often by consultants;
- strength of the environmental movement is such that it has a large
  influence over competitive ability to the extent that it is no longer
  economically viable to work against it;
- can see how maintenance of a healthy environment is critical to overall
  regional economy, e.g. revenue generated by tourist industry, water
  quality and fisheries
- can assist in the identification/protection of development land for
  industrial/port uses that require a coastal location; help to guard against
  being priced out of the market by other commercial uses, with higher land
  values (e.g. housing development such as waterfront gentrification);
- increased professionalism within the environmental movement – more
  of a force to be reckoned with – with increased recognition of this professionalism; and
- assistance with ensuring that access to the waterfront is managed more
  strategically, without ad hoc requests that potentially can disturb and/or
  have safety implications for port operations.

initiatives. Clearly private companies which depend on coastal resources have a keen interest in how the coastal initiative is managed and planned. The traditional adversarial role between government and industry is beginning to be broken down in some parts of the world, with government and industry forming partnerships for coastal management initiatives. A good example of this comes from the Thames Estuary described in Box 3.10 (Kennedy, 1996a,b).

The partnership approach shown in the case of the Thames Estuary has been extended in some parts of the world to full community and industry participation in coastal management decision making. Also, management arrangements for jointly managing resources, called collaborative management, are becoming increasingly important in coastal management. These initiatives are discussed in more detail in Chapter 4.

3.4.3 Guiding statements for coastal programmes

Many governments chose to clarify their administrative arrangements for coastal management by articulating what a coastal programme is attempting to achieve. As described above, there are a variety of ways to formalize this process – for example, legislation and the production of various types of documents to guide networked approaches. The structure of these documents generally begins with the philosophy underlying the coastal programme, followed by a list of guiding statements, issues to be addressed, and steps to be taken to tackle those issues.

Statements which guide coastal programmes are usually separated into a hierarchy. Section 3.1.3 covered a number of terms used to describe the various statements in this hierarchy, and developed the following standardized terminology: overall goal, objectives, and actions, guided by statements of principle. How various governments and international organizations have worded these guiding statements is looked at in the following sections.

There are significant advantages in producing formal written statements of programme philosophy and guidance. As will be demonstrated later in this section, in most cases the advantages outweigh any disadvantages. Nevertheless, it is important to weigh up the pros and cons of formalizing a programme's goals, principles, objectives and actions, both for the organizations involved in the programme and for the stakeholders (who are often the key to a programme's ultimate success or failure) charged with its implementation (Steers *et al.*, 1985). While these issues are discussed throughout the following sections, it is worth summarizing these pros and cons at this point (Table 3.8), if only to focus attention on the realities of putting the various coastal zone management concepts into practice.

Table 3.8 Advantages and disadvantages of formalizing organizational objectives (adapted from Steers et al., 1985)

	For organizations	For individuals
Advantages	• Focuses attention to common goals • Rationale for organizing and prioritizing programmes • Provides a set of standards for assessing programme effectiveness • Source of legitimization • Recruitment of staff through identification of programme priorities	• Focuses attention • Rationale for working with an organization • Vehicle for personal goal attainment • Personal job security • Self-identification and status with an organization
Disadvantages	• Means to an end can become the real goals • Measurement stressed quantitative goals at expense of qualitative • Goal specificity problem (ambiguous goals fail to provide direction; highly specific goals may constrain action and creativity)	• Rewards may not be tied to goal attainment • Difficulty in determining relevant performance evaluation criteria • Inability of individuals to identify with abstract goals • Organizational goals may be incongruent with personal goals

(a) Coastal programme principles

Statements of principle within a coastal programme describe the programme's overall philosophy (section 3.1.3; Figure 3.1; Box 3.11). The most pervasive coastal management principle in use today is that of sustainability, in many coastal programmes being the primary principle.

(b) Overall goal in coastal management programmes

The concept of balance between development and conservation pressures has been used by many coastal nations as the centrepiece of their coastal zone management efforts. Sustainable development principles in coastal programmes (section 3.2) essentially grew out of earlier notions of balance, but the balance concept remains in use by some coastal nations, as shown by the examples in Box 3.12.

The vision statement chosen for the Sulawesi Selatan Province (Indonesia) Coastal Strategy forms part of a hierarchy of actions (Figure 3.7) (Bangda, 1996): it is guided by national strategies and policies and it

---

Box 3.11

Example guiding principles for coastal management programmes

Principles of integrated coastal zone management in the Pacific Islands Region Project Proposal (SPREP, 1993)

- The needs of present generations must be met without compromising the ability of future generations to meet their own needs.
- Equity in participation must be promoted in sustainable development.
- Adverse environmental impacts of economic development must be minimized.
- The precautionary principle must be taken into account.
- Resource use and development planning policies must integrate environmental considerations with economic and sectoral planning policies.
- International responsibilities in the Pacific islands region must be met.

Main headings of the Guiding Principles for the Australian Federal Government's Coastal Strategy (Commonwealth of Australia, 1995)

- Sustainable Resource Use:

  - integrated assessment;
  - the precautionary approach;
  - resource allocation;
  - the user-pays principle;

- Resource Conservation; and
- Public Participation.

---

also guides regional and local coastal initiatives. Similarly, the Central Coast Regional strategy in Western Australia developed a set of Founding Principles guided by an overall statement of purpose (Box 3.13).

The Indonesian and Western Australian case studies shown above represent the geographic, cultural and administrative spectrums. At one end Indonesia represents a tropical archaepelagic nation of 185 million people with a diversity of cultures which intensively use their coast and a dominant 'top down' government. The Central Coast is a length of sub-tropical coast in sparsely populated developed country (Australia) where the major issues are linked to growth management and conservation. Nevertheless, all the case studies contain similar planning goals based on concepts of sustainable development. Both planning systems are described at greater length in Chapter 5.

Box 3.12

Examples of overall coastal management goals

Western Australian State Government goal for coastal planning and management (Western Australian Government, 1983)

- To achieve a balance between the protection of environmental quality and provision for the social and economic needs of the community.

Purpose of the Coastal Resources Management Program of Equador (Robadue, 1995)

- The preservation and development of the coastal resources in the provinces of Esmeraldas, Manabí, Guyas, El Oro and Galápagos.

Draft Vision for the Sulawesi Selatan Province (Indonesia) Coastal Strategy

- Coastal resources of Sulawesi Selatan will be managed on an integrated basis so as to ensure optimization of the economic, social and ecological benefits they provide.

Draft Overriding Coastal Zone Management Goal of the Western Australian State Government

- To sustainably manage the coastal zone for the long-term benefit of the people of Western Australia, the nation and international visitors by balancing the protection of environmental quality and features of cultural significance with the provision for social and economic needs.

(c) Coastal programme objectives

The objectives of a coastal zone programme are, in many ways, the programme's cornerstone. Badly framed objectives can fundamentally weaken the status of a programme, and seriously hamper its implementation. Well defined objectives, on the other hand, can bolster a coastal programme and significantly enhance its chances of succeeding. However, defining programme objectives to strike the right balance between clarity of purpose on the one hand, and becoming overly rigid on the other, is often not straightforward. The issues which influence this balance are described in this section.

The objectives of coastal programmes are important for a number of interrelated reasons: programme development, implementation and evaluation, and the role of programme stakeholders. The most important issues are discussed below:

Figure 3.7 Sulawesi Selatan Province (Indonesia) coastal planning system (Bangda, 1996).

- whether or not to develop quantifiable objectives;
- multiple objectives;
- conflicting objectives; and
- methods of objective setting (including implied versus stated objectives).

Choosing whether or not objectives should be quantifiable depends on their intended purpose. Coastal programmes focused on achieving outcomes generally develop quantifiable objectives which can help to measure whether the programme has been successful. Other programmes may decide to develop a small hierarchy of objectives statements – called operative and operational objectives. These are defined by Steers *et al.* (1985) as:

- Operative objectives represent the real intentions of an organization. They reflect what an organization is actually trying to do, regardless of what it claims to be doing.
- Operational objectives are those with built in standards that can be used to determine if objectives are being met.

Whether the objectives of a coastal initiative are operative or operational largely depends on its geographic coverage and focus. Water quality

---

Box 3.13

Central Coast (Western Australia) Regional Strategy Purpose and Founding Principles (Western Australian Planning Commission, 1996a)

The Central Coast Regional Strategy (Western Australia) – Purpose and Principles

The background to the Central Coast Strategy is given in Box 2.4 and its planning context and the approach to its development shown in Box 5.12.

Primary Purpose:

- To provide a link between State and local planning which is based on a balance of economic, social and environmental considerations.

Broad Founding Principles

These relate to:

- ecologically sustainable development;
- regional identity;
- managing natural resources;
- facilitating the development of community facilities and social services;
- fostering economic development and promoting diversification; and
- coordinating and integrating regional planning and development.

---

management plans may, for example, have very specific operational objectives relating to the attainment of certain levels of pollutant loads in the water column, or the extent of marine ecosystem damage from changes in water quality. Most other types of coastal plans and programmes generally employ operative objectives, because of their wide-reaching aims. Examples of both types of statements are shown in Box 3.14.

In general, the greater the geographic coverage of a coastal initiative, the higher is the likelihood that its objectives will be operative in nature. The broad coverage of national coastal programmes or regional plans or policies means that specific targets are difficult to write into objective statements. Specific targets are often left to a lower level in the hierarchy of guiding statements. For example, the objectives statements in the Thames Estuary Management Plan stated a set of general objectives (aims), and left the development of targets to achieve those objectives to specific actions listed within an action plan (Figure 3.8). These actions

---

Box 3.14

Example of operative and operational objectives for coastal management programmes

Operative objectives

Goals (Objectives) for the Sulawesi Selatan (SulSel) Province (Indonesia) Coastal Strategy

- To provide for the present and future interests and aspirations of SulSel residents with respect to coastal resource use, access and enjoyment.
- To protect, restore and enhance coastal ecosystems of SulSel and the ecological processes which sustain them.
- To encourage optimal, efficient and sustainable use of coastal resources.
- To ensure that coastal management and planning activities are undertaken on an integrated basis and that management resources are used efficiently and effectively.

Operational objectives

The Western Australian Government Department of Environmental Protection has developed a series of recommended water quality criteria to meet environmental quality objectives.

| Recommended Environmental Quality Objective | Example recommended environmental quality criteria (for example, inorganic | | |
toxicants) $(\mu g l^{-1})$			
	Arsenic	Cadmium	Manganese
Maintenance of biodiversity			
Maintenance of ecosystem integrity			
Class I – Conservation Zone			
Class II – Multiple Use Zone	50.0	2.0	100.0
Class III – Industrial Buffer Zone	50.0	2.0	100.0
Maintenance of aquatic life for human			
consumption	0.02	0.2	100.0
Maintenance of recreational values			
Maintenance of aesthetic values			

---

specify a time-scale for their achievement, and list the 'partners and players' required to implement each action. As such the actions specified by the Thames management plan contain the built-in standards (time frame and responsible agencies) required for operational objectives.

Localized coastal initiatives are more likely than those at national or regional level to contain specific operational objectives. However, if

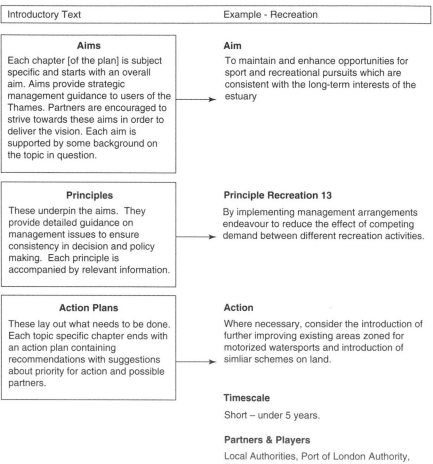

Our Vision
A shared vision for the Thames

Introductory Text   Example - Recreation

Aims

Each chapter [of the plan] is subject specific and starts with an overall aim. Aims provide strategic management guidance to users of the Thames. Partners are encouraged to strive towards these aims in order to deliver the vision. Each aim is supported by some background on the topic in question.

Aim

To maintain and enhance opportunities for sport and recreational pursuits which are consistent with the long-term interests of the estuary

Principles

These underpin the aims. They provide detailed guidance on management issues to ensure consistency in decision and policy making. Each principle is accompanied by relevant information.

Principle Recreation 13

By implementing management arrangements endeavour to reduce the effect of competing demand between different recreation activities.

Action Plans

These lay out what needs to be done. Each topic specific chapter ends with an action plan containing recommendations with suggestions about priority for action and possible partners.

Action

Where necessary, consider the introduction of further improving existing areas zoned for motorized watersports and introduction of simliar schemes on land.

Timescale

Short – under 5 years.

Partners & Players

Local Authorities, Port of London Authority, Environment Agency, Marine Safety Agency, British Marine Industries Federation, Governing Bodies, user groups

Figure 3.8 Thames Estuary Management Plan: explanation of guiding statements and an example (English Nature, 1996a).

coastal plans or programmes are wide-ranging, aimed at assisting in integrated coastal management, their objectives are likely to be too broad to be directly implementable. In these cases too, the standards required for ensuring objective implementation are written at the action statement level (Figure 3.1).

More often than not coastal management programmes have multiple objectives, as shown previously in this chapter. One reason for this is the complex nature of the exercise, as neatly summarized by Owens (1992, p. 144):

> ... from the outset the purpose of coastal zone management in the United States has not been simple or straightforward. It includes multiple goals, both preserving and developing coastal resources, some of which are invariably conflicting.

The objectives of coastal programmes generally fall into four groups:

- environmental;
- economic;
- social and/or cultural; and
- administrative.

These groups, and the objective statements within them, will depend on the overall goal of the particular programme. The wording of the overall goal will also affect whether subsequent objectives conflict with each other. In cases where either balance or sustainability is a key part, multiple and conflicting objectives will be inevitable. This is because balance and sustainability goals promote both conservation and development, and objective statements will contain objectives that encourage both the development of coastal resources and their conservation. Clearly, both objectives cannot be met in the same place at the same time. Resolving this inherent complexity is commonly one of the most important roles of coastal planning and management activities. Addressing the issue of conflicting objectives is often central to overall programme design, and to the various programme components, especially in coastal management plans. It is consequently a topic which is discussed throughout this book.

(d) Coastal programme action statements

Action statements are the 'doing' part of coastal programmes. They specify what tasks and activities will occur in order to meet the programme's objectives, and ultimately its overall goal.

The form of action statements will depend to a large extent on the geographic coverage of the particular coastal programme (Box 3.15). For example, if a coastal programme covers an entire nation, action statements are most likely to be oriented towards the development of new plans, or the implementation of broad management initiatives, such as a dune rehabilitation programme. In contrast, a more localized coastal programme may include action statements defining specific on-the-ground/water coastal management actions, such as the provision

Box 3.15

Example action statements from the Mandurah (Western Australia) Coastal Strategy (City of Mandurah, 1993)

Recommended Action	Responsibility					Timeframe			Priority		
	State Govt	Local Govt	Local Govt, Coastal Committee	Coastcare	Community	Short	Med	Long	High	Med	Low
Delegate appropriate management and maintenance activities to Coast Care Groups	•		•			•				•	
In association with appropriate user groups, institute a priority-based program for the widening – and relocation where necessary – of existing beach access paths . . .			•	•	•	•				•	

Note: Each of the selected action statements listed above accords with a higher level strategic objective in the coastal plan. Coastcare groups are community based.

of access ways in certain locations or the installation of public boat moorings. Again, depending on the coverage of a coastal programme, action statements can define matters such as those responsible for carrying out the action, the time within which an action should be completed, the required budget, and so on.

(e) Ownership of guiding statements in coastal programmes

It is tempting to think, after reading this chapter or leafing through the various publications of national governments and international organizations, that writing a set of guiding statements for a coastal management programme is relatively easy. Indeed, it would not take much effort to put together a set of statements. In most cases, however, this would at best be likely to result in a short-term improvement in the management of the coast, and at worst both a short- and long-term reduction in the quality and quantity of a coastal programme. Poorly defined objectives may contribute to the actions of those involved in a coastal programme being unproductive, or even counter-productive. The reason is simply to do with ownership.

Because most coastal management programmes attempt to tackle a multitude of issues which occur over different spatial and temporal scales, simply developing a set of guiding statements – and then telling stakeholders what they are – can do more harm than good. This is often referred to as the 'top down' management approach; that is, government agencies impose their ideas on those affected by the decisions of government. There can also be disadvantages in developing guiding statements purely from the 'bottom up', due to local biases and problems in how non-expert opinions are formed.

Successful coastal management programmes use a systematic process to develop guiding statements. These attempt to integrate the views of both the top and bottom decision-making levels with those of stakeholders in the management of the coast. There are various ways to achieve this, such as public workshops, seminars, enquiries and the public release of draft statements for discussion. Importantly, it is not what particular techniques are used, but that a conscious effort is made when deciding that a process be established for developing guiding statements which actively facilitate the meaningful involvement of stakeholders. Hence, a sense of ownership and commitment to achieving the objectives of the programme is developed. This decision is often made within the context of reforming a nation's coastal management administrative system, during the development of a coastal planning strategy, or during its evaluation. In all cases the factors which can influence the development of guiding statements, described in previous sections, should be taken into consideration.

3.5 Evaluating and monitoring coastal management programmes

Imagine a recently elected Government Minister responsible for coastal planning and management asking Government officials to 'tell me if the coastal management programme is successful or not'. Assume that these officials have included monitoring and evaluation criteria as part of the coastal programme's design. The Minister's question could be answered through the monitoring and evaluation results and how management objectives are being met throughout the lifetime of the programme (King *et al.*, 1987).

This hypothetical example highlights both the practical and political importance of programme evaluation and monitoring. The emphasis on 'proving' what coastal programmes have achieved is becoming increasingly important as a means to maintain programme funding, and to secure the commitment of all programme stakeholders.

Monitoring and evaluation are not simple tasks, but a systematic approach to them throughout its lifetime is a vital part of any successful programme (Herman *et al.*, 1987; Rossi and Freeman, 1989).

Anyone contemplating programme evaluation might usefully start by asking the set of questions framed by Owens (1993) (a characterizing key word for each question is bracketed):

- What is the ultimate reason for undertaking an evaluation (Orientation)?
- What is the state of implementation of the programme (State)?
- What aspect(s) of the programme should the evaluation be concentrated on (Focus)?
- What is the temporal relationship between the evaluation and programme development and delivery (Timing)?
- What is the appropriate underlying evaluation approach, and what are acceptable methods of collecting and analysing relevant information consistent with this approach (Approach)?

There are two main bases for programme evaluation – objectives-based, and needs-based (Owens, 1993). As the name suggests, objectives-based evaluation examines how a programme has performed in relation to its stated objectives. This is the most common type of programme evaluation (Cronbach, 1992). In an alternative approach, assessment is used to determine if a programme meets identifiable needs; for example, whether the needs of the beneficiaries of the programme are catered for. This type of evaluation has been called 'goal-free' evaluation (Owens, 1993), because the evaluation is of all programme effects and not necessarily restricted to those outcomes linked to programme objectives. This depends to a large extent on what aspects of the coast are being evaluated, and how the programme was designed. For example, the statements of programme objectives may have been written with evaluation in mind (termed 'operational objectives' – section 3 4.3). Such a programme is 'objectives led'. In this case, an objectives-based evaluation would be the most appropriate measure of that programme's performance.

The right type of objectives-based evaluation will depend on what exactly the evaluation is trying to achieve, the status of the programme and how the results of the evaluation will be used. Owens (1993) divides the many different types of evaluation into five 'evaluation forms'. The important features of each type of evaluation form, in relation to the major components of any evaluation programme, are shown in Table 3.9 and discussed below.

A common perception is that evaluation can only take place after a programme has been completed. As shown in Table 3.9, this is not the case; only one form of evaluation – impact evaluation – occurs after programme completion. The other four forms are undertaken either before the programme has started or during its operation. Thus, the evaluation of coastal programmes should be viewed as a process which occurs before the programme is initiated, continues throughout its life and is completed after the programme is finished. A wide range of techniques can be used to undertake the different forms of evaluation depending on the orientation, focus and timing of the evaluation. Detailed information on various approaches to evaluation can be obtained from specific texts on evaluation techniques (eg. Owens, 1993; Cronbach, 1992).

The problem of evaluating coastal programmes has recently been high-lighted in the United States, where coastal management is undertaken primarily by State and local governments with Federal financial and other assistance (Box 3.8). Major problems were experienced in evaluating the State coastal management programmes for four main reasons (Box 3.16) (Knecht et al., 1996).

In order to overcome the evaluation problems listed in Box 3.16, Knecht et al. (1996) used a surrogate measure of coastal programme performance, namely the opinions and perceptions of those knowledgeable about a programme. The results found considerable variation between sample groups (academics, coastal programme managers and coastal interest groups), but no systematic difference between the performance of different programme structures. The latter issue is discussed in more detail in section 3.4. Perhaps the key point from this study is that considerable difficulties remain in evaluating the performance of United States state coastal programmes, and considerably more work needs to be done before a thorough picture of state-by-state performance can be built up.

The study by Knecht et al. (1996), supported by previous evaluative studies of the coastal management systems (e.g. University of North

Table 3.9 Forms of programme evaluation (adapted from Owens, 1993, p. 22)

	Evaluation forms				
Factor	Impact evaluation	Evaluation in programme management	Process evaluation	Design evaluation	Evaluation for development
Orientation	Justification	Accountability	Improvement	Clarification	Synthesis
State (of programme)	Settled	Settled	Development	Development	None
Focus	Outcomes/delivery	Outcomes/delivery	Delivery	Design	Context
Timing	After	During	During	During	Before
Typical approaches	• Objectives-based • Needs-based	• Component evaluation • Systems approach	• Implementation studies • Action research • Responsive evaluation	• Evaluability assessment • Programme logic • Accreditation	• Needs assessment • Review of practice • Research synthesis

---

Box 3.16

Problems in evaluating US state coastal management programmes (Knecht et al., 1996)

1. The 'federalist' problem: are state coastal programmes expected to achieve national goals, state goals, or both?
2. The 'process versus substance' problem: should state coastal programmes be evaluated in terms of process-related goals (i.e. number of new coastal zone regulations) or in terms of substantive (on the ground) outcomes of specific problems found in each state?
3. Problems related to the general lack of outcome-related information and data: how can programme performance be evaluated without information on results actually being achieved? Furthermore, state coastal programmes often lack clearly articulated goals and objectives, which compounds the problem of evaluation.
4. The 'attribution' problem: a state coastal programme is not the only programme seeking to protect and enhance the coastal zone and its resources. Consequently, an outside observer would face difficulty in knowing which programme to credit with success or failure. In addition, factors external to any management programme can also affect what occurs at the coast, such as social and economic trends.

Carolina Center for Urban and Regional Studies, 1991; Born and Miller, 1988; Donaldson *et al.*, 1994), suggests that there should be a well thought-out process for evaluating and monitoring how a coastal programme is performing. Further, using the work of Owens (1993) shown in Table 3.9, this process should include consideration of evaluation at different stages of the programme, including the programme design stage, during its implementation and after its completion (Box 3.17).

These evaluation themes will be returned to throughout the book, especially in relation to the planning of coastal programmes. It is important to emphasize that programme evaluation is often the last component to be thought about in the design and implementation of a coastal initiative. Indeed, as will be shown in Chapter 5, 'monitor and evaluate' is often stated as the last step in the planning process. However, experience from the United States, which can look back at 25 years of its coastal programmes, has shown that evaluation should be the first issue that should be thought about. Perhaps the most pragmatic reason for this is to justify future funding by governments and international donor agencies who want to know their money is being well spent.

There are also some political issues related to monitoring and evaluation which should be borne in mind by potential evaluators. Coastal management programmes are essentially those of government, with varying

---

Box 3.17

Programme evaluation stages (Owens, 1993)

- Consideration of evaluation procedures should occur before the programme starts (Evaluation for Development). This pre-assessment should include an emphasis on the wording of statements of objectives and actions.
- Early in the implementation of the programme a preliminary evaluation should be undertaken in order to check on the overall design of the programme and how easy it will be to evaluate its performance once it is in full operation (Design Evaluation). Modifications to the programme can be made to ensure it performs as expected.
- During the implementation of the programme, periodic evaluations should be undertaken both for purposes of accountability and in order to highlight possible areas for improvement (Process Evaluation and Evaluation in Programme Management).
- Finally, once the programme has matured or has been completed, the justification of the programme should be tested (Impact Evaluation). This evaluation could focus on either objectives- or needs-based evaluation. This will depend on how the programme was designed, and how previous evaluations have been carried out.

---

degrees of involvement of private industry and the community (section 3.4.2). Thus any evaluation of the successes and failures of a coastal programme can be viewed as essentially an evaluation of the performance of government itself. This can be particularly so in nations with a strong coastal focus, where much of the nations' infrastructure is located.

Thus, coastal programme evaluations can rapidly take on a high level of importance, attracting the attention of politicians and senior government officers. As a result, there may be an impression formed that it is better not to do an evaluation, simply because its results may not reflect well on the government of the day. This impression may be enhanced by key interest groups who may be benefiting from current coastal management arrangements, and who may feel they would lose out from any actions which flow from the evaluation's results. This view of evaluations may be shared by some government officials working on a coastal programme. A negative evaluation, they may feel, could reflect poorly on a career.

There are also forces working in favour of coastal programme evaluations, most notably within governments and international donor organizations, that borrow much of their philosophy and practice from the business world. Within such environments there is an increased awareness of setting and attaining targets, performance measurement and accountability. However, although this encourages evaluations to be carried out, there can be disadvantages in focusing on outcomes which are easy to measure. This has the potential to downgrade the importance of other important issues, such as coastal environmental quality or scenic beauty which may require more concerted efforts to yield meaningful evaluation criteria.

3.6 Summary

The major themes of this chapter are neatly summarized by Kenchington (1993) who states that, whatever approach is taken, an effective coastal programme has the following characteristics:

- a dynamic goal or vision of the coastal zone for the next 25 to 30 years;
- national objectives which guide the development of regional and local objectives and plans;
- a strategy, commitment and resources to meet the objectives;
- legally based authority, precedence and accountability for achievement of objectives;
- monitoring and review processes; and
- political, administrative and stakeholder commitment to implement the strategy.

The issues not covered in this chapter, but highlighted by Kenchington (1993) as central to an effective coastal management programme, namely strategy development and legal instruments in coastal programmes, are described in the following two chapters. These descriptions are supplemented by analysis of the major tools available to coastal managers, including the range of coastal planning approaches.

Chapter 4

Major coastal management and planning techniques

A wide range of techniques is commonly used in coastal management and planning. They can be used individually to address specific problems, combined to address more complex issues, or used as part of a coastal management plan. The number is enormous, and effectively covers all the techniques available for the management of the natural environment, urban centres and systems of government.

In order to narrow down the range of choice we have selected the coastal planning and management techniques which are the most common and/or important to assist in the sustainable development of coastal areas. They include those used today, such as policy, and Environmental Impact Assessment, and those emerging techniques which are being used in some coastal nations and whose application we believe will expand in the future. These techniques include the application of customary (traditional and indigenous) management practices and visual analysis techniques.

Though we have chosen to focus on the most important techniques, the number is still relatively large, meaning that the description of each will be necessarily broad. Nevertheless, each section describing a technique is structured to allow an introduction to the main factors important in its application to coastal planning and management, and is illustrated through the use of case studies. Sources of further reference are given throughout to enable additional detail on each technique to be readily obtained.

The major techniques are grouped into administrative, social and technical. This grouping is undertaken to highlight the similarity between some techniques, while showing the differences between others. This grouping is useful if at times somewhat artificial in that there are techniques which contain elements of more than one group. For example, Environmental Impact Assessment is a government process, a technical procedure, and also involves social components.

As in previous chapters, case studies are used to demonstrate the application of each technique to actual coastal management problems and issues.

4.1 Administrative

Governments can assist in improving the management of coastal areas in a variety of ways: by encouragement, through force or through the use of research and information. Approaches include the use of policies or general guidelines, or much more targeted means such as the enforcement of regulations or the issuing of permits and licenses. Increasingly, a softer, less authoritarian approach than emphasizing coastal management problems is being taken using education and training programmes.

4.1.1 Policy and legislation

'Policy' and 'legislation' are two words easily recognized by the public. When managers or politicians announce the passing of new policy or a new piece of legislation it is a visible sign that the coast has a high priority for decision makers. And depending on their implementation and enforcement powers, policy and legislation can be powerful tools for managing the coast.

Policy and legislation as described in this section are used by most coastal nations, but in different combinations and to varying degrees. To a large extent this reflects economic, cultural and political circumstances and also the length of time coastal programmes have been active. In some cases it reflects the maturity of a nation's coastal planning initiatives. As will be shown through case studies, coastal programmes, especially in developed countries, have tended to evolve through early controlling stages founded on policy or legislative control (government dominated) into communicative and participatory stages where education and other techniques dominate. Indeed, such evolution in coastal programmes in many cases cannot take place without first establishing a clear set of operating parameters, often established through policy and/or legislation.

(a) Policy

Politicians, administrators and managers often cite 'policy' as a basis for decision making. But what exactly is policy? A useful generic definition is 'purposive course of action followed by an actor . . . in dealing with a problem' (Anderson *et al.*, 1984). Policy is about guiding decisions (Figure 4.1), specifically about decisions regarding choices between alternative courses of action (Colebatch, 1993). Policy therefore is deeply rooted in decision-making processes and hence is interwoven within the mechanics of organizational behaviour – public and private, large and small. Consequently, there is a risk that analysis of policies in coastal planning and management becomes no more than sweeping generalizations for looking at the way in which decision-making processes operate. As described by Davis *et al.* (1993, p. 7) in the Australian governmental context:

> The idea of 'public policy' works on a range of levels. It can simply mean a written document expressing intent on a particular issue, or imply a whole process in which values, interests and resources compete through institutions to influence government action.

Nevertheless, the importance of policy to the effective management of the coast is so important that such an analysis must be undertaken here. In this section policy will be linked wherever possible to other chapters where government processes are discussed, most notably Chapter 3.

Policies important in the management of the coast can broadly be divided into public policy (that is, the policies of government agencies and their staff) and non-public policy. The latter refers to the polices of all organizations not part of the public sector, and their staff – including private businesses, non-governmental organizations and community groups. In practice, there is little or no difference between the concepts of policy development and implementation between the public and the non-public, but the distinction allows the extensive literature on public policy, most notably from the United States (e.g. House and Shull, 1988; Considine, 1994), to be divided from that on policies in the private sector (Christensen, 1982).

The broad notion of policy described above shares common elements with the general definition of planning adopted in Chapter 3, the most important being that both planning and policy assist in setting some conscious course of action. There is no distinct boundary between planning and policy formulation; indeed, in some cases coastal plans may be considered as spatially oriented policies. Policies attempt to steer a course of action by deliberately affecting decision making; planning

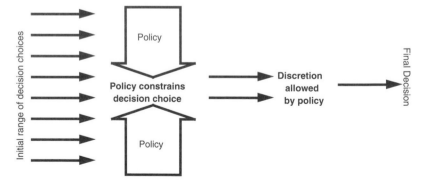

Figure 4.1 Policy and discretion in guiding decision making (adapted from Mukhi *et al.*, 1988).

Box 4.1

National-level coastal policy and planning in Australia and New Zealand

An interesting contrast between the use of 'policy' and 'plan' in developing national actions on coastal management is shown by the difference between Australia and New Zealand. Both nations have developed national approaches; Australia between 1993 and 1995 and New Zealand between 1991 and 1994. The Australian Federal Government chose to describe its policy as 'Living on the Coast: The Commonwealth Coastal Policy (1995)' but to describe its implementation jointly with State and Territory Governments as the 'National Coastal Action Plan'. In New Zealand the 'New Zealand Coastal Policy Statement' (1994) contained a number of well defined policy statements and expanded on the requirement of a framework of regional coastal plans (see Box 5.14).

In both Australia and New Zealand, national-level coastal policy statements were used to establish a national coastal planning framework. Again, in each case the policy statements use many planning elements, such as the use of guiding statements.

Examples of regional coastal planning initiatives in both Australia and New Zealand are described in Chapter 5.

---

attempts to do the same. Both attempt to produce structured, deliberate and consistent decisions by first clearly stating objectives, then actions in order to achieve those objectives.

In practice, the similarities between policy and planning increase as the geographic coverage of each increases. At the national and international level especially, coastal management plans and policies provide guidance as to how decisions are made – generally there is discretion to allow decisions to be made at regional and/or local level. At this level of planning the difference between planning and policy can become merely semantic, and does not necessarily reflect true differences in approach. This language difference is shown by the terminology chosen by the neighbouring countries of Australia and New Zealand shown in Box 4.1.

A useful way of describing policy in coastal management is through the terms 'expressed' and 'implied' policy used in business management (Mukhi *et al.*, 1988):

Expressed policies are written or oral statements that provide decision makers with information that helps them choose among alternatives.

Implied polices are not directly voiced or written. They lie within the established pattern of decisions.

Table 4.1 Management techniques used in the Sri Lankan Coastal Management Strategy (White and Samarakoon, 1994; Coast Conservation Department, 1996)

Policy	Management technique*
Erosion Control	
Regulate development suitability at specific sites	Education, Permit
Insure proper location in relation to the shoreline	Setback, Education
Regulate amount, location and timing of sand mining	Permit, Devolution
Build coast protection structures at appropriate locations	Master Plan for Coast Erosion Management
Regulate private construction of groynes, revetments	Permit
Limit construction in erosion prone areas	No-build zones
Habitat Protection	
Regulate location/use of development activities relative to valued habitats	Education, Permit Special Area Management
Regulate discharges from development which may affect habitats	Permit, Education
Reduce resource use conflicts	Special Area Management
Coastal Pollution	
Regulate effluent discharge of new development activities	Permit
Archaeological, historical, cultural and scenic sites	
Regulation of development activities in relation to valued sites	Permit, Education

* More than one management technique is normally used to implement a given policy; only primary techniques are listed.

The use of expressed policies in coastal management is widespread. Coastal programmes, for example, may choose to specify a set of general statements of policy (Box 4.1). Such policies may operate at a range of geographic scales, from international to local. They can have a broad range of applications, and degrees of prescriptiveness. Examples or policies developed for the Sri Lankan coastal management programme (Table 4.1) demonstrate one possible range of application. A further example of expressed policies is taken from the New Zealand Coastal

Box 4.2

Expressed policies for coastal hazard management in New Zealand

The New Zealand Coastal Policy Statement (Box 4.1) lists a number of specific requirements of New Zealand governments. For example, for the management of the impacts of coastal hazards and potential sea-level rise the New Zealand policy contains six specific policies:

POLICY 3.4.1

Local authority policy statements and plans should identify areas in the coastal environment where natural hazards exist.

POLICY 3.4.2

Policy statements and plans should recognize the possibility of a rise in sea level, and should identify areas which would as a consequence be subject to erosion or inundation. Natural systems which are a natural defence to erosion and/or inundation should be identified and their integrity protected.

POLICY 3.4.3

The ability of natural features such as beaches, sand dunes, mangroves, wetlands and barrier islands, to protect subdivision, use, or development should be recognized and maintained and, where appropriate, steps should be required to enhance that ability.

POLICY 3.4.4

In relation to future subdivision, use and development, policy statements and plans should recognize that some natural features may migrate inland as the result of dynamic coastal processes (including sea-level rise).

POLICY 3.4.5

New subdivision, use and development should be so located and designed that the need for hazard protection works is avoided.

POLICY 3.4.6

Where existing subdivision, use or development is threatened by a coastal hazard, coastal protection works should be permitted only where they are the best practicable option for the future. The abandonment or relocation of existing structures should be considered among the options. Where coastal protection works are the best practicable option, they should be located and designed so as to avoid adverse environment effects to the extent practicable.

Policy Statement (Box 4.1) which lists the policies developed for the management of coastal hazards (Box 4.2).

The vast majority of expressed policies allow a degree of discretion in decision making. Allowing the professional staff of organizations to make decisions within the broad confines of expressed policies is one of the underlying principles of many organizations. Within governments discretion has been described as an 'inevitable, inescapable characteristic' (Bryner, 1987, p. 3). One way of visualizing the role of policy and discretion in decision making is shown in Figure 4.1, which highlights the role of policies containing the range of possible decision-making choices. Figure 4.1 shows a policy acting to reduce the range of possible decisions. In this visualization the degree of discretion narrows as the width of the gap constrained by policy reduces.

In many cases the link between expressed and implied policy is blurred with the discretionary powers of an organization's staff intertwined with that organization's culture or unwritten rules. The result can be a substantial grey area between expressed and implied policies. The grey area often occurs in cases where decision-making authorities are required to make individual decisions in the absence of expressed policy. Such situations can occur where formal expressions of policy have not yet occurred in newly established authorities, where decision-making powers have extended beyond the boundaries of existing policies, or where day-to-day decisions have been made with the assumption that expressed policies existed because 'that is how things have always been done'.

For example, a permitting authority is developing 'policy on the run', because once a decision is made to allow a particular activity at a particular location policy has been set to allow others to undertake the same activity. However, this is not an expressed policy, unless there is a process to document that decision formally as a precedent that will be applied uniformly to all subsequent permit decisions.

There are significant advantages and disadvantages of implied policies (Table 4.2). Their major disadvantages include being hidden from public scrutiny, and hence the communication of them to stakeholders involved in decision-making processes possibly being poor. Implied policies can also lead to ad-hoc and sometimes inconsistent decisions. This can be exacerbated if informal policy formulation is undertaken by a few individuals without consideration of their flow-on effects.

In conclusion, policy-making is one of the central components of many coastal programmes around the world. The expression of formal policies can act as a guide to decision makers by helping them to choose between actions. In addition, many coastal initiatives contain unwritten (implied) policies which can be a critical part of how programmes operate in practice. The interaction between these different types of policy with legislation for coastal management is described in the next section.

Table 4.2 Advantages and disadvantages of implied policy-making in coastal management

Advantages	Disadvantages
Provides flexible decision-making	Open to decision-maker bias
Can be quick	Decision making may be inconsistent
Can be used to test future expressed policy	Lacks documentation and transparency Not discussed with the broader community

(b) Legislation

Legislation is the government of the time's response to community demands for government action or management of particular issues, areas or activities. Legislation or law is defined through a parliamentary or legislative process and the outcome is often expressed as an Act or Law and associated regulations. Before the assenting/passing of an Act or Law considerable debate in parliament and the community usually takes place. The government and community view legislation as a long-term approach to management of issues, areas or activities irrespective of the ruling political party. Because the formulation, passing and amending of legislation consumes considerable staff and financial resources, changing the law is often avoided.

Legislation has a number of functions in coastal planning and management, especially in translating concepts, as discussed in Chapter 3, to plans and management actions. Most importantly it sets out the broad purpose for managing the coast and the guiding principles for planning and management. It enables governments to incorporate sustainable development principles, including the precautionary principle and intergenerational equity, into a formal management framework, thereby establishing a basis for sustainable use of the coast while meeting international and national obligations. Also, in some countries legislation is used to define the coast spatially (Chapter 1).

Legislation can define or clarify institutional arrangements; or, if a new agency is required, it can specify how that agency will be formed, resourced and operated. If a new agency is not formed, legislation can specify the linkages and interactions of the various institutions. Kenchington (1990) suggests using existing institutions where possible and to use inter-agency agreements to effect management. Legislation also specifies the basis, scope and nature of planning and management. It can detail the steps undertaken to declare a planning area and to formulate the plan, including the requirements for public involvement. It can include the type of plans that can be produced, such as zoning plans, and make provisions so that plans also have the force of law.

An Act or Law can make provisions for the basis for management; it can also facilitate the use of specific mechanisms for management such as permits, licences, enforcement, education, monitoring and evaluation; and it can specify how the Act or Law will be enforced and who will enforce it. Similarly, legislation can facilitate the formulation of regulations so that provisions in the Act or Law can be implemented and that day-to-day management activities in the coast can be undertaken as highlighted in Chapter 5. Finally, legislation can specify the resourcing of planning and management activities.

4.1.2 Guidelines

The term 'guidelines' is used here to describe a group of documents which are less prescriptive and/or forceful than formal legislation, policies or regulations, but nevertheless guide the actions of decision makers. Clearly, there are many ways to 'guide' decisions, such as using advertising campaigns. This section does not focus on these, but rather examines the informal, yet structured, approaches used by governments for the production of guidance documents.

A useful way to consider the range of ways decisions may be guided was developed by Kay *et al.* (1996a) for examining the variety of approaches available to guide the examination by governments of potential future coastal vulnerability to climate change and sea-level rise (Figure 4.2).

The concept in Figure 4.2 is a spectrum of guidance which varies according to levels of prescriptiveness, direct applicability, flexibility and extent of required local knowledge. The practical outcome from the

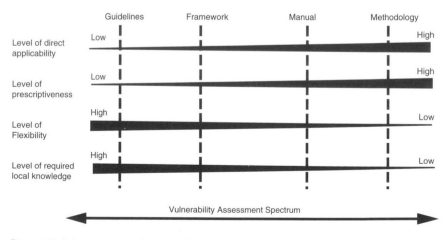

Figure 4.2 Schematic coastal vulnerability assessement guidance spectrum (from Kay *et al.*, 1996a).

consideration of such a spectrum is that the form of guidance could range from guidelines, through broadly structured frameworks and manuals, to methodologies.

At one end of this guidance spectrum are very broad, flexible and non-prescriptive guidelines. For example, sea-level rise vulnerability assessment guidelines could describe the range of possible assessment techniques and approaches for different biophysical, governmental, social, economic and cultural settings. Such guidelines would have to be interpreted according to need. Although the degree of flexibility is high, the level of direct applicability is low (Figure 4.2). At the other end of the guidance spectrum are highly prescriptive methodologies which aim to be directly applicable, but by their very nature are inflexible and require little local knowledge for their implementation.

Midway in the vulnerability assessment guidance spectrum are documents which allow some degree of flexibility while maintaining some direct applicability. Such documents include 'frameworks' and manuals.

Manuals are becoming increasingly important in Australian coastal management efforts (New South Wales Government Department of Public Works, 1990; Oma et al., 1992). They are designed to describe clearly the range of approaches available to coastal managers, and to discuss their strengths and weaknesses. Manuals can also be designed to include case study materials, as well as technical appendices as required.

The choice of guidance document types will be determined in part by the advantages and disadvantages shown in Figure 4.2, and in part by the way they are intended to fit within the broader coastal management system. In some cases the use of a manual will simply be explaining a range of techniques which may be available to implement a particular policy, legislative requirement or coastal management plan; in which case the manual is being used as an implementation tool that may supplement, or replace, the need for more detailed site-level planning. In other circumstances an education programme may require additional material which explains things such as the approach of governments in their coastal management efforts.

4.1.3 Zoning

Zoning is one of the simplest and most commonly tools in coastal planning and management. It is also one of the most powerful. Zoning, which is based of the concept of spatially separating and controlling incompatible uses, is a tool which can be applied in a range of situations and which can be modified to suit varying social, economic and political environments.

Zoning grew from the 'nuisance' crisis in urban management in newly industrialized cities in Europe and North America, especially in relation to health, sanitation and transportation problems. These problems were exacerbated early in the 20th century by the advent of the new technologies of the motor car, electricity, telephones and elevators; and the new construction methods, most notably steel-framed modular construction, which allowed high-rise buildings for the first time (Leung, 1989; Campbell and Fainstein, 1996). Zoning was promoted in the United States as a form of 'scientific management' for urban areas (Cullingworth, 1993). The result was that zoning became one of the founding principles of land-use planning systems in Europe and the United States. For the latter country, Haar (1977, cited in Cullingworth, 1993) described zoning as 'the workhorse of the planning movement'. According to Hall *et al.* (1993):

> In Britain as elsewhere, town planning had grown up as a local system of zoning control designed to avoid bad neighbour problems and to hold down municipal costs.

The use of zoning in land-use planning in the United States is summarized by Cullingworth (1993, p. 34) as:

> The division of an area into zones within which uses are permitted as set out in the zoning ordinance. The ordinance also details the restrictions and conditions which apply in each zone.

Thus zoning provides a simple mechanism for urban planners to integrate complex and often competing demands and land uses on to a single plan or map; and zoning plans provide an effective tool for communicating implicit and often complicated management objectives to the community in an easily understood form.

The widespread use of zoning schemes in urban planning has spread into larger scales of regional planning, where broad-scale land use zones can be identified. Use of zoning has broadened considerably from urban planning through its use in ecological conservation, especially in protected area management where the 'biosphere' model of core, buffer and utilization zones is used to manage and protect biodiversity (Gubbay, 1995). Zoning is also used extensively in the management of ocean space under international maritime regulations, which ensure the spatial separation of marine traffic in order to avoid collisions at sea. The use of zoning in urban planning, described above, has expanded greatly past the restriction through the issuing of permits being the primary land-use control mechanism. Zoning in many coastal management schemes now involves the three categories of 'allowed', 'permitted' and 'restricted use'.

(a) The mechanics of zoning

Zoning manages an area (land or marine) using management prescriptions which apply to spatially defined zones. Activities within a zone are managed by either specifying which activities are:

- allowed, or allowed with permission; and if an activity is not specified it is assumed not allowed unless permission is given; or
- prohibited, or allowed with permission, and if an activity is not specified it is assumed to be allowed.

It is worth noting these two approaches since they will influence how activities will be managed. In the first, and more common approach, new activities can be managed since a permit will only be issued if that activity meets management objectives. In addition, the permit may contain conditions which minimize the impacts of the new activities. Under the second approach new activities are allowed unless management can demonstrate that they are inconsistent with management objectives or have adverse environmental impacts. This approach is not used very often since it is costly and time consuming for managers to demonstrate the inconsistencies associated with each new activity.

Zoning as a concept can be applied at varying planning scales. Zoning plans can be formulated for broad geographical areas spanning political boundaries, or for a small area of only a few hundred square metres. The types of zones, the management objectives within the zone and the types of activities managed within these zones will, however, vary with scale. Zones such as 'tourism', 'agricultural' and 'industrial' are effective for broad management of a region or district, but are ineffective in managing conflicting recreational uses along a narrow beach.

There are a number of discrete steps in developing a zoning scheme in coastal management. The application of these steps depends on the existence of legislation to give effect to the zoning plan. In some cases, such as that governing the management of the Great Barrier Reef Marine Park (see Box 4.3), the legislation specifies the types of zones and the purposes for which they can be used. Such legislative prescriptions are more common for the land component of coastal areas, enforced through land-use planning legislation. Where land-use zoning legislation applies there may be very detailed zoning requirements in place which prescribe details of permitted and/or excluded activities.

The scale of management and the objectives for each zone underpin the formulation of a zoning plan. Again these objectives may be pre-determined by legislation, policy, or policies. In cases where the objectives are not predetermined, there is scope for clearly stating why a particular zone is being developed (see Chapter 3 for details on objective setting in coastal management). Where management is at the broad regional scale, zones will be defined to manage a range of uses, will have broad management objectives, and will cover broad areas. As the scale decreases, the range of uses is likely to decrease, and management objectives usually become more specific and operate at a fine scale in order to simplify the community's understanding of zoning provisions.

Existing environmental, social and economic information combined with community input on the current and future use of the area forms the base information for the establishment of a zoning plan. The complexity of the information required varies according to the intensity of use of an area and complexity of the zoning plan.

Finally, zones generally define the appropriate uses within a given area. Where possible, issues, activities or uses which can be differentiated into separate spatial areas should be allocated to appropriate zones. For example, if the risk of an accident between water skiers and windsurfers is an issue for a particular area, motorized and non-motorized water sports zones may be an option for managing the area. The non-motorized vessel area may also protect areas of higher conservation value because the damage caused by propellers is reduced.

When zoning is used to manage an area, the zoning scheme should be as simple as possible and the number of zones should be kept to a minimum. With more complex zoning and more numerous zones, the difficulty in implementing the plan increases and the community's understanding and support for the plan may decrease. An example of a zoning scheme applied to the management of the Australian Great Barrier Reef (GBR) is shown in Box 4.3.

The GBR approach to zoning integrates the Biosphere Model to the zoning of protected areas. In this model a core zone is used to give a high degree of protection to a specific area. The core zone is then surrounded by a buffer zone which allows limited use of the area while providing some protection. This buffer zone is surrounded by a utilization zone where there is limited or no protection (Figure 4.3).

The Biosphere Model is one of the simplest zoning plans and because of its simplicity is used by many agencies for protected area management (e.g. Indonesia, where it forms the basis of all protected area zoning plans, including marine protected areas – MPAs). The definitions of core zones which consist of a network of research and reference sites, buffer zones which manage human impacts for sustainable use and ecological function, and utilization zones to manage conflicting uses by spatial separation are simple. These broad definitions enable planners to use the broad objectives without modification, to redefine the objectives in light of local needs and to use these zones as a basis for a more detailed zoning scheme. The details of developing a zoning scheme, oriented towards the management of marine protected areas, are provided by Kenchington (1990) and Gubbay (1995).

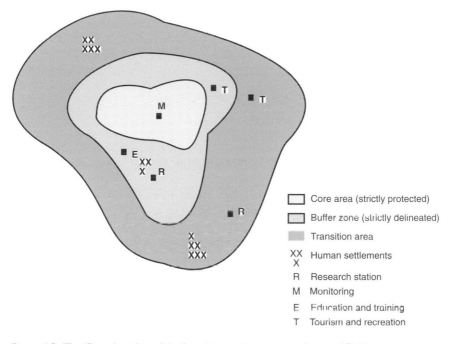

Figure 4.3 The 'Biosphere' model of zoning marine protected areas (Gubbay, 1995).

An area is 'zoned' using criteria which the planning team has developed in consultation with the community. The criteria are based on a range of ecological, social and economic values including: conservation and the presence of threatened or endangered species; access, recreation, traditional use and proximity to urban centres; existing use and potential and current commercial and industry development in such areas as tourism, fishing, mining, port development, mariculture or aquaculture.

Zoning boundaries should be clear and consistent. Setting the boundaries of the zones must also be considered, especially where zones extend into the marine environment. Zone boundaries can be precisely defined using geodetic reference points, but this may be of limited use to users who do not have the equipment or skills to locate these points. Geophysical features may be used, such as depth, high/low water mark, streets, depth/elevation, vegetation line, etc. The disadvantage of many of these features is that they are subject to change. Often the two are combined, with geophysical features the preferred method and with reference points used when features are not available. This approach is used in the establishment of the zones on the Great Barrier Reef (Boxes 4.3 and 4.4).

Box 4.3

The broad-scale zoning scheme of the Australian Great Barrier Reef

Zoning of the Great Barrier Reef Marine Park

Zoning at varying scales is used in managing the Great Barrier Reef Marine Park (GBRMP). The GBRMP is large (348 700 km²) and to undertake operational management on a park-wide scale is difficult. To overcome the problem of size, the park is divided into Sections which have the capacity to manage or regulate impacts, and to buffer the more highly protected areas from impacts originating outside the Marine Park (Kenchington, 1990). Within a Section of the park, a zoning system is used:

Original and modified zones within the Cairns Section of the Great Barrier Reef Marine Park

Original zones	Date defined	Current zones	Date defined
General Use 'A'	1981	General Use	1992
General Use 'B'	1981	Habitat Protection	1992
		Estuarine Protection	1992
Marine National Park 'A'	1981	Conservation Park	1992
Marine National Park Buffer	1982	Buffer	1983
Marine National Park 'B'	1981	National Park	1992
Scientific	1981	Preservation	1992
Preservation	1981	Preservation	1981

The initial zoning scheme was based primarily on extractive uses and minimizing these uses while providing for reasonable use (Kenchington, 1993). As issues, uses and community expectations and perceptions of the reef's management have changed, zoning has changed accordingly. The table shows how the names of the current zones have less focus on use, but greater emphasis on using other zones for habitat and resource protection to ensure general use zones are sustainable. In turn this reflects the evolution of management objectives.

Zoning which manages uses over a broad area may not be suitable for managing activities at a specific site. For example, tourism is allowed by permit in a number of zones, and the zoning plan does not specify the nature and intensity of tourism throughout the park or within a specific zone. As a consequence, zoning alone cannot manage the tourism at a specific site; it has the potential to allow nearly every site to be intensively

continued . . .

developed for tourism in an ad hoc manner. Permits issued to tourist operators give some degree of flexibility in managing the impacts of these activities, but do not provide much scope for managing at the site level. Area plans, which encompass a large area within the Section, and reef-use plans are two options for managing at a smaller scale.

How the broad-scale zoning provisions outlined above relate to the zoning plan for Green Island in the Great Barrier Reef region is discussed in Box 4.4.

Box 4.4

Reef activities zoning plan of Green Island, Great Barrier Reef (Zigterman and De Campo, 1993)

Within the Cairns Section of the Great Barrier Reef Marine Park, a management plan for Green Island and Reef, a popular tourist destination, is used to intensively manage tourism at the site (Zigterman and De Campo, 1993). The site is zoned National Park and the overall purpose of this zone is to provide for the protection of areas in a natural state while allowing for public appreciation of natural features which are relatively undisturbed;

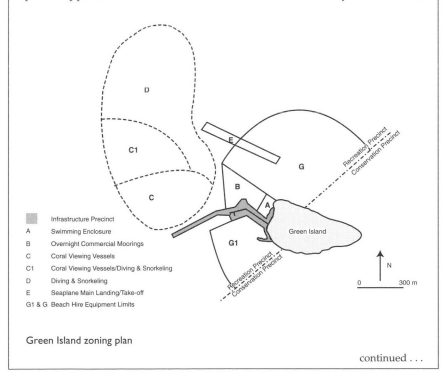

	Infrastructure Precinct
A	Swimming Enclosure
B	Overnight Commercial Moorings
C	Coral Viewing Vessels
C1	Coral Viewing Vessels/Diving & Snorkeling
D	Diving & Snorkeling
E	Seaplane Main Landing/Take-off
G1 & G	Beach Hire Equipment Limits

Green Island zoning plan

continued . . .

and to provide for traditional fishing, hunting and gathering (GBRMPA – Cairns Section Zoning Plan). Within this zone tourism is an acceptable use, but the zoning system does not make any provisions for determining the level, form and intensity of tourism.

In the site plan a number of strategies are used to manage tourism: restriction of the amount and types of use through limiting the number of day visitors to the site to a daily maximum of 2025; limiting the number of permitted operators at the site, and a form of tourism facility zoning; reduction of the impacts of uses which are allowed; hardening of the site; and monitoring. The management plan for the site includes the use of precincts (zones) to separate conflicting uses. Three precincts are used: conservation, recreation and infrastructure (see figure and table). These precincts complement or reflect the purpose and use of the National Park zoning. Implementation of the Green Island Plan commenced in 1993 and the use of zoning appears to have addressed many of the issues associated with conflicting use.

Precinct	Purpose
Infrastructure	To provide for the development of permanently fixed structures associated with access and use of the site for tourism and recreation
Recreation	To provide for intensive recreational use and enjoyment of the reef, with structures limited to re-locatable facilities
Conservation	To maintain a large portion of the reef free from built facilities

Where possible the pattern of zones should form a series of transitions in terms of restrictions or access (e.g. avoid placing a conservation zone beside a heavy industry zone: if possible try to separate the two with a buffer zone or recreation/commercial zone).

(b) Linking zoning with other coastal planning and management tools

Once zones have been established through a zoning plan, a number of related forms of management can be used in conjunction with the zones (Table 4.3). These other forms of management can overlay the zoning plan so that management can be fine-tuned for a particular area or resource.

The effectiveness of a zoning plan will ultimately rely on the community's acceptance of this plan and the government's commitment to provide the resources to implement it. Studies have shown that where the public has been actively and meaningfully involved in the planning process there is a greater acceptance of the plan, its regulations and their implementation

Table 4.3 Coastal management tools linked with zoning

Tool	Purpose
Time partitioning	Restricting access to an area or resource to specific times of the year
Facility/infrastructure restrictions	Specifying the type of gear which is used or what type of infrastructure can be constructed
Permit/licence quotas	Restricting the number of users accessing an area, using a resource or being allowed to undertake an activity
Production quotas	Restricting the amount of the resource which can be harvested
Licence–quota combinations	The number of users accessing the area is limited as well as the level of harvesting controlled

(Savina and White, 1986; Stone, 1988; Ehler and Basta, 1993; Kelleher, 1993). Techniques for involving the community in planning and management are discussed in Chapter 5.

A number of activities are undertaken to implement a zoning plan, with communication, education, Environmental Impact Assessment and enforcement playing major roles. These activities are discussed in this section. The implementation of zoning plans is similar to other plans and is discussed in Chapter 5.

4.1.4 Regulation and enforcement

Regulation and enforcement are often perceived by the community as simple and easy options for achieving compliance with mangement initiatives. The basis for this simplistic view is that the majority of the community by its very nature tends to comply with the law and assumes that the rest of the community is the same. Clearly there is a sector of the community which, for a number of reasons, including a lack of understanding of the purposes of management initiatives, blatant disagreement with them, or economic motives, does not comply. For this sector of the community, regulation supported by enforcement is used along with other mechanisms such as awareness and monitoring.

(a) Regulations, permits and licences

Acts of parliament provide the broad legislative basis for managing particular resources and activities, but often do not provide detailed

Table 4.4 Permitted activities and examples on the Great Barrier Reef Marine Park (Alder, 1993)

Permitted activity	Example
Exception to normal activities	Harbour works
Variable by their nature and need to be addressed on a case-by-case basis	Tourist programmes
Subject to potential conflict between allowed uses	Mariculture ventures
New activities with unknown impacts – – once their impacts are understood they can be classified as either allowed or prohibited	Establishing structures on the reef

prescriptions which can be used to implement an Act's provisions. Regulations, permits and licences commonly provide implementation mechanisms by specifying what actions are acceptable under the Act, and the penalties for breaching it. Because regulations, permits and licences are usually easier to amend than an Act, they provide a flexible mechanism for managing the coast. However, as will be shown below, regulations, permits and licences only remain effective when sufficient resources are provided to enforce them and, in the long term, when implemented in combination with education and communication programmes.

Permits and licences are written approvals from government to conduct specified activities in specified areas. Commonly permits are used in conjunction with zoning plans as a means of enacting a zone's specifications and/or restrictions. The processes and criteria for issuing permits are generally controlled by either policy directions or regulations, or are specified in legislation.

Permits can be used in a range of activities to assist in day-to-day coastal management activities, as shown by their use for the management of the Great Barrier Reef Marine Park (Alder, 1993) (Table 4.4).

(b) Enforcement

Enforcement is a management tool used to effect compliance with Acts, regulations, permits, licences, policies or plans with a legislative basis. Enforcement is a management activity that is highly visible, and generally outcomes are achieved in a relatively short time when compared with other management mechanisms such as education programmes. As a consequence, the public and politicians often perceive enforcement as 'the answer' to compliance. Enforcement is one of many mechanisms available to managers to encourage compliance with legislated management

Box 4.5

Enforcement of a marine reserve in the Philippines

In 1980 a marine reserve was established around Apo Island, Philippines. The marine reserve was established to assist in enhancing and maintaining fisheries resources for the local community of about 700 persons. The initial management of the reserve, however, was constrained by outside fishers who entered the area and not only over-harvested fish resources but also used destructive fishing methods such as illegal nets and explosives. Then, in 1985, an intensive community-based conservation programme started on Apo Island under the guidance of Silliman University (a Negros Island based institution with a history of community outreach programmes). This two-year programme formally established a fish sanctuary on one side of the island and assisted the community to develop a management committee for full-time surveillance and protection of the sanctuary and reserve surrounding the island. This community-based enforcement combined with an extensive education programme and other initiatives have resulted in a significant increase in fish catch to island residents over the last 12 years. Today, the Apo Island coral reef and community groups are the focus of numerous educational field trips from communities with similar interests in other parts of the Philippines. Today there are over 100 community-based coastal resource management projects (targeting fisheries, mangroves and coral reef resources) in the Philippines (Pomeroy et al., 1997).

provisions, but it is generally temporary and short term. Research has shown that as long as the 'big stick' of enforcement is applied by an enforcement agency having a high profile in the community and actively patrolling the area, there will be compliance. Once the big stick is removed, however, many members of the community will revert back to their undesirable activities. But research has also found that when enforcement is used in combination with other management tools, long-term compliance can be realized (Box 4.5).

The various regulations, licences, permits and legislative tools used in coastal management are sometimes not worth the paper they are written on because they are not enforced. Of course, there can be a myriad of reasons for the non-enforcement – a lack of resources (not just financial but also staff); staff may lack the expertise needed to undertake various enforcement activities, or it may be culturally difficult to act as an enforcer; there may be a lack of political support to prosecute offenders and previous efforts to prosecute may have been unsuccessful, resulting in a reluctance to undertake further enforcement activities. The most common reason is simply a poor understanding of what it actually takes to effectively enforce the various 'rules' imposed by governments.

Box 4.6

Enforcement programme of the Great Barrier Reef Marine Park

Between 1985 and 1991 the Cairns Section enforcement programme consisted of air surveillance and vessel patrols. Air surveillance was designed on an annual basis to survey particular areas of the Section at certain frequencies based on a stratified random sampling scheme. Vessel patrols were also designed to cover specific areas at a certain frequency, but weather and staffing constraints limited the statistical basis for the patrols. In either programme, breaches of the Great Barrier Reef Marine Park Act, regulations or Section zoning plan were recorded; these records were then used to examine changes over the six years of the study.

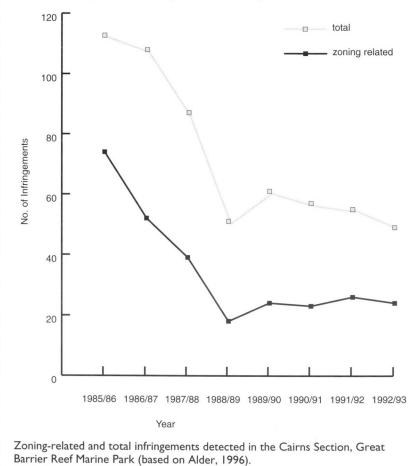

Zoning-related and total infringements detected in the Cairns Section, Great Barrier Reef Marine Park (based on Alder, 1996).

continued . . .

As the figure shows, total infringements declined steadily until 1988/89 and then remained constant; this pattern was also evident for infringements related to zoning compliance, which declined from 74 in the 1985/86 financial year to 18 in 1988/89 and remained at that level. Other types of infringements, however, were variable over the same time. The total number of infringements detected and zoning plan infringements were not significantly correlated ($P > 0.05$) to the amount of staff time or funds spent annually on enforcement (Alder, 1994).

In the corresponding time frame an extensive awareness and communication programme was implemented. The programme focused on raising user awareness of the Park and that there were areas (zones) where certain activities were not allowed. To simplify users' understanding of zoning, all visual material for each zone was colour-coded, e.g. green was a National Park zone which meant 'look but don't take'; blue was General Use zone which allowed fishing; and pink was preservation – 'no-go'. Offices within the management agencies would also refer to the colour system when they explained the zoning system. A subsequent survey of the effectiveness of awareness and communication programmes indicated that the zoning information was disseminated throughout the community and that there was support for management of the Park. It would appear that awareness programmes contributed to reducing zoning infringements (Alder, 1994, 1996) (see Box 4.8).

Enforcement programmes can also be very expensive and time consuming, and can be stressful for the enforcers. The constant reinforcement of an essentially negative message ('you are not allowed to do that') by enforcement officers can erode their morale and also lead to long-term inefficiencies in programme delivery. Hence, the trend in the effective compliance of coastal programmes is to integrate enforcement with communication strategies aimed at pointing out to those who breach the rules what the consequences of their actions are, and more importantly why the rules were established to begin with. Communication and enforcement are now seen to go hand-in-hand, acting to support each other.

Experience with enforcement programmes in marine parks has shown that most people in the community want to comply with regulations, permits and licences (Alder, 1994). For this sector, compliance is quickly gained once they are aware of the rules. The various regulations used in the management of the Great Barrier Reef Marine Park require an active enforcement programme. The effectiveness of this programme is described in Box 4.6.

The case studies shown in Boxes 4.5 and 4.6 highlight the need to include enforcement as a component of any coastal management planning programme. Enforcement programmes can be undertaken in a number of ways. Staff within an organization can be designated as inspectors/ officers and therefore have the power to enforce the provisions of an Act, or a plan if it has a legislative basis. Although one organization may have responsibility for management, it may delegate enforcement activities to other organizations, as is the case in the Great Barrier Reef (Box 4.6). If the expertise does not exist within an organization or affiliated institutions the use of private security officers or subcontracting out the programme is an option. This option is sometimes used in American national parks (Christensin, 1987). Which option to use depends on a number of factors such as funding, expertise, support from politicians and support from the community.

In summary, whatever option is used to enforce permits, licences policies or plans, the long-term effectiveness of enforcement programmes is enhanced when they are designed and integrated into other programmes. This is especially so when enforcement is integrated with communication and education programmes.

4.2 Social

The social dimension of coastal planning and management is often dealt with as an afterthought. Technical and scientific aspects can be emphasized, sometimes because it is easy to hide behind their 'objectivity'. The emotional and spiritual links and community values (aspects which are much more difficult since they are dealing with human nature, which is not predictable) are easier to avoid or to be given cursory consideration. As emphasized throughout this book, managing the coast is inextricably linked with managing society's use of the coast and therefore the social aspects must be an integral part of any management or planning programme.

4.2.1 Customary (traditional) practices

> Traditional knowledge is being lost very rapidly as its possessors die. Recording it is thus a truly urgent matter. Allowing it to vanish amounts to throwing away centuries of priceless practical experience. To record it with care and in the interest of its possessors – not just for the economic benefit of industrialised societies – is essential.
>
> (Johannes, 1989, p. 9)

This section outlines traditional resource management practices of non-western cultures and discusses how they relate to the planning and management of the coast. Customary resource management practices as they relate to the coast are introduced first, drawing on general literature in the area (McCay and Acheson, 1987; Johannes, 1989) as well as some excellent texts written specifically about the coast (Ruddle and Johannes, 1983; Johannes, 1984; Smyth, 1991, 1993). How these factors relate to the development of formal coastal programmes is then discussed.

Cultural factors play a central, if not the central, role in the successful management of coastal areas. As described in Chapter 3, the cultural norms of a coastal nation will shape the boundaries of a coastal programme, often long before notions of the exact details of programme design have been considered. Much of the content of this book focuses on the development and implementation of coastal planning and management systems which are essentially founded on the cultural norms of western developed countries. These western norms include the basic rules of data collection and analysis, and consideration of alternatives within essentially Christian values of the relationship of humans with their environment.

However, much of the global coastline is inhabited by people of cultural groups having their own cultural values and religious beliefs. Often these do not conform to western Christian values. The result can be that these non-western views of the relationship between people and the coastal environment can be viewed as somehow diverging from the western 'norm'. Of course, this view is misleading – all cultural settings require unique management and planning solutions, including western cultures.

Consideration of cultural factors in coastal management is driven to a large extent by the re-vitalization of indigenous cultures since the reduction of colonial powers over the last 100 years or so. The gradual withdrawal of European and North American influence from Asia, Africa, South America and the Pacific has seen a re-emergence and formalization within government systems of indigenous cultures. This is coupled with attempts to reconcile colonial and indigenous cultures in the 'new world' of North America, Australasia, southern Africa and South America.

Like the other tools described in this chapter, using traditional knowledge and practices to assist in coastal management is a specialized activity. As such, relevant experts, such as sociologists and anthropologists trained in culturally appropriate communication techniques, should ideally be used. The authors have both witnessed attempts to elicit traditional knowledge in clumsy, inappropriate ways. This can often lead to those engaged in traditional practices to tell outside researchers what they think the researchers want to hear. Sometimes, locals can be mischievous, deliberately misleading outsiders who do not go about things in the right way, or can refuse to grant access or interviews to subsequent researchers.

(a) Types of traditional knowledge and practice in coastal management

Traditional knowledge and practice in coastal management can be broadly divided into knowledge of the biophysical characteristics of the coast, and of the various management practices developed to manage the resource. The former focuses on traditional understanding of elements of the coastal environment of direct use to local populations, including an understanding of local oceanographic factors (tides, wave refraction patterns) for navigation and to help predict the movement of fishery resources; and knowledge of biological resources, most commonly linked in the coastal environment with the exploitation of fish, crustaceans and other marine fauna. An understanding of the schooling habits of a particular species of fish, for example, may be used to design more efficient ways of catching those fish with available technology. The use of so-called traditional ecological knowledge has been documented in hunter-gatherer cultures from the Inuit of northern Canada to Australian Aborigines (Smyth, 1991).

Interwoven with traditional knowledge of the biophysical factors in the exploitation of coastal resources are customary rules and decision-making hierarchies. The social structure of traditional groups, such as extended families and tribal groups, determines to a large extent how traditional knowledge of the biophysical environment is applied. For example, Cornforth (1992) demonstrated the importance of customary decision making in Western Samoa to day-to-day coastal management. In Western Samoa, and many other Pacific nations, villages 'hold tenure' over coastal lands and waters, including lagoons and nearshore reefs. The traditional basis of this is that villages communally gain access to all the potential resources on an island, from hilltops to the ocean (Crocombe, 1995). Indeed, traditional customs include the use of management tools described elsewhere in this chapter, including zoning, quotas on fish catches, development of regulations and policy (rules) and enforcement mechanisms (punishment and shaming). The use and application of these techniques in the Pacific is well documented (for example Zann, 1984).

The third important factor in traditional coastal management is the role of religious or spiritual beliefs. In many cases these beliefs are intimately linked with cultural systems and decision making, so that for all practical purposes they are one and the same.

Examples of traditional cultural values being followed, but with assistance on introduced technologies, are fairly common. Again with reference to the Pacific, religious ceremonies or visits from high-ranking members of neighbouring families may require the presentation of 'sacred' foods, such as a turtle or prized reef-fish. The importance of such occasions can outweigh day-to-day resource management considerations to the extent that dynamite, poisons or other destructive actions may be used in order to satisfy the cultural protocols.

Spiritual beliefs may also extend to restrictions on the taking of certain species of marine life , such as where they may be within the 'totem' of a family group; while other species may have special significance to particular age groups or genders. In the Gilbert Islands of the Pacific, for example, no clan (extended family) member is allowed to eat its totem; thus 'porpoise callers' cannot eat any crustacean, eel, octopus or scorpion fish (Grimble, 1972, cited in Zann, 1984).

(b) Balancing traditional and western approaches to coastal management

The prevailing view of the use of traditional approaches to coastal management is that it should be viewed in the same analytical way as any other approach (Johannes, 1989), a view that has evolved from opposing positions on the efficiency of customary practice. Some view customary approaches as being the most efficient and equitable methods of exploiting natural resources, being honed over hundreds (and sometimes thousands) of years. Others point to the view that such practices were only sustainable due to low population densities in the past, and are now inefficient and unsustainable. Both views point to examples drawn from around the world. However, these views are used here to describe two ends of a spectrum (which has considerable 'grey' areas) which balances traditional and western approaches, as summarized by Johannes (1989, p. 7).

> The truth lies between these extremes; wise and unwise practices coexist in many, if not most, cultures. The existence of the latter practices does not diminish the importance of the former.

Achieving a balanced view between the use of traditional and 'outside' approaches is one of the biggest challenges to effective coastal management in many nations today, especially in light of recent decisions to recognize indigenous rights over resources in coastal areas. Tensions between traditional and introduced management techniques may reflect larger tensions related to colonial influences and/or long-standing cultural differences. Nevertheless, the potential for harnessing traditional knowledge and integrating this with western approaches is enormous. Again, Johannes (1989) states with reference to biological information:

> The potential for the application of traditional environmental knowledge . . . is quite simply, vast. Such information must not only be collected and verified. It must be balanced with more technical

forms of biological research – population dynamics, pollution genetics . . . , before it can be put to use.

(c) Integrating traditional knowledge, practice and beliefs into coastal management programmes

How, then, can traditional knowledge, practice and beliefs be integrated into some form of structured coastal management programme? As has been alluded to above, the answer will depend on the scale and intensity of coastal management problems and the respective opinions and power of traditional groups and formal government organizations. The interplay of these factors can lead to a range of programme types. For example, where coastal problems are not severe, and there is joint desire by governments and traditional groups to retain traditional customary management, a decision may be taken to develop a 'minimum intervention' strategy. Thus, the coastal programme simply formalizes customary coastal management practice.

In cases where coastal resource degradation is significant, there is often the requirement for government intervention to employ western techniques to assist and/or overarch traditional approaches. In many cases the use of outside techniques is required because of the accelerated damage to coastal resources through the integration of western technologies with traditional practices. For example, the use of outboard motors on fishing boats has extended their range and speed, while using nylon fishing lines, nets and imported hooks has increased the fishers' efficiency, leading to overfishing.

The degree of traditional and government integration will depend to a large extent on the degree of local decision making and empowerment agreed to by those within the central and the traditional systems of governance (see Chapter 3). For example, governments may wish to formally recognize major parts of customary practice and management through the development of community management programmes. A way to work out the relative use of western and traditional management approaches is through consideration of them in coastal management planning. Through the use of the participative management planning process (described in Chapter 5) the customary importance of an area to its stakeholders can be discussed, and the various roles, responsibilities and management actions required agreed upon. The result can be the clarification of the use of customary knowledge, practices and spiritual values.

4.2.2 Collaborative and community-based management

Collaborative and community-based management are powerful tools which have the potential to help address coastal problems at the local level. Both are capable of effecting socioeconomic changes, modifying people's activities at the source of the problem in a way which can ultimately help to meet management objectives. Poverty, for example, is often the reason for environmentally inappropriate fishing practices in many coastal areas. Managers will therefore often focus on improving the people's income, and in doing so will bring about a shift from inappropriate to appropriate methods.

Collaborative and community-based management can also assist in integrating environmental and resource management activities into people's everyday lives: where a community makes some resource management decisions that affect their activities, management becomes a part of their lives. Furthermore, this type of management contributes to the socioeconomic development of the community. As mentioned above, problems are not just environmental and therefore all aspects of the community context must be addressed. Partnerships between people and nature can be strengthened by actively involving the community in management. A sense of stewardship and responsibility for managing resources is often an outcome of collaborative and community-based management (Drijver and Sajise, 1993). Various governments are aware of the benefits of collaborative and community-based management; the challenge for managers is to facilitate these forms of management. The next section describes collaborative and community-based management, and their role in planning and managing of coastal areas.

(a) Background to the development of collaborative and community-based management

Collaborative and community-based management in marine and coastal areas evolved from a convergence of several advances in protected area management, rural development and fisheries development during the 1980s. The 1980 World Conservation Strategy and 1982 Bali World Congress on National Parks emphasized the linking of protected area management with local area economic activity (Wells and Brandon, 1992). This concept was further developed in the late 1980s to link conservation with sustainable development, and led to the establishment of Integrated Conservation and Development Plans (ICDP). These plans focused on balancing the conservation needs of an area with the socioeconomic development of the community which is dependent on the area. The ICDP approach has been developed in agricultural and forestry projects, which have advanced community involvement in the management of land-based protected areas. Community involvement in managing marine and coastal areas has, however, lagged behind land areas due to the issue of managing shared resources in multiple-use areas.

The role of the community in coastal management is wide ranging and depends on a number of factors such a geographic scale, issues to be addressed, governance context, community motivation and capacity, and policy processes (Zeitlin-Hale, 1996). The community has several potentially important roles which contribute to planning and managing in coastal areas.

PARTICIPATORY COASTAL RESOURCE ASSESSMENT (PCRA)

Coastal dwellers and users are knowledgeable about local resources and can provide some of the biophysical information needed to make appropriate resource allocation decisions. Similarly, users can provide socioeconomic information more efficiently and effectively than most agencies. Through this PCRA maps and environmental profiles can be produced. Management costs (time, staff and funds) can be substantially reduced as a result.

PARTICIPATION

Stakeholders within a collaborative or community-based management programme are generally more accessible if communities are organized. This provides more opportunities for managers and stakeholders to discuss key issues and to interact with each other. It also ensures prompt feedback from both groups which leads to more efficient resolution of issues and faster integration of stakeholders in the planning and management of an area.

DECISION MAKING

Stakeholders bring ideas, judgements and perspectives which can lead to substantive results and a final product of high quality (Baines, 1985). This is particularly important since stakeholders are usually the groups who bear the majority of impacts related to access and resource use within an area. They are the users who generally have further restrictions enforced on their use of the area's resources and must bear the financial and social consequences. The design and implementation of programmes and management prescriptions are more readily supported by the stake-holders and the general community when they play a major role in decision making than in the absence of participation.

INITIATING ACTION

Stakeholders can readily identify needed management actions; this provides a better incentive to suggest, initiate and implement or support the needed actions. Again, this can make efficient use of limited resources.

PROGRAMME EVALUATION

As discussed in Chapters 3 and 5, stakeholders have a vital role to play in the formulation and establishment of evaluation criteria for management, and to be active participants in implementation of programme evaluation studies. Stakeholders can provide valuable insights and lessons about the design and implementation of a management programme. This information is otherwise likely to remain unknown (Wells and Brandon, 1992).

Managing agencies are also aware of the role the community has in planning and managing the coast, and many are shifting towards greater community involvement. This shift is increasingly being linked to broader trends in resource management toward a greater awareness of the relative roles of the community and lobby (or special interest) groups (Smith *et al.*, 1997). Collaborative and community-based management are two approaches available to managers to increase the level of community and interest group representation in decision making, and are described below.

(b) Making the choice: collaborative or community-based management?

Collaborative and community-based management are the two major forms of effective community participation in coastal management programmes. Which to pursue depends on the factors which affect the community's role, as discussed above. As an example, collaborative management is better suited to Sri Lanka's form of government and social structure, while in the Philippines community-based management is more of a possibility since local authorities have jurisdiction in coastal waters (White and Samarakoon, 1994). The differences between the two forms of management are discussed below.

Collaborative management, as the name implies, involves all stakeholders in the management of resources. In this form of management the aim is to achieve mutual agreement among the majority of stakeholders on the available options. White *et al.* (1994) note that collaborative management has a number of common elements: all stakeholders have a say in the management of resources; sharing of management responsibility varies according to specific conditions but government assumes responsibility for overall policy and coordination; and socioeconomic and cultural objectives are an integral part of management. Collaborative management is well developed in fisheries management (Jentoft, 1989; Lim *et al.*, 1995) and a set of common characteristics of is emerging (Table 4.5).

Community-based management uses a holistic approach to management by incorporating environmental, socioeconomic and cultural considerations

Table 4.5 Characteristics of collaborative and community-based management (based on Jentoft, 1989)

Characteristics	Collaborative management	Community-based management
Initiative	Decentral	Local
Organization	Formal	Informal
Leadership	Participant	Mutual Adjustment
Control	Decentral	Decentral
Autonomy	Some	Yes
Participant	Yes	Yes

in decision making by stakeholders. It is based on the concept of people empowered with responsibility to manage their resources. That is, the community together with government, business and other interested parties share an interest in co-managing resources with some decision making devolved to the community. The characteristics of community-based management are listed in Table 4.5.

In both approaches, consensual planning as discussed in Chapters 3 and 5 is the ideal process to formulate a plan of management. Community-based management, however, is rarely achieved since governments are reluctant to devolve power, communities are often viewed as unqualified or unskilled to take on responsibility for managing, or communities are reluctant to take responsibility for decision making. Nevertheless, community-based management represents a set of ideals that many communities and their managers might usefully adopt.

Collaborative and community-based management represent the bottom of Arnstein's ladder of citizen participation (see Figure 3.6). Collaborative management itself is not at the bottom of the ladder because it retains an element of government decision making. In a well developed community-based management programme, local decision making is undertaken by community representatives (Figure 4.4). This form of management represents the bottom of the ladder. Examples of community-based management from the Philippines (Buhat, 1994; Christie and White, 1994) and the Caribbean (Smith and Homer, 1994) demonstrate the effectiveness of this form of management in meeting management objectives. In the Philippines many islands and their surrounding reefs are planned and managed by the local community with the assistance of non-governmental organizations (NGOs). Because the community, especially the fishers, have determined the management regimes, there has been a wider acceptance and compliance resulting in improved fisheries resources. In many similar cases community-based management is intimately linked with government and traditional cultural groups joining to develop culturally appropriate coastal management systems.

Figure 4.4 Marine management workshop participants, Seychelles.

Collaborative and community-based management are not just a developing country phenomenon; they are also being developed in other countries such as Australia, Japan, Norway and the United States. Collaborative management has developed more widely than community-based, with a number of partnerships being established with resource management agencies. Collaborative arrangements can be based on either a sector or a geographic basis. Queensland Fisheries management (see Box 5.25), for example, includes working through advisory committees based on the various fishing sectors or geographical locations along the Queensland coast. Collaborative and community-based management can also be used to approach other management challenges such as incorporating traditional management practices, involving peripheral interest groups, and facilitating participation in planning.

Either of these management approaches provides a framework for governments to work with indigenous cultures in the joint management of coastal resources. A good example is the Great Barrier Reef Marine Park Authority, where aboriginal communities and the Queensland Fish Management Authority have agreed to ban gill-netting in the southern section of the park to address the problem of declining dugong populations (Anon., 1997). Community-based management is effective in involving urban and urban-fringe residents in on-the-ground management activities. Coastcare is a federally funded coastal management initiative in Australia which includes a major component focused on involving communities in on-the-ground management. Under this initiative, community groups are encouraged to assist in dune, reef, mangrove and beach management, through activities such as the construction of dune access-ways.

(c) Developing collaborative and community-based management programmes

Community participation usually begins with a bottom-up approach involving major stakeholder groups. The process is initiated through a government commitment to devolve some power to the community, and the community's recognition of the need to manage local areas. If the commitment is made and stakeholders are aware of the need to manage, then community-based management begins to evolve in the community. Subsequent actions and developments by government and the community determine the progress towards full empowerment. The development and implementation of community-based management programmes has been rapid in some countries such as the Philippines, and slow in other countries. In countries where there has been a strong paternalistic or government dominated approach to management, collaborative management is more likely to be a possibility, with slow progress towards greater involvement and empowerment following.

There are five common principles in developing community-based management as identified by Drijver and Sajise (1993):

- Process approach (similar to a bottom-up approach): managers and stakeholders agree on overall objectives, and then develop ideas and activities step-by-step towards achieving these objectives.
- Participation: all participants have some form of power in all phases of planning and management.
- Conservation and sustainable use: developed in partnership with all sectors of the community so that sustainable use programmes are socioeconomically acceptable.
- Linkages: between local management prescriptions, and regional or national level policies and strategies.
- Incentive packages (or readily observed tangible benefits – social or economic): these are an integral part of any community-based management programme. Stakeholders must perceive some benefit from participating in the planning and management of an area.

A community-based management programme has a number of components: community organization, education, NGO involvement, social benefits, government support and institution building. Initially, community organization is undertaken. It involves creating committees with

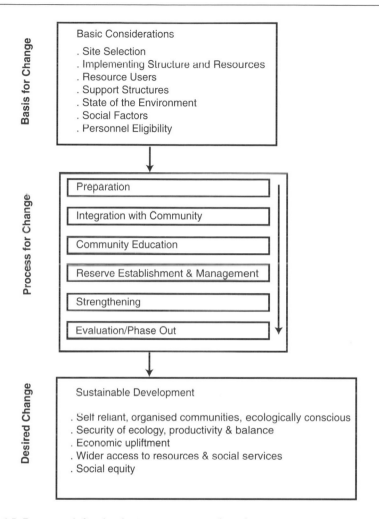

Figure 4.5 Framework for developing a community-based management programme (based on White *et al.*, 1994).

representatives from various sectors of the community so that particular issues can be discussed and programmes planned and implemented. Here the NGO component is included since they can assist in community organization and education. Education is an important component; it informs the community on the resources they are using, their value, and how they can be managed through a community-based management programme. Education programmes also explain how management of the area or resources will benefit the community, and how they can be a part of the planning and management of resources.

Box 4.7

Community-based fisheries management in the Philippines

Many Philippine islands were suffering the same problems: deteriorating marine environments due to destructive fishing practices. The consequences of these activities were reflected in declining fish catches and a corresponding loss of disposable income derived from sales of valuable fish. Increasing poverty was forcing people to use more efficient and destructive fishing methods (Savina and White, 1986).

The Philippines government recognized the need to actively involve the community early in the development of management programmes of marine and coastal areas to stem the decline of coastal resources. This recognition, combined with legislative changes giving local governments jurisdiction 15 km seaward of the low water mark (Rolden and Sievert, 1993), and the early work of academics in the area of community development, set the scene for effective community-based management of marine and coastal areas in the Philippines (Ferrer, 1992).

The community-based management programme was part of the Marine and Coastal Development Programme (MCDP) which was designed to enable local communities to protect and/or enhance their marine resources. The focus of the two-year programme (1985–1987) was the initiation of local marine management programmes in the form of marine reserves and sanctuaries (White, 1986; White and Savina, 1987).

Apo Island was one of the sites selected for the MCDP and symbolized an experiment in coastal management which has proved effective for coral reefs surrounding small islands and along some large island shorelines. The reserve model included limited protection for the coral reef and fishery surrounding the entire island and strict protection from all extraction or damaging activities in a small sanctuary normally covering up to 20% of the coral reef area (White, 1988a,b). This reserve and sanctuary approach is providing real benefits to local fishing communities through increased or stable fish yields from coral reefs which are maintained and protected (White, 1989; Alcala and Russ, 1990).

The objectives of the MCDP programme included the following.

1. Institutional development at Silliman University by raising awareness of resource management methods and community development skills.
2. Implementation of marine resource management programmes through the establishment of marine reserves, fish sanctuaries and buffer zones to increase diversity and fish abundance, and increase long-term fish yields.
3. Community development programmes to facilitate community marine resource management; alternative livelihood projects; community education centre.
4. Agroforestry and water development.

continued . . .

5. Outreach and replication to extend programmes to neighbouring communities and to establish a network of local and national organizations concerned with marine conservation and management.

The approach used to facilitate community based management was composed of five major activities:

1. Integration into the community: field workers live and work in the community so that they can be made aware of the island's culture and its problems, collect baseline data (environmental, socioeconomic, demographic and resource knowledge). This information enables further planning of the project.
2. Education is a continuous activity on a non-formal basis using small groups or one-on-one contact.
3. Core group building using existing community groups or facilitating the formation of new groups: core groups provide guidance on how the project should be implemented and suggest potential solutions. Groups often reflect the interests of community members; for example, the marine management committee (MMC) was formed by members interested in the reserve.
4. Formalizing and strengthening organizations providing ongoing support to the core group and its management efforts: assistance was given to the groups to identify new projects such as mariculture, training and tourist initiatives. Apo Island is now a training centre, which has strengthened the core group and solidified support for the marine reserve among the community.

The results of the MCDP at Apo Island are substantial:

- municipal legal support for the reserve exists;
- demarcation of the reserve using buoys and signs;
- reserves are managed by island resident committees which also patrol the area;
- municipal ordinances for the area are posted in the local language;
- moral support from the Philippine police is sometimes given;
- community education centre is established and is the focus for meetings and training programmes;
- dive tourism has increased;
- reef fishery resources have increased as well as diversity; and
- increased community satisfaction with management.

The success of Apo Island has been realized at two other islands in the Philippines which have followed the Apo Island approach. These case studies highlight the need to combine community, environmental and legal approaches for a particular site with long-term institutional support from government, non-government groups and academia to set the framework for effective community based management.

Real or perceived personal or social benefits, including ownership of resources or the management of those resources for sustained use, must be integral to the programme. This can only be achieved if there is government support which ensures that the legal mechanisms allow for some of the management responsibility to be given to the community and financial support for particular development programmes. Once community-based management is initiated, institutional development is another component. It is focused on supporting and training community groups so that they are given the skills and resources for long-term management. Support is often maintained through networking with other communities so that they have a support system to call upon.

The framework for community-based management is described well in Figure 4.5. Application of the framework and associated processes in the Philippines is illustrated in Box 4.7.

(d) Conclusion

It is interesting to compare the success of Apo island in community-based management with the example of the Sumilon islands, also in the Philippines. Apo Island has maintained a strong community-based management programme of reef resources since 1985: management included reducing destructive fishing methods and closing a section of the reef (up to 20%) to extraction and other damaging activities (White, 1996). The community agreed to the closure. Sumilon Island management commenced in 1974; however, it has had a weak, intermittent programme and less municipal support. Comparing fishery stocks of the two reefs shows that the island with a strong community management programme has increased its stocks significantly, while the other community has seen only a minor increase (Russ and Alcala, 1994).

The successes of collaborative and community-based management bear many similar features to consensus planning and implementation of such plans. Indeed, community-based management is one form of consensus management. Both are flexible management tools which can be applied to a range of social and cultural environments. They are flexible enough to meet the legislative requirements set by government as well as incorporating traditional practices within the same management programme. Collaborative and community-based management are recent planning and management tools which are being embraced by many nations. The challenge facing planners and managers is to improve the effectiveness of these tools and to broaden the scope of their use on the coast.

4.2.3 Capacity building

The ancient Chinese philosopher Lao Tse said:

> Give a man a fish and he will eat for a day.
> Teach a man to fish and he will eat for a lifetime.

The fishermen in the Central Visayas region of the Philippines taking part in a community-based fisheries management programme (Alix, 1989) modernized this proverb to:

> Give a man a fish and he will eat for a day.
> Teach a man to fish and he will eat until the resource is depleted.
> Teach a community to manage its fishery resources and it will prosper for generations to come.

'Capacity building' is a term used to describe initiatives which aim to increase the capability of those charged with managing the coast to make sound planning and management decisions (Crawford et al., 1993). The term is used commonly by international organizations, especially the United Nations in its various programmes. Capacity building is also increasingly used by national governments when new programmes or initiatives are introduced and there is recognition that relevant expertise among the participants is limited. This rather sweeping term, then, can be used to encompass a great number of apparently different activities, all of which are focused on supporting and improving coastal management decisions. The focus of these activities is on the 'human capacity' of individual decision makers and coastal managers as well as the 'institutional capacity' (Crawford et al., 1993). The latter refers to the coastal management capacity of businesses, governments, non-governmental groups and communities.

The distinction between human and institutional capacity is a useful one: human capacity building is centred on training and professional development, while the other aims to improve institutional arrangements for coastal management. There is a blurring of the boundaries between the two in the discussion of research and data management, which require the building of both human and institutional capacity, as is discussed in Table 4.6. Institutional arrangements were discussed and analysed in Chapter 3, and hence are not discussed further here.

Human capacity building can include anything from providing written training material, videos, facilitated meetings or workshops to extensive long-term formal education programmes, partnerships and mentoring schemes. The common theme of all these activities is on training, professional development and improved expertise. They are not just restricted

Table 4.6 Example components of a capacity-building programme

Capacity-building group	Example tools
Human capacity	• Education
	• Training
	• Professional development
Institutional capacity	• Policy development
	• Legislative change
	• Restructuring
	• Communication
	• Marketing
	• Education
	• Training
	• Information dissemination
	• Database management
Research	• Data collection and analysis
	• Database management
	• Results dissemination

to administrative types of activities but apply to other areas such as strengthening the research capabilities of individuals or organizations. Collectively these various activities contribute to strengthening individual or institutional capability to plan and manage the coast efficiently and effectively.

An emerging component of the way in which coastal programmes are developed and implemented is through the use of communication and marketing tools. These tools are increasingly fulfilling a variety of roles in coastal programmes, such as promoting the use of a particular policy, law, plan, management tool, or the application of a particular institutional design (Table 4.6).

The nature and scope of capacity building programmes will vary with the range of staff functions of the organization. If the organization's primary functions are administrative, then a capacity building programme will focus, for example, on improving skills in various administrative functions, policy formulation and strategic planning. An organization which is technically or operationally focused will have a capacity building programme to strengthen field operations to improve their surveillance and enforcement capability, or develop skills in resource assessment and community development. Similarly, if participants are experienced bureaucrats from either government or industry, a capacity building programme will be very different from one which is used to improve the community-based management skills of local coastal residents who have had limited exposure to decision making. Irrespective of the administrative level or management focus, individuals and institutions need the knowledge, skills and confidence to participate in decision making. Capacity building programmes play a critical role in providing this.

(a) Communication, education and training

This section introduces the use of communication, education and training techniques to assist in coastal planning and management initiatives. Communication is used here to describe the general act of imparting information in such a way that understanding is achieved and ultimately behaviour and attitudes change. Within this broad umbrella is a range of approaches including programmes in education, training and corporate-style communication. The term 'communication' is used to describe these, unless specifically referred to otherwise.

Five strategies can be used alone or in combination to influence behaviours and attitudes to ultimately achieve compliance with coastal plans and strategies (Global Vision, 1996):

- technological, employing new methods or equipment such as the use of moorings rather than anchors; or economic incentives or dis-incentives as discussed in section 4.3.4;
- enforcement, as discussed in section 4.1.4b;
- social marketing, which draws on marketing and communication techniques; and
- education, to raise awareness and understanding.

Communication has several functions in coastal management, including:

- reduction of social conflicts and resource impacts;
- gaining support for management practices;
- reduction of management costs;
- the potential for increasing users' experiences of the coast; and
- contributing to the development of community-based management.

In contrast to the use of regulations and enforcement, implementation of communication programmes can be inexpensive. In Australia, for example, it was estimated that an effective education programme targeting fishers could be implemented for 2% of the cost of enforcement (Bergin, 1993). Alcock (1991) also noted that education costs less money and effort than enforcement. Communication programmes take time and require a long-term commitment of staff and funds before benefits are evident; but communication can effect long-term behaviour changes, thereby reducing management costs over time.

A major factor limiting the funding and support of communication programmes is the time taken for their benefits to be realized. The impact

Box 4.8

Changing awareness and attitudes of Cairns (Australia) residents toward management of the Ciarns Section of the Great Barrier Reef Marine Park

A long-term study (Alder, 1996) of changing awareness and attitudes of Cairns (Australia) residents towards management of the Cairns Section of the Great Barrier Reef Marine Park has demonstrated the effectiveness of education programmes in several areas. Their value in changing awareness and attitudes was evaluated using face-to-face surveys of Cairns residents in 1985 when management of the Section began, and in 1991 prior to the review of the zoning plan. The results of the first survey were used to focus education and awareness programmes on informing the community of the existence of the Great Barrier Reef Marine Park, its values, issues and management regimes.

The six-year study highlighted changes in community awareness and attitudes. Awareness of the park's existence increased significantly; although the understanding of zoning (the basis for park management) increased, it was not significant. A detailed knowledge of zoning, however, decreased (see table). A total awareness score was formulated for the 1985 survey. The median score increased from 3 to 4 in the period 1985 to 1991. In addition, support for restricting or encouraging specific activities in the park such as resort development, shell collecting, and commercial, spear and recreational fishing remained high for both surveys. Support for encouraging fishing competitions and island camping remained unchanged and support for floating hotel development declined significantly. Park Management support remained high, but most respondents (46%) were undecided about how effective management was.

The education and awareness programmes contributed to improving community participation in the formulation of zoning plans. Although it did not increase the level of participation significantly, it enabled participants to focus on specific issues rather than broad general concepts.

Changes in % responses in community awareness to the Great Barrier Reef Marine Park

Variable	1985		1991	
	%	N	%	N
Park's existence	10	348	45	454
Zoning concept	65	34	70	201
Zoning details	46	22	19	24

of enforcement activities is immediate and publicly visible, while the effects of communication programmes are less obvious to the community and politicians; and managers are reluctant to assign adequate funding for them since it is difficult to measure the benefits. This issue was studied in relation to the management of the Great Barrier Reef Marine Park in Australia by Alder (1994) (Box 4.8).

Communication programmes can be developed to involve stakeholders in aspects of coastal management ranging from facilitating participation in the management planning process (including defining goals and objectives) and developing policy and drafting action plans, to involvement in monitoring programmes. Motivation and involvement of stakeholders is maximized when they can perceive the relevance of their participation. Again, communication programmes can address this issue (Box 4.9). In the Caribbean, the recreational diving community is involved in monitoring coral reefs using simple methods that require a minimum of training (Smith and Homer, 1994).

Other examples of communication strategies which can be easily understood by those targeted by a particular message in subsistence fishing communities in Indonesia and Papua New Guinea are shown in Box 4.10.

Corporations use marketing strategies to develop products or services that will satisfy wants. They communicate the benefits of the products or services on offer to existing and potential customers, ensuring that demands are fulfilled to the satisfaction of the customer and the business (Armstrong, 1986). This concept also applies to communication programmes used in coastal management. Managers may wish to develop communication materials (products) and programmes (services) which will alter specific behaviours or change awareness amongst users, which will satisfy management needs and users' wants. This focus on satisfying management objectives in the short term in order to benefit users and management in the long term distinguishes the use of education programmes in coastal management from marketing in the business environments. Nevertheless, marketing concepts are becoming increasingly important in the development and implementation of communication programmes in coastal planning and management. Examples of the use of marketing techniques that can be used in the management of marine parks are shown in Table 4.7.

Specialist education and training programmes are becoming increasingly used as an integral part of coastal management initiatives. Such programmes are offered both in-house as part of the on-going professional development of staff and by international organizations and tertiary training institutions. In recent years the fostering of regional centres of expertise in coastal management and planning has contributed substantially to the local delivery of education materials. For example, in

Box 4.9

Indonesian communication strategies for coastal management

Designing and delivering communication programmes in Indonesia, as in any developing country, is not an easy task. Several constraints, other than the chronic ones of limited resources and expertise, need to be addressed in the development of any communication programme at the national or regional level. Indonesia has 583 languages and dialects (Department of Information, 1992) and a diversity of cultures. Although Bahasa Indonesia is the national language, only those people who have completed high school studies understand and use it. Coastal dwellers, the most intensive users of marine resources, mostly speak their own local dialect and therefore any communication programme must include native speakers. Similarly, the literacy rate for coastal residents is considered low (Ministry for Population and Environment, 1992); consequently communication programmes must use alternatives to print-based media. Cultural and religious differences should also be incorporated at the area level. Particular attention is given to the different status of women since they are often the major exploiters of near-shore coastal environments.

Act No. 5, Article 37 of the Conservation of Living Resources and Their Ecosystems Act (Republic of Indonesia, 1990) specifies that education is a part of the management of protected areas in Indonesia. Clearly the Government of Indonesia recognizes and supports the role of communication in protected area management, but it has not historically provided the resources needed to use this management tool. Reviews of publications and reports on the development and progress of Marine Protected Areas (MPAs) (Soegiarto, 1981; Haeruman Js, 1988) do not indicate the use of communication programmes in their development, suggesting that communication has had a low profile in MPA management, until recently.

In 1992, World Wide Fund for Nature Indonesian Program (WWF-IP) developed a communication strategy (1991–1995) which focuses on raising awareness among key agencies to address marine conservation issues and strengthen information, education and communication at the park level (Schoen and Djohani, 1992).

Communication programmes at the national and MPA level are underway in Indonesia. Current initiatives in MPA communication in Indonesia are focused on either specific issues or areas. Specific issues include dugong and turtle conservation, coral reef management and mangrove management. Outputs from such programmes include posters, brochures and comic books. These media are usually inexpensive to produce and easy to distribute. Their effectiveness, however, depends on the education level of the recipients and how relevant the messages are to them. Although these programmes are nationwide, usually a portion of the resources is used to

continued . . .

undertake a case study to demonstrate the potential success of the programme. There are no published studies of the evaluation of these case studies, and few programmes are subsequently funded.

Major communication programmes are currently under way at specific sites across Indonesia, including Taka Bone Rate (South Sulawesi). This is part of the area's management planning programme and development of community-based management. The project is part of a WWF communication strategy.

At Taka Bone Rate, communication programmes are focused on a number of areas (Alder *et al.*, 1995a):

- awareness of marine resources, especially giant clams and turtles, and the impacts of destructive fishing methods;
- organizational skills amongst community members;
- business management; and
- basic planning and management methods.

The full impact of Indonesian communication programmes will not be realized for a number of years. However, there are some tangible results beginning to appear. For example, Taka Bone Rate residents are now aware of the impacts of destructive fishing practices and are investigating alternative income generating activities. A clam aquaculture pilot project has also commenced with the support of the community and funded by the World Wide Fund for Nature. These initiatives suggest that communication has an important part to play in developing sustainable management practices for the atoll.

the Asia-Pacific region the United Nations Environment Programme has established an education and training programme based on a distributed network of centres of excellence and country nodes (Hay, 1994) (Box 4.11).

(b) Research and data management

Many coastal management decisions focus on complex issues of resource allocation and are therefore made with a degree of uncertainty. Managers attempt to deal with this uncertainty by basing their decisions on an analysis of the best available sources of information, including the opinions and perceptions of stakeholders. What, then, does a manager do if he or she judges that a large degree of scientific uncertainty remains?

Some existing information may have come from previous or current research programmes. Often the planning process identifies information gaps or highlights the need for better or more appropriate information. If resources are available then research programmes are undertaken to obtain the necessary information. If resources are scarce, the plan may

Box 4.10

Example cartoon books for communicating impacts of coastal dynamite fishing

Indonesia

Papua New Guinea

continued . . .

Cartoon books for public education can be an effective way to communicate a message, especially in areas where literacy rates may be low. Examples of the use of such material to help reduce the use of explosives for fishing in Indonesia (Bason, 1996) and Papua New Guinea (Hershey and Wilson, 1991) are shown above. The first cartoon strip from Indonesia, using colloquial Indonesian, shows 'that evening' he 'makes his first bomb'. This strip also shows the common technique for making such bombs, which is filling a bottle with explosives, lighting the fuse and then launching. Primitive fuses are used, often resulting in severe personal injury, as graphically shown by the second example from Papua New Guinea.

recommend a range of research programmes to provide that information, or it may recommend research programmes (e.g. a coastal processes research programme) to answer specific questions or issues. Whatever factors initiate the research programmes, it is increasingly recognized that their outcomes, including the data and the management of the data, have an important part to play in reducing the level of uncertainty in any coastal management programme (National Research Council, 1995b):

> This need (scientific information) is becoming more evident as the complexity of the relationships among the environment, resources and the economic and social well-being of human populations is fully recognized and as changes and long-term threats are discovered.

Before the 1990s there were few coast-specific research programmes (e.g. the GESAMP – Joint Group of Experts on the Scientific Aspects of Marine Environmental Protection) developed to answer specific coastal management questions (National Research Council, 1995b; GESAMP, 1996). Much of the research on the coast was focused on ecological or science questions, or to provide information for engineering projects or EIA programmes at the site-specific level. However, the passing of environmental legislation, including coastal, in the 1960s and 1970s (section 4.3.1) highlighted the need for scientific information for decision making (National Research Council, 1995b). In turn, this spawned a belief by decision makers that research programmes were an essential prerequisite for decision making, and the views of 'uninformed' non-scientists were secondary. The result was a mountain of literature on various coastal research programmes (scientific and social), including some very comprehensive analyses of the techniques for developing such programmes and ways of managing the resultant data. The results were often collated and used opportunistically to provide scientific justification to reduce uncertainty in decision making. Despite this mechanistic view

Table 4.7 Marketing activities and their application to coastal communication programmes (based on Armstrong, 1986)

Activity	Description	Application in coastal communication programmes
Market planning	Setting of targets and markets based on corporate objectives and formulation of action plans	Establishing what changes are required in user behaviour and awareness to meet management objectives (e.g. reducing uncontrolled access to beaches by 50%)
Product development and planning	Developing new ideas and concepts and testing the products to ensure they meet customer needs	Developing new education material and ensuring that it will work with intended audience before distribution (e.g. a TV advert to inform the community of the impacts of uncontrolled access)
Sales planning	Defining fields or sales outlets	Defining the groups which will receive the program (e.g. surfers, fishers and local residents)
Marketing research	Collating information on actual and potential markets and users of goods and services	Determine the information wants and needs of the target audiences (e.g. where are the designated access points?)
Sales forecasting	Assessment of potential sales and market trends	Assessment of the potential short and long term impacts of the education programme on the target audience (e.g. how long will it take to see a measurable reduction in uncontrolled access?)
Analysis	Analysis of the product life cycle	How long will the programme be effective for? (e.g. how long will the TV advert have an impact on the target audience?)

Table 4.7 continued

Activity	Description	Application in coastal communication programmes
Target marketing	Formulating a more detailed definition of different groups that make up the market (segmentation) and determining where efforts should be targeted	A more detailed definition of the target audiences (e.g. surfers who belong to a club, surfers who are local residents, etc.)
Developing the market mix	Setting the blend of product, price, place and promotion to generate the responses the organization wants in the target market	Balancing the available funding (price) for the programme, with the intended messages (product), the most appropriate media (promotion), and target audiences (place) (e.g. balancing the cost of a TV advert and target audience within a limited budget)
Marketing and sales operations	Implementation of the marketing plan	Implementation of the education programme
Marketing and sales control	Monitoring performance to ensure targets are achieved within the budget	Monitoring and evaluating the effectiveness of the education programme
Feedback	Amending the plan as necessary	Revising the programme to improve its effectiveness

of research, it has contributed to improved decision making in the coast. The outcome of coastal processes programmes, greatly improving our knowledge and ability to define appropriate shoreline management strategies, and to improve the definition of coastal buffer zones, is a good example.

Global initiatives such as the Rio Summit emphasized that decision making needs to be supported by a range of information, including social, from a variety of sources; and the need to link scientists and managers initiated a change in information requirements. As discussed earlier, purely research-based systems evolved during the 1990s into a process by which the opinions of local or traditional knowledge and feelings of stakeholders are combined with scientific research through the principle

Box 4.11

Coastal management training in Asia and the Pacific

NETTLAP is an innovative networked approach to building the capacity of coastal managers, amongst others, in the Asia-Pacific region. NETTLAP was established by the United Nations Environment Programme (UNEP) in 1992, in response to a UNEP Governing Council decision and a subsequent regional meeting of experts convened to develop a programme of action for environmental education and training in the Asia-Pacifc region.

NETTLAP was established to:

- enhance the environmental expertise of decision makers, policy formulators and tertiary-level educators;
- increase the environmental skills and awareness of tertiary-level students;
- enhance environmental technologies and capacities for their use; and
- strengthen the overall environmental expertise in the region at technical, management and policy levels.

The activities of the network are to (Bandara, 1995):

- develop and apply innovative methods in environmental training;
- identify regional training needs and share knowledge through ongoing interaction amongst network partners;
- prepare and disseminate curriculum guidelines, resource materials, learning aids and packages for environmental training; and
- implement targeted technical Training and Resources Development Workshops.

Coastal zone management is one of NETTLAP's thematic networks, established because of the strong coast dependence of many of the island and continental coastal nations in the region. Assessments of the styles of coastal management training in the Asia-Pacific region showed that traditional sectoral (discipline) based education was dominant, while an interdisciplinary educational approach was more appropriate to stress the requirement for integration in the region's coastal management approaches (Chua, 1991; Hay *et al.*, 1994).

Currently NETTLAP links over 200 key tertiary institutions and more than 2000 staff members who are active in environmental education and training. Governments from 36 countries in the Asia-Pacific region have designated National Focal Points for the NETTLAP Network.

Staff of tertiary institutions (i.e. universities, technical institutes, training institutes and teacher training colleges) were chosen as key targets for

continued . . .

NETTLAP. The 'training the trainers' approach produces a large multiplier effect inherent in training such staff, this arising from the immediate transfer to colleagues and students (Yodmani, 1995). The role of tertiary institutions is particularly important in the region's developing countries where the staff are usually highly respected for their expertise.

The implementation of network programmes is organized by a NETTLAP Regional Coordinating Unit based at UNEP's Regional Office for Asia and the Pacific (ROAP) in Bangkok. The activities under each of the themes, namely Toxic Chemicals and Hazardous Waste Management, Environmental Economics and Coastal Zone Management, are organized by a Thematic Network Coordinator, and subregional Thematic Network Nodes with a proven academic record in the respective fields are identified and entrusted with relevant tasks.

The Coastal Zone Management component of NETTLAP has undertaken a number of activities, including the publication of five reports and the staging of a number of training workshops and discussion forums – for example, workshops in Ch-am, Thailand, Kandy, Sri Lanka and Manila, Philippines (Chou, 1994; Hay and Ming, 1993).

NETTLAP is focusing on innovative training approaches rather than depending on traditional discipline-bound training methods. Role playing, participatory field mapping, and the use of the Internet and databases are currently being explored as additions to curriculum development of tertiary-level courses.

The operation of NETTLAP for the past five years has shown that it 'has simply been spread too thinly across 35 participating countries' (Yodmani, 1995). NETTLAP is moving forward by helping to build national-based programmes. The first such project was recently developed in the Philippines. The aim of this approach was to bring together non-governmental and community-based organizations as well as industry and government. In addition, NETTLAP has recently catalysed substantial funding for national-level activities and partnerships in Thailand and Malaysia for the establishment and operation of Inter-University Networks on Training and Research on Environmental Management.

of precautionary decision making. This evolution effectively recognized that research programmes remain a crucial part of effective coastal management, albeit in a modified form.

Despite current research efforts, our understanding of coastal eco-systems and processes, social features and economic value remains poor at best. As a consequence, environmental decision making in the coast, as in most environments, is characterized by uncertainty. For researchers, uncertainty points to further inquiry, description and explanation; for policy makers, a concern with adequately reflecting societal values (National Research Council, 1995b). Decision makers, however, bound by ethical and practical considerations, do not have the luxury of suspending their decisions until all the scientific information is collected and analysed (Latin, 1993). Indeed, some have suggested that, from a management perspective (Welch, 1991 p.205):

> While good science and information are important, inter-agency co-operation is the first prerequisite to sound management.

The need to further evolve the linking of science and management, and heightened awareness of the importance of science, remains despite the advances made to date (National Research Council, 1995b). Latin (1993) suggests that conventional scientific norms may impede rather than promote reef conservation, because the response of science to uncertainty, in the absence of considerable knowledge and reliability, is no decision. This 'no decision' response by scientists also applies in the coast. The consequence of a no decision is significant since it maintains the status quo and does not contribute to problem resolution; in some situations existing problems may exacerbated. However, constraints in the links between researchers and decision makers can be reduced if researchers relax their decision-making norms and managers involve researchers in all stages of decision making. Latin (1993) further suggests that scientific norms of knowledge and reliability must be relaxed if scientists want to

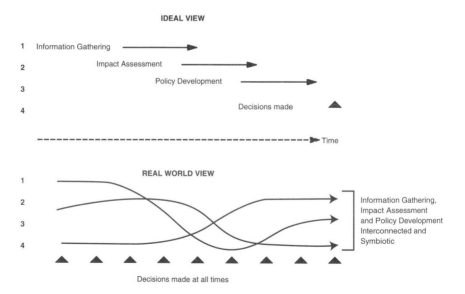

Figure 4.6 Ideal and 'real world' views of the interaction of information gathering, impact assessment and policy development for decision making (adapted from Kay *et al.*, 1996b).

facilitate better environmental management. The nexus between the gathering of information, the development of policy and decision making is shown with reference to Figure 4.6. An ideal view of this interaction is that decsisions are made after a well ordered sequence, whereas in 'real world' decisions information gathering, policy development and impact assessment are all occurring at the same time, thus influencing each other (Kay *et al.*, 1996b). The result is that information gathering, including science, policy and decision making become interconnected and symbiotic (Feldman, 1989).

The National Research Council (1995b) has identified areas where researchers can be involved in policy formulation, including provision of internal advice in the form of a report, or through an internal advisory group using researchers within the agency or contracted services. Advice can be obtained quickly this way, and can be targeted. Using advisory groups external to policy-making agencies can provide an independent evaluation of information, an approach which is useful when agencies require an independent review of internal mechanisms, and when it is cost effective to obtain the information outside of the organization. Workshops are another forum to provide advice, but it is important that workshop participants include policy makers and scientists. Another approach is the use of informal policy advisory groups composed of a range of internal and external researchers, stakeholders and decision makers to review published information and advise the decision-making agency.

The question which arises from the above discussion is: 'what coastal information do decision makers need which will also facilitate cooperation between researchers and managers?' The information needs will vary with the scale of planning. Information at the international level will be focused on large areas, summarized by country, highly qualitative and with limited precision. Information relevant to this level can include remotely sensed data, summarised demographic statistics and national economic analysis. At the site level, however, information will be very detailed and precise. Detailed site maps indicating individual plants, species lists with possible densities, and detailed geology and geo-morphological characteristics would be typical information needed at this scale of planning. A survey of coastal managers in Australia highlighted 28 types of information needed. The 10 most important types provide an insight into the scope and nature of the information (Table 4.8). The table indicates that environmental information is a high priority, but information on social factors and other planning initiatives is also considered important.

The analysis of information for decision making is critical to coastal management, but it is often inadequately performed due to reasons such as limited time, funding and expertise (Bower, 1992). Just as important is

Table 4.8 Ten most important information needs for Australian coastal managers (Brown, 1995)

Top ten (of 28) information needs for coastal managers	Rank order of availability of top ten information needs (1 = best; 28 = worst)	Top ten gaps in information as identified by coastal managers	Barriers to information transfer (% frequency of respondents suggestions)
1. Ecosystems, habitats and species	14	Ecosystems, habitats and species	Inadequate information services (48%)
2. Environmental Impact Assessment	12	Developmental benefits and losses	Unclear locus of responsibility (15%)
3. Condition of rivers, estuaries, etc.	25	Recreation and tourism	Need for research and investigation (15%)
4. Recreation and tourism	6	Condition of soils and beaches	Absence of coordination and integration (6%)
5. Community priorities for coastal areas	24	Condition of rivers, estuaries, oceans	No access to local information (6%)
6. Strategic plans	17	Regulations and by-laws	Inadequate resources – financial, human (5%)
7. Condition of soils and beaches	16	Pollution indicators	Lack of clear policies on coastal management (2%)
8. Integrated resource management	26	Community priorities for coastal areas	
9. Public participation	15	Integrated resource management	
10. Coastal hazards	13	Public participation	

the reporting, since it must be in a form that is understandable to decision makers and others; and the methods of analysis must be well presented for peer review (Bower, 1992). Therefore it is important to include data analysis and reporting as part of the planning process and to include other interested parties who can assist in defining what information is needed and in prioritizing information needs. Information needs will be guided by the issues identified, goals and objectives, programme evaluation criteria and planning scales. The analysis techniques available will also influence the choice of information to collect and store.

Once the research component of a coastal management programme is under way, it is imperative that the outcomes of research, including data, be adequately captured, stored, retrieved and reported (National Research Council, 1993). This raises a number of issues regarding coastal data such as ownership, consistency and access. Data management options used will depend on how these issues are resolved and the sources of data. Large environmental data sets should be structured so as to be transparent, reliable, scalable, and distributed; where possible, data entry should be automated (Malafant and Radke, 1995).

The use of Geographic Information Systems (GIS) to undertake these data management tasks is becoming increasingly widespread, especially when linked to the use of satellite remote sensing techology (Asian Development Bank, 1991b). GIS can be extremely useful for coastal management, and especially coastal planning, because of the ability of such systems to store and analyse spatial data captured at a range of scales. For example, GIS was used extensively to produce much of the background information for the regional coastal planning exercise described in Boxes 3.13 and 5.12. GIS technology is becoming cheaper, easier to use, and more reliable (Huxhold and Levinsohn, 1995). The greatest constraint to the use of such techniques is rapidly becoming the quality of the data, and how it can be updated and improved, rather than computing limitations (Wegener and Masser, 1996; National Research Council, 1997). In this regard, some countries are moving rapidly to develop stringent data quality, archiving and retrieval systems, known as 'data warehouses' (Inmon and Hackathorn, 1994). This notion, introduced recently in the United States, aims to ensure the rapid access to all US Federal spatial data sets through Internet technologies (National Research Council, 1995a).

Whatever research agenda and data management system eventuates it is worth remembering Bower's (1992) 'Four Facts of Life' with respect to information for decision making:

1. No analysis for integrated coastal management can include all the information and analyse all the alternatives.
2. There are physical and psychological limits to a human's capacity as an information processor and decision maker. Too much information

obscures the various trade-offs that are involved, which are the heart of the political process.

3. Only a limited amount of data relating to any given analysis can be presented at one time due to the complexities of coastal ecosystems, the complexities of decision making in the coast and the multiple use nature of the coast.

4. The format used to present the results will affect the amount of data that can be presented and will affect the extent to which the results are understood.

Management decisions must be made, and dealing with uncertainty is part of the decision-making process. Research plays a critical role in reducing uncertainty and providing advice on a range of environmental, social and economic factors. Research, however, can only make a significant contribution when information sources, processes and outcomes are efficiently managed. When scientists balance their strictly scientific norms with pragmatic considerations the effectiveness of management decisions is usually enhanced (Bower, 1992). Recent initiatives such as the Earth Summit have introduced some of the needed changes – changes which are continuing through a range of workshops and training programmes (e.g. the Asia/Pacific Network described in Box 4.11) which can only further develop constructive working relationships between researchers and decision makers.

Finally, the use of the Internet is likely to bring significant benefits to coastal managers and planners in coming years, although initially this will be concentrated in developed countries where communications infastrucutre is better developed and more reliable. The ability to quickly find and download information on coastal problems, experiences and techniques from around the world will add to the Internet's ability to bring like-minded people together. Innovative education and training initiatives of NETTLAP in the Asia/Pacific region (Box 4.11), or the International Coral Reef Initiative (Box 5.4) are examples of what is possible with Internet communication. However, searching through the enormous range of information available on the Internet is increasingly requiring the use of specialized sites to help 'navigate' through the apparent morass of data. (The first such site dedicated solely to this purpose can be found at http://www.coastalmanagement.com and is managed by the principal author.)

(c) Section summary

This section has demonstrated the diversity of approaches to building the skills and professional and organizational infrastructure now acknowledged as essential for an effective coastal management programme.

Unlike other sections of this chapter which have outlined relatively clearly defined tools and techniques, capacity building remains an area of endeavour which requires extreme flexibility and cultural sensitivity.

As many coastal programmes evolve from 'rule based' to 'participatory based', there is likely to be increasing demand for capacity building to play a central role in helping to deliver acceptable results; and perhaps for a corresponding evolution of coastal managers towards Olsen's (1995) 'ideal' coastal manager who 'besides being a good strategist and leader' will be equipped with the skills and knowledge for:

- conflict resolution;
- managing group processes;
- design and administration of transdisciplinary research programmes;
- design and administration of public education and public participation programmes; and
- programme evaluation;

and perhaps as well will be skilled in the delivery of the coastal plans, tools and initiatives described in this book, including recreation and tourism management outlined in the next section.

4.2.4 Recreation and tourism management

The significance of tourism and recreation is often most evident in the coast. In fact coastal tourism is the most significant form of tourism, with domestic and international tourist flows in many countries dominated by visitors seeking the sun and the sea (Pearce, 1987). The coast, with its beaches, dunes, coral reefs, estuaries and other coastal waters, has always been a natural playground. Coastal environments provide open space, the opportunity for leisure, relaxation, contemplation and physical activity. Changing recreation-oriented life styles in developed countries and the rapid expansion of tourism facilities in developing countries has placed considerable strain on coastal resources and in many cases intensified conflicting pressures on them.

Recreation and tourism are growth industries worldwide. For many countries tourism is now a significant part of the economy (see Chapter 2). Indeed, in many coastal nations around the world, tourism is the most important single industry (Miller and Auyomg, 1991; Stronge, 1994) (Figure 4.7). The indications are that the growth in tourism will continue into the next century, with tourism in coastal zones being the major focus of that growth (Miller, 1993).

This section includes a brief consideration of recreation and tourism planning principles; concepts such as recreational carrying capacity, tourism succession, and recreational planning methodologies focusing on

Figure 4.7 Recreational pressures, Green Island, Great Barrier Reef (credit: John DeCampo).

the Recreation Opportunity Spectrum. This 'toolkit', described using case study examples, can be used in recreation and tourism planning at a range of spatial scales.

The terms 'leisure' and 'recreation' are used here in the sense of Patmore (1983) cited in Veal (1992):

> Leisure related to time, and the whole of non-work time in particular, and . . . recreation related to the specific activities pursued in that leisure time. But the distinction is a convention, and its rigid application can occasionally stifle a full exploration of the values and satisfactions of the leisure experience.

Following these definitions, the difference between tourism and recreation is defined by Kenchington (1993, p. 2) as:

> tourism is the business of trading recreational opportunities for economic gain.

Tourism can be generally considered to be the 'business' of recreation (Figure 4.8); but the distinction between tourism and recreation becomes less clear when it is acknowledged that the provision of recreational facilities on the coast by private industry and governments requires funding, and government funding is increasingly being gained through

Figure 4.8 Green Island beach hire, Green Island, Great Barrier Reef (credit: John DeCampo).

user-pays charges, such as park entrance fees. Given this trend, the distinction between recreation and tourism is judged here to be sufficiently blurred that the two terms are used interchangeably.

Recreational management and planning aims to enhance users' recreation experience of the coast while protecting and upgrading the coast as a recreation resource; in other words making coastal recreation more enjoyable and safe, without changing the coast in a way which actually reduces its attractiveness.

Recreation planning for a coastal area often aims to produce strategies which identify the appropriate degree of naturalness, levels of access, type and extent of facilities, intensity of management, and level and type of recreational use.

Encouragement of a tourism industry by government usually has multiple objectives which can be outlined according to scale. At the national level the aim of tourism may be to facilitate broad-scale economic development within the nation's sustainable development strategy. At the site level the aims may be to improve the local economy, maintaining the area's cultural assests, improve local social conditions, and protect local coastal environments. Early tourism planning was focused on physical or promotional planning for the growth of tourism, but it has now evolved to using a balanced approach recognizing the needs and views of tourists and developers as well as the wider community (Pearce, 1989). This change in approaches to tourism management has also seen a call for tourism planning to be integrated with other forms of planning, and not to rely on tourism sector planning. Such integration is shown in the approach taken for developing tourism in the Shark Bay Region of Western Australia (Box 5.2).

There has often been an absence of broader-scale national and regional planning for the growth of tourism developments, with much of the focus of tourism planning being on managing the development of the industry within a defined area (Agardy, 1990). In many cases around the world this local focus has produced a short-term economic gain, but long-term environmental degradation and resulting economic decline (Coccossis and Nijkamp, 1995). Patterns of resort evolution have been described by Butler (1980) who outlines a six-stage evolutionary process: exploration, involvement, development, consolidation, stagnation and rejuvenation or decline (Figure 4.9). In this model of tourism it may be assumed that decision makers are seeking to reach the upper outcome shown in Figure 4.9: rejuvenation or at least stabilization.

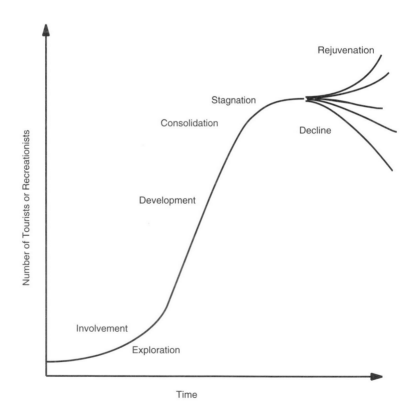

Figure 4.9 Butler's model of the hypothetical evolution of a tourist area (Butler, 1980, as adapted by Pearce, 1989).

Planning for tourism development also seeks to address issues in the coast unique to the industry, such as visitors 'loving the environment to death', conflicts with other industries such as aquaculture and sand mining, and the strain on existing resources and infrastructure with the community (Mathieson and Wall, 1982; Kenchington, 1993).

The result has been the development of specialist recreational and tourism management tools and techniques (e.g. Lieber and Fesenmaier, 1983; Kraus and Curtis, 1986; Jubenville *et al.*, 1987; Torkildsen, 1992), some of which have been applied directly to coastal areas (eg. Fabbri, 1990; Wong, 1993; Goodhead and Johnson, 1996). These tools can complement or be included within specific coastal recreation or tourism management plans, or incorporated in broader integrated management plans as is demonstrated in Chapter 5. The most important of these management tools, and their underlying concepts, are described in the following sections.

(a) Concepts of recreation and tourism management

People choose to recreate. They choose to go surfing on weekends, walk on the beach after work, or go on holiday to coastal tourist resorts. Recreational choices include the type of recreational experience or activity sought, with whom to recreate, and recreational time and location. The notion of choice permeates the concepts of recreation and tourism management. By altering choice, through the provision of 'things to do' or 'things to see', or deliberately limiting choice by restricting the provision of such choices, people's recreational experiences are being managed (McCool *et al.*, 1985).

Recreational management concepts have taken the notion of choice and linked it to the relative impacts of different intensities of recreational uses on the environment, and on the recreational experiences of tourists themselves. One of the driving concerns of recreation managers after the Second World War (especially in North America) was an awareness of the problem of recreation 'succession' (Jubenville *et al.*, 1987), referring to the evolution of a recreation site as more people become aware of its attractions. As visitor numbers grow at a particular site, facilities are upgraded, attracting even more visitors. The danger, then, is that this positive feedback spirals out of control, with high intensity recreational uses inevitable and a predictable 'sameness' of recreational choices. The succession model is similar to the models of tourist resort developments outlined above (e.g. Figure 4.9)

Concerns such as these are encapsulated in the concept of recreational carrying capacity, which focuses on the notion that there is a finite number of people who can visit an area before its capacity to absorb them diminishes. Once capacity is reached, a degradation of the environment or a reduction in the users' recreational experience occurs (Wager, 1964). While this notion is instinctively appealing it has been of 'little utility for the manager looking for some rational reason for limiting use' (Jubenville *et al.*, 1987, p. 29). Practical difficulties such as these led to a 'deceptively simple restatement of the problem' (Prosser, 1986), the purpose of which was to examine explicitly the desired social and biophysical attributes of an area and how those attributes are to be effectively managed. Called the Limits of Acceptable Change (LAC) concept, it focuses on the environmental and social conditions that are deemed to be acceptable, and the management actions required to achieve those conditions (Prosser, 1986).

There are four main stages in the Limits of Acceptable Change planning process (Prosser, 1986, p. 6):

1. Specify acceptable and achievable environmental and social conditions and define those conditions by a set of measurable indicators.
2. Analyse the relationship between existing conditions and those judged acceptable.
3. Identify management actions needed to achieve acceptable environmental and social conditions.
4. Monitor the indicators of condition of an area and evaluate the effectiveness of management actions.

In practice the LAC concept is closely linked to a complementary recreational planning concept – the Recreation Opportunity Spectrum (ROS) (Clark and Stankey, 1979), which considers recreation in terms of various settings and experiences available to different users. The ROS assists in recreational planning at a range of scales, but is commonly used at local and regional levels.

ROS recognizes that different people look for different types and intensities of recreation, and that through the provision of a range of 'recreation opportunities' most users are accommodated. A recreation opportunity is defined as 'a chance for a person to participate in a specific recreational activity in a specific setting in order to realize a predictable recreational experience' (Stankey and Wood, 1982, p. 7), which translates into planning for combinations of activities, settings and probable experience opportunities across a spectrum ranging from 'primitive' to 'modern'.

Thus, a recreational opportunity setting is made up of the combination of social, physical, biological and managerial conditions that give value to a place (Clark and Stankey, 1979). This value can include those qualities provided by nature (e.g. vegetation and topography), those qualities associated with recreational use (e.g. use types and levels), and those conditions provided by management (e.g. facilities, roads and regulations). By varying these conditions management can offer recreationists a wide range of recreational settings and hence experiences ranging from modern holiday resorts to primitive 'back to nature' wilderness settings.

The value of the ROS as a planning framework is that it offers a conceptual tool for considering recreation as something more than simply different activities or areas. Instead, ROS highlights the issue of recreation and tourism management as being more than solely the provision of physical developments, such as resorts, campsites and walktrails, but rather providing a diverse set of recreation opportunities (Clark and Stankey, 1979; Schmidt, 1996). Beyond this value the ROS has specific application for (Clark and Stankey, 1979):

- making inventories of, allocating and planning recreation resources;
- estimating the consequences of management decisions on recreation opportunities; and
- matching experiences people desire with available opportunities.

An example of the application of the ROS and LAC concepts is their combination for the purposes of national park management in the south coast region of Western Australia (Box 4.12).

Once criteria for management have been established for recreational management under the ROS and LAC system, tangible management steps can be designed. Examples of such steps for the management of the national parks shown in Box 4.12 are listed in Table 4.9.

(b) Recreation and tourism planning

Planning for recreation and tourism can be carried out as a 'recreation and tourism only' exercise through sector-specific subject-plans, or through integration with other sectors. Often both are carried out, with broad-scale national or regional tourist development plans concentrating on the requirements for the promotion of viable industries. There is an increasing trend for tourism planning to be incorporated into integrated planning initiatives: to recognize environmental thresholds that reflect the concept of carrying capacity and its application; to acknowledge the constraints for siting facilities (not every facility has to be on the coast); to incorporate sustainable principles into the design and construction of developments (e.g. silt curtains for marine construction); and to integrate social values into tourism developments, which is often done through community participation. Examples of such initiatives are described in more detail in Chapter 5, using the Gascoyne Tourism Development Strategy and the Shark Bay Region Plan in Western Australia, and local tourism development planning in Sri Lanka.

Often the needs of the recreation and tourist industry, such as access to coastal land and good transport links, are planned for through integrated

Box 4.12

The Recreation Opportunity Spectrum used for national park planning in the south coast area of Western Australia (adapted from Western Australian Department of CALM, 1991, and Jubenville et al., 1987)

The Recreation Opportunity Spectrum has been used by the Western Australian State Government Department of Conservation and Land Management (CALM) for planning ongoing management of national parks on the State's south coast (Western Australian Department of Conservation and Land Management, 1991). CALM manage around 70% of this extensive (approximately 1500 km), sparsely populated and biologically rich section of coastline (Donaldson et al., 1995). The ROS concept has been applied to this section of coast to provide a management framework for recreation and tourism management.

National parks in the South Coast region, including those abutting the coast, were placed along the spectrum, as shown in the figure.

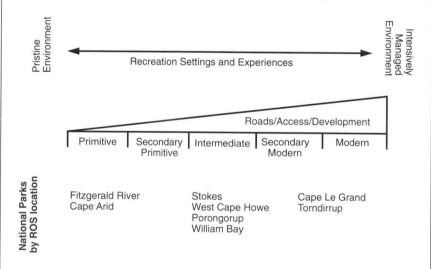

Recreation opportunity spectrum for the management of coastal national parks on the south coast of Western Australia.

Management prescriptions and recreational opportunities for parks at each position on the spectrum are in the process of being developed. These management prescriptions will cover the range of factors listed in the table.

CALM is working towards using each of the above management guidelines prescriptions for each setting on the ROS (from primitive to modern, as shown in the figure). For example, the management actions for the factor of

continued . . .

proximity of parking to key features for each ROS class could mean that users are expected to walk over 2 km from their cars in 'primitive' parks; whereas in the 'modern' parks they may be required to walk much shorter distances, in the order of 50–100m, to reach recreational opportunities.

Recreational opportunities	Management issues
Access	How to get there; distance from nearest town; road types; access, proximity of parking to key features; parking capacity
Other non-recreational resource uses	Presence of western features (buildings, power lines etc.)
On-site modification	Visual impact; complexity; facilities; disabled access; walks – with or without signs
Social interaction	Groups; availability of on-site information interpretation; appropriate use
Acceptability of visitor impact	Visitor impact
Acceptable regimentation	Visitor management; safety signs; management presence

regional planning. As is shown in Chapter 5, these plans aim to bring together the various competing demands for coastal resources, including tourism demands, into an overall planning framework. Examples of recreation and tourism issues which may be addressed at various scales of integrated coastal planning are also discussed in Chapter 5.

The steps common to most recreation planning initiatives follow the generic planning stages described in Chapter 5, especially as they relate to each scale of planning. However, recreation or tourism planning may require specific information and analysis techniques as part of integrated plans. The most commonly required recreation-specific requirement is an assessment of potential recreation demand. Recreation demand analysis requires the identification and analysis of existing patterns of use (i.e. how many people use each area, who uses which areas and facilities, when, and for what activities) and prediction of possible future changes in use patterns. A key issue in demand analysis is the difference between actual use of recreational opportunities versus the 'latent' or untapped demand.

An important issue in tourism and recreation planning is planning for the evolution of recreational opportunities and tourism products over time. A realisation that without deliberate planning a 'sameness' would creep into the style of recreational uses and opportunities was one of the

Table 4.9 Some measures to control the character of intensity of recreational use to meet desired management objectives in coastal parks (Schmidt, 1996)

Type of control	Method	Specific control techniques
Site management (emphasis on site design, landscaping and engineering)	Harden site	• Install durable surfaces • Irrigate • Fertilize • Revegetate • Convert to more hardy species • Thin ground cover and overstorey
	Channel use	• Erect barriers • Construct paths, roads, trails, walkways, bridges, etc. • Landscape
	Develop facilities	• Provide access to under-used and/or unused areas • Provide sanitation facilities • Provide overnight accommodation • Provide activity-oriented facilities (e.g. camping, boating, swimming platforms) • Provide interpretive facilities
Direct regulation of use (emphasis on regulation of behaviour; individual choice restricted; high degree of control)	Increase policy enforcement	• Impose fines • Increase surveillance of area

Table 4.9 continued

Type of control	Method	Specific control techniques
	Zone use	• Zone incompatible uses (e.g. hiker only zones, prohibit motor use)
	Restrict use intensity	• Rotate use (e.g. open or close roads, trails, campsites) • Require reservations • Limit usage via access points • Limit size of groups, number of horses, vehicles etc. • Limit permitted duration of stay in an area (maximum and/or minimum)
	Restrict activities	• Restrict building campfires • Restrict fishing or hunting
Indirect regulation of use (emphasis on influencing or modifying behaviour, individual retains freedom to choose; control less complete, more variation in use possible)	Alter physical facilities	• Improve (or not) access roads, trails • Improve (or not) campsites and other concentrated use areas • Improve (or not) fish or wildlife populations (e.g. stock or allow to die out) • Advertise specific attributes of the area • Identify range of recreational opportunities in surrounding area • Educate users to basic concepts of ecology • Advertise under-used areas and general patterns of use
	Set eligibility requirements	• Charge constant entrance fee • Charge differential fees by trail, zone, season, etc. • Require proof of ecological knowledge and recreational activity skills

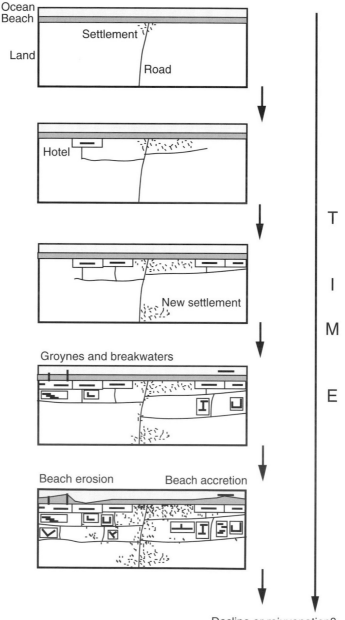

Figure 4.10 Tentative beach resort model (adapted from Smith, 1992).

driving influences in the development of the Recreational Opportunity Spectrum, described above. The same holds true for tourism goods and services. In a similar vein the evolution of tourist resorts and resort towns has become an important planning topic. Miossec's (Miossec, 1976, cited in Pearce, 1989) model of tourist development, which conceptualizes the growth of tourist regions through space and time, describes the staged growth of tourist areas and the resultant environmental and social impacts. This model was developed further and applied at Pattaya, Thailand by Smith (1992) (Figure 4.10).

Sustainability of the tourism industry is also reliant on local support, best obtained by encouraging local people to participate in all phases of tourist development. Without this support the industry may experience a succession of declining community attitudes towards tourism as described in Figure 4.11 (Doxey, 1975). Again, that such issues of local community attitudes are now realized as important in tourism planning stresses the requirement for tourism to be managed through integrated area management programmes (Pearce, 1989; Smith, 1992).

Miossec's model links the number of tourists with sustainable

(1) Euphoria
• Enthusiasm for tourist development
• Mutual feeling of satisfaction
• Opportunities for local participation
• Flows of money and interesting contacts

(2) Apathy
• Industry expands
• Tourists taken for granted
• More interest in profit making
• Personal contact becomes more formal

(3) Irritation
• Industry nearing saturation point
• Expansion of facilities required
• Encroachment into local way of life

(4) Antagonism
• Irritations become more overt
• The tourist is seen as the harbinger of all that is bad
• Mutual politeness gives way to antagonism

(5) Final level
• Environment has changed irreversibly
• The resource base has changed and the type of tourist has also changed
• If the destination is large enough to cope with mass tourism it will continue to thrive

T
I
M
E

Figure 4.11 Example of succession of community attitudes towards tourism (Doxey, 1975, cited in Mercer, 1995).

Box 4.13

Tourism growth and management of the Red Sea coast

Nestling between the desert sands of Africa and Arabia, the Red Sea has lured adventurous explorers for hundreds of years. Now it attracts thousands of tourists (Hawkins and Roberts, 1993).

The exotic and previously remote northern shores of the Red Sea within Egypt, Israel and Jordan are now the focus of coastal tourism developments. Development has been rapid over the past 20 years since early tourist centres of Hurghada (Egypt), Sharm el Sheikh (Egypt) and Eilat (Israel) were established. There are now numerous coastal tourist centres, and these are expected to increase by 1300% from 1993 to 2000 as mainly European holiday makers seek new destinations, especially those that offer a wilderness experience. Countries, and residents in the region, seek to diversify their income base and to maximize the economic benefits of tourism (Hawkins and Roberts, 1993).

A key draw for tourists is the well developed fringing coral reefs set in clear, warm waters. SCUBA diving and snorkelling are the two main recreational uses of these reefs. For example, of the 200 000 visitors to Sharm el Sheikh in 1992, an estimated 50 000 were divers; there are plans for diver numbers to grow to 300 000 per year from a projected 1.2 million visitors (Hawkins and Roberts, 1992). The actual growth of tourist numbers and their direct impact on the growth of marine tourist numbers is illustrated in the table.

The impacts of the growth of coastal and marine tourism on coastal and marine environments have varied significantly around the Red Sea. In some areas, degradation of fringing reefs has been significant due to the direct

Growth of numbers of hotel beds and dive boats in the Sharm el Sheikh (Egypt) Region 1988–1995 (from Anthias, 1994)

Year	No. hotel rooms	No. dive boats
1988	1,030	23
1989	1,276	25
1990	1,358	47
1991	2,906	60
1992	3,306	89
1993	5,190	120
1994	8,234	200
1995	11,384	240

continued . . .

effects of construction (infilling, sedimentation), indirect effects of tourist development (sewage, desalinization, irrigation and rubbish) and the direct effects of tourist boats and tourists themselves (Hawkins and Roberts, 1993). The latter effects are due to both the effect of divers and snorkellers damaging coral, anchor and mooring damage and the impact on local fisheries from increased demand for seafood. Future issues which need to be considered and managed include living-aboard dive boats which traverse maritime boundaries and are therefore more difficult to regulate.

There is an increasing recognition in the region that the current rate of tourism expansion can significantly damage the very coastal ecosystems that draw visitors to the Red Sea. A range of planning and environmental studies have recently been completed or are underway. A key component of these strategies is the expansion of the region's marine protected area network. An example of an existing protected area is the Ra's Mohamed Marine Park at the southern tip of the Sinai peninsula (Sharm el Sheikh, Egypt). The park is a significant area in the Sinai; typically, marine areas on the Gulf of Suez are shallow but at Ra's Mohamed the reefs are visually spectacular with deep drop offs, steep walls and large numbers of pelagic fishes. The coral fauna of Ra's Mohamed is also quite different from that found on the reefs in the Gulf of Suez side of the Sinai. The park, first declared in 1983, was expanded in 1989, 1991 and 1992 (Anthias, 1994). Management is based on a multi-faceted approach of a zoning plan, park-wide provisions, scientific research, site improvements, pollution controls and awareness.

The zoning plan aims to maintain areas of high conservation or biodiversity by designating closed areas and allowing other areas (open) to be used for a range of activities including tourism. Park-wide laws include prohibitions on the collecting of marine and terrestrial resources, any form of hunting, spearfishing, fish feeding, anchoring on reef areas and bottom line fishing (except in designated areas). Site improvements include the construction of floating jetties using perspex to reduce the impact on coral of shading by the structures. Pollution controls are enforced by using Egypt's Environmental Impact Assessment legislation and placing restrictions on effluent discharges, such as brine from desalination units. Awareness programmes are also under way to encourage visitor appreciation of, and responsibility towards, the protection of natural resources as well as community and local business cooperation and participation in conservation planning (Anthias, 1994). Community cooperation and participation is enhanced with the involvement of local Bedouin fishermen as park rangers.

The aim of management at Ra's Mohamed and other initiatives in the area is to provide a strategic long-term view of tourist development and attempt to mix conservation and development to provide a sustainable future for both the reef ecosystems and the tourism industry.

development. A decline in tourist numbers can often be caused by significant environmental degradation from poorly planned tourist developments, and from the impacts of tourist themselves on natural and social environments (Coccossis and Nijkamp, 1995). Tourist developments built too close to sandy coastlines can, for example, cause chronic erosion problems, resulting in the loss of the beach that most of the tourists came for in the first place! An example of such conflicts is clearly demonstrated by the rapid growth of tourist development on the Red Sea coast discussed in Box 4.13.

The incorporation of tourism development (including the Red Sea case study introduced in Box 4.13) into integrated coastal management planning is shown through a number of case studies in Chapter 5; for example, how the promotion of a locally sustainable coastal and marine tourism industry in Sri Lanka is integrated with other development goals is demonstrated in the analysis of the Hikkaduwa Special Area Management Plan (see Box 5.16).

This section on recreation and tourism planning has demonstrated the subtle interactions between coastal tourists or recreationists and their natural and social environment. The degree of this subtlety is being increasingly realized as the importance of providing a diverse range of recreational and tourism opportunities is understood. In addition, the use of management and planning tools, including environmental impact assessment, economic analysis and risk management, is becoming increasingly widespread in the tourism industry. The most important of these tools are described in the next section.

4.3 Technical

Coastal planners and managers can choose to use a number of 'technical' approaches to plan and manage the coast. Many of these approaches are not specific to the coast: in many cases these approaches were developed for land-based systems and then modified where necessary for use in the coast. The full range of technical techniques which can be applied in the coast is far too wide to include all of them here. Instead, this section discusses both the major and most common tools currently used in coastal management, and some which are not in widespread use but we judge to be on the brink of the mainstream.

4.3.1 Environmental impact assessment

Environmental impact assessment (EIA) is one of the most frequently used tools in coastal management. EIA is globally widespread, and is used in a variety of planning and management contexts, each of which is outlined in this section. Environmental impact analysis is another term for environmental impact assessment. Environmental impact statements (EIS) are generally the written reports required of an EIA process.

The following clear working definition of EIA was developed in New Zealand during the reform of their resource management (Ministry for the Environment, 1988, p. 6):

> Environmental Impact Assessment (EIA) is a process by which the impacts of a proposal (whether as a policy, management plan or intended development) are identified early on in the decision-making process, so that these considerations are taken into account in the design and approval of the proposal.

Within the EIA context, 'environment' refers not only to biophysical aspects, but also social and economic aspects. The general aim of the EIA process is to provide decision makers with the best available information which will help to minimize the costs (environmental and financial) and maximize the benefits of the proposed actions. Minimizing environmental costs is often associated with managing environmental risk, as discussed in section 4.3.2. EIA is now an integral part of environmental planning and management of the coastal and marine environments of many coastal nations (Sorensen and West, 1992).

The purpose of this section is not to detail the EIA process but to outline its use as a tool in planning and managing the coast. For more information and/or details on EIA there are several good texts, of which those of Glasson *et al.* (1994), Gilpin (1995) and Thomas (1996) are drawn on in the following section. A number of international organizations have also produced EIA guidelines or manuals, including many international donor agencies (e.g. Asian Development Bank, 1991a). Each of these manuals and guidelines has recently been listed and summarized by Gilpin (1995, pp. 74–87).

There is no coast-specific EIA, but the legislation or policies used throughout the world are either generic in their application, or 'the environment' is defined in terms such that it applies in the coastal zone. However, the special attributes of the coastal environment have been recognized through the publication of a number of guidelines and manuals written to assist in undertaking coastal EIA (e.g. Sorensen and West, 1992; SPREP, 1992; Vestal *et al.*, 1995).

Some managers view EIA as both a political and a technical process (Gilpin, 1995; Thomas, 1996). EIA might be seen as a political process because it is based on society's value judgements, therefore decisions on whether a development should proceed are influenced by social politics. Furthermore, most governments have legislated for EIA, meaning that EIA decisions are also political judgements. Both social and political types of value judgements take place throughout the EIA process. Most governments have found that the best way to manage these judgements is through public participation. Depending on local legislative requirements, public participation can be an important element of many of the steps in the EIA process. The integral role of public participation in EIA also reflects its relatively recent evolution, and the simultaneous rise of citizen interest and subsequent involvement in decision making.

Initially EIA was used to assess the impacts of a single development or set of actions. Often these assessments were independent of each other and issues such as cumulative impacts were not addressed. Recent initiatives, however, have extended the use of the EIA process to strategic planning, addressing cumulative impacts and formulating environmental management systems to International Standards Organization (ISO) 14000 standards. There have also been a number of extensions and modifications to the EIA process, such as landscape and visual assessment (section 4.3.3), economic assessment (section 4.3.4), and health and social impact assessments. In addition, some detailed EIAs have merged risk analysis through the use of quantitative probabilistic analysis of potential impacts (section 4.3.2). These processes can be a substitute for an EIA in specific circumstances (described elsewhere in this chapter), but more often they are used in addition to the EIA process.

EIA became part of US mainstream coastal management in 1970 with

---

Box 4.14

Requirements of an EIA specified by Section 102 of the US Federal National Environmental Policy Act (1969)

Section 102(C) of the US Federal National Environmental Policy Act (1969) states that all agencies of the Federal Government shall:

Include in every recommendation or report on proposals for legislation and other major Federal actions significantly affecting the quality of the human environment, a detailed statement by the responsible official on:

I.   The environmental impact of the proposed action,
II.   Any adverse environmental effects which cannot be avoided should the proposal be implemented,
III. Alternatives to the proposed action,
IV.   The relationship between local short-term uses of man's environment and the maintenance and enhancement of long-term productivity, and
V.   Any irreversible and irretrievable commitments of resources which would be involved in the proposed action should it be implemented.

the passage of the Federal National Environmental Policy Act (NEPA) (1969) (Black, 1981). The NEPA laid the foundations for all subsequent EIAs, as shown in Box 4.14.

The purpose of an EIA, as specified by Section 2 the NEPA (1969), is to embody the 'practicable means' whereby 'Federal plans, functions, programmes and resources (are) improved and evaluate(d)' so as 'to promote efforts which will prevent or eliminate damage to the environment and the biosphere and stimulate the health and welfare of man' (Black, 1981, p. 22; Bregman and Mackenthun, 1992).

Passage of the NEPA is viewed as a logical step in the increasing environmental awareness in the United States in the 1960s (Black, 1981, p. 1; Glasson et al., 1994). Thus, the development of EIA parallels the formulation of the Coastal Zone Management Act in the same country. This paralleling of EIA and coastal management policy was subsequently followed in many other coastal nations (Sorensen and West, 1992), although EIA was generally distinct from coastal management programmes during the 1970s and 1980s.

Since 1970 governments in both developed and developing countries have initiated legislation and policy for assessing the impacts of development or issues (Glasson et al., 1994)

(a) The need for an EIA

The requirements for undertaking an EIA can be specified in a number of ways, including through legislation and guidelines and through specific needs, such as those specified by funding bodies (Table 4.10).

Another method of distinguishing between the requirements of EIA (as shown in Table 4.10) is whether they are defined in an Act – for example,

Table 4.10 Forms of establishing EIA systems (adapted from Glasson et al., 1994)

Requirement for EIA	Comments
Mandatory regulations, acts of statutes	Generally enforced by requiring the preparation of an adequate EIA before permission is given for a project to proceed
EIA guidelines	Non-enforceable, but generally impose obligations on the administering agency
Legislation outlining discretionary EIA requirements	Enforcement dependent on discretionary requirements
Ad hoc preparation	May be required for specific reasons; for example by funding bodies such as the World Bank

the NEPA (1969) definition shown in Box 4.14 – or through the specification of a list of projects, or by defining whether an EIA is required for government projects only (such as in the NEPA), private projects only, or both (Glasson *et al.*, 1994). Various EIA systems around the world use different combinations of these approaches.

(b) Steps in the EIA process

All EIAs share a number of common components, carried out in sequence, as neatly summarized by Smith (1993):

- scoping;
- prediction;
- significance assessment;
- evaluation;
- monitoring; and
- mitigation.

In operation, these common elements are incorporated into the 'classical' generic model of EIA, as a systematic process, shown in Figure 4.12 and discussed step by step below.

In undertaking an EIA, negotiations and discussions usually take place between those who may be required to prepare an environmental impact statement (EIS) (usually called the 'proponent') and key government agencies (Figure 4.12). Such agencies are often those who will be involved in the process for assessing the EIS and/or those who would be involved in implementing the management commitments required by it.

The negotiating process before the proposal's application is lodged can be a very important part of the EIA process, depending on constraints placed by legislation, guidelines or day-to-day practice of administering agencies. For example, proponents may discuss at length what modifications are required to reduce the impacts of the proposal, resulting either in a reduction in the level of assessment or completely exempting the proposal from further assessment. In some EIA processes, such as the NEPA requirements in the United States, the pre-lodgement negotiating phase has become very important, effectively becoming a form of pre-EIA. This has been one of the major reasons for the dramatic reduction in the number of formal EIAs required in the United States (O'Riordan and Sewell, 1981).

Following lodgement of the proposal, the administering agency decides whether it warrants an EIA. The decision is based either on a list of designated projects or on criteria which consider the nature and scale of the project, depending on the relevant legislative constraints; or it may be at the discretion of the assessor, which may in turn be constrained by

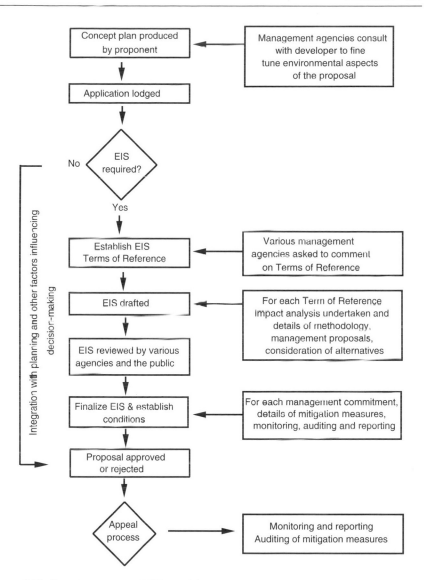

Figure 4.12 Generic operational EIA model.

guidelines or formal criteria. The constraints direct the issues which need to be addressed, including biophysical, social and economic factors. In addition, the environment which is at risk under the proposed project is assessed to help identify potential risks. Once these risks are identified, alternatives to avoid, mitigate or remedy impacts are investigated. A key component of EIAs is the consideration of alternative actions which may be taken to mitigate environmental impact.

The decision whether or not the project should proceed is based on an evaluation of the results of the previous stages in the EIA (Figure 4.12). If the decision is to allow the project to proceed, then a permit or licence is often the legislative outcome. The approval can contain conditions, which are often linked to performance, monitoring and evaluation criteria. A decision on whether or not to proceed with the project may also be subject to appeal in some countries. Once the approval is given, the EIA process does not end; depending on requirements, monitoring and other processes may continue for several years. In some countries there are provisions for revoking approvals if the monitoring requirements of an EIA are breached.

Techniques which can be used in an EIA to consider the environmental impact, and assess alternatives, are extensive. The body of literature which has developed in the EIA area provides a thorough exposition of these techniques. Listed in increasing level of sophistication, the five major types of impact identification and summarization are checklists, interaction matrices, overlay mapping, networks, and simulation modelling (Westman, 1985; Smith, 1993). These methods are described in detail in most EIA texts (see list above). In more complex EIAs there is also often a close relationship with the risk assessment techniques described in section 4.3.2.

(c) Operational EIA

The range of EIA uses is shown in Box 4.15 where EIA is an integral part of the management of the Great Barrier Reef. Box 4.15 shows how EIA is used to manage of range of activities, from the establishment of pontoon-based tourism on specific reefs (Figure 4.13) to the dumping of dredged material.

(d) Integrating EIA with planning

Close linkages between EIA and planning occur at a number of points in the EIA process. In general these links are becoming closer as the various forms of coastal planning described in this book (especially in Chapter 5) generally include consideration of environmental impacts.

In this sense, the classical EIA step-by-step practice has been linked to planning theories, most notably the synoptic (comprehensive or rational) theory (Smith, 1993). Synoptic planning is based on the development of plans through a series of steps undertaken in a purely rational manner (Chapter 3). The synoptic model can be used to integrate impact assessment with planning. Thus, EIA can be used as method for looking at the impacts of 'planning' and plans on the environment by using a series of steps similar to synoptic planning (Smith, 1993):

Box 4.15

EIA and the management of the Great Barrier Reef

The use of pontoons as tourist facilities was first proposed in the early 1980s for a reef location, Agincourt Reef in the northern Great Barrier Reef (GBR). The pontoon was designed as an aluminium platform permanently anchored to the seabed and would cater for approximately 350 day visitors. Prior to the installation of the pontoon a comprehensive EIA process was undertaken. Of particular concern were the effects of shading on coral, whether the pontoon would act as a fish aggregation device, and the performance of the pontoon under cyclonic conditions. Since very little was known about the effects of shading, the permit to establish the facilities included the conduct of a monitoring programme for the impacts of shading. Subsequent to this installation, other pontoons have been installed and similar monitoring programmes established. As a consequence a substantial information base has been established on the impacts of pontoons on coral reefs. What managers have found is that shading by the pontoons has minimal impact. Engineers have also used monitoring information to assist in refining the design of pontoons to improve their performance under cyclonic conditions

Disposal of dredged material

Throughout the world many port facilities require regular dredging of the seabed to enable large vessels to utilize port facilities. The Cairns Port in north Queensland requires regular dredging of built-up sediment so that deep draft vessels can enter the Port. It is an industrial port which is also used for recreational boating and fishing. Several marine-related industries are also based on adjacent land.

The EIA focused on two major issues: the toxicity of the dredged material, since industries adjacent to the Port had historically disposed of their wastes directly into the Port; and the location of the spoils site for the dredged material. The analysis of sediment to be dredged indicated low toxicity levels and therefore this subject was not an issue, but other issues such as turbidity and possible sediment drift to upstream beaches needed to be addressed.

Historically, the material dredged from the port was disposed of in an area offshore to its north. This area was selected because of its closeness to the port (reducing dredging operational costs), but the area used is within the GBR World Heritage Area. The regular disposal of dredged material in the World Heritage Area was questioned as part of the EIA. As a consequence, proponents were asked to investigate alternative disposal sites. As an interim measure, strict conditions on disposal of the material were imposed. These included limiting the amount of material dredged,

continued . . .

specifying the type of dredge used, the time of the year dredging could take place, and the sea conditions in which the material could be dumped; and requiring an aerial surveillance programme to monitor sediment plumes. Imposing these conditions as part of the EIA approval process assisted in managing the impacts of disposing the dredged material on adjacent marine and coastal areas.

- identification of problems;
- defining goals and objectives in planning (corresponds to the processes which assist analysing the potential impacts of goals and objectives);
- identifying opportunities and constraints in planning (similar to analysing the current environment and predicting the effects of planning actions);
- defining alternatives (similar in both planning and EIA);
- making a choice and implementing that choice (also similar in both processes).

At each of these steps, impact analysis (similar to the generic EIA process discussed above) is undertaken. The advantages and disadvantages of rational/synoptic planning, as discussed in Chapter 3, also apply here. That is, the model usually only applies in the early stages of the process.

A 'manual planning' model is based on assessing the impact of 'the plan' using a process similar to the generic EIA described above or the use

Figure 4.13 Tourist pontoon, Great Barrier Reef (credit: GBRMPA).

of a set of guidelines specific to assessing plans. In the manual planning model, the outcome of the planning process ('the plan') is subject to an single EIA process. If a synoptic approach is used, then an EIA is conducted at each step of the planning process. With either process, the potential impacts of a plan as part of the approval planning process are investigated.

The assessment of plans can be a statutory or non-statutory process depending on the nature of the EIA legislation or policy. If there are no statutory requirements for assessing the impact of a plan, governments will nevertheless assess the plan. Depending on the legislation and prevailing political climate, the public may or may not be consulted.

A project-by-project approach to EIA dominates the way that many developments and activities are managed around the world. This approach dominates primarily due to the legislative and administrative arrangements which generally apply to projects, and generally does not make provisions for how the proposal which is the subject of an EIA fits into the broader context. This limitation on the range of environmental measures that managers can use within the EIA process has been criticized as a major failing of EIA (O'Riordan, 1981; Smith, 1993). Individual projects may not have a significant impact, but several projects may in combination produce a 'cumulative impact'. Linear impacts can occur with developments that are spatially linear, such as strip development of continuous urban residential land along coastlines (Court et al., 1994; Vestal et al., 1995).

Planning processes, especially the production of plans with a broad strategic focus, can enable the incorporation of measures to address some of the legislative and administrative constraints to the production of project-by-project EIAs. Strategic planning documents can be used to manage potentially cumulative environmental impacts. Planning processes may also trigger EIAs, through either statutory land-use planning systems or in the identification of potential impacts during strategic planning.

Plans or policies can limit the number or levels of cumulative impacts in several ways, ranging from defining the number of facilities that discharge or contribute to environmental degradation, to defining the total discharge into the environment for specific pollutants such as sulphur emissions, and specifying/setting the environmental standards for activities or structures that are allowed or permitted in a particular area. Plans and policies can also provide a mechanism for consulting the public on decisions regarding strategic land or water use, and other developments or facilities with the potential to add to the environmental problems of an area.

Similarly, cumulative impacts which occur in a linear fashion can be managed through planning by specifying where activities can and cannot

Box 4.16

Linking EIA and strategic planning on the Great Barrier Reef

The Cairns Section Zoning Plan for the Great Barrier Reef Marine Park is an example of how planning was used to address the issue of linear impacts. In the rezoning of the Cairns Section, the community expressed concern with

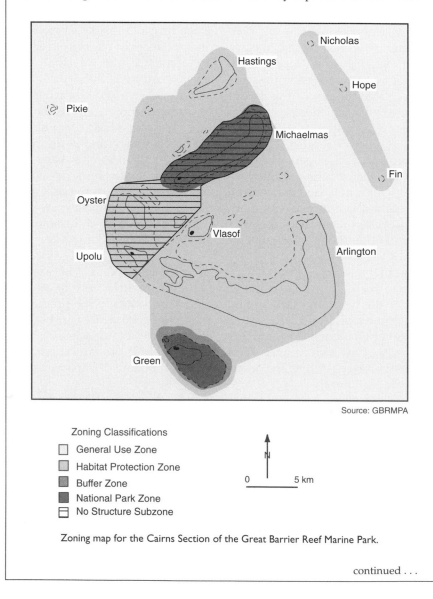

Source: GBRMPA

Zoning Classifications

☐ General Use Zone
▨ Habitat Protection Zone
■ Buffer Zone
■ National Park Zone
⊟ No Structure Subzone

0   5 km

N

Zoning map for the Cairns Section of the Great Barrier Reef Marine Park.

continued . . .

number of applications for developing pontoons and other similar permanent structures on the outer barrier reefs of the Section. These outer reefs form a line/chain of reefs and are noted for their aesthetics and good coral cover. To minimize risks, the zoning plan was modified to include a structure/no-structure subzone (see figure). The no-structure subzone did not allow or permit the establishment of permanent structures on reefs. This ensured that linear developments along the outer reefs would be avoided.

take place. An example of how EIA is linked with regional strategic planning in the Great Barrier Reef is shown in Box 4.16 (see also Box 4.15).

The inclusion of EIA principles in coastal planning can ensure that the outcomes of planning decisions are similar to the desired outcomes for EIA. Similarly EIA can be used to ensure that plans will cause the minimum impact on the environment. Hence good coastal planning reduces the need for EIA, and when EIA is required, its scope can readily be defined by pre-existing planning (Brown and McDonald, 1995).

(e) Strategic environmental assessment

The focus of 'classical' EIAs on project-by-project environmental impacts, and their resultant inability to adequately address strategic environmental impact issues, as described above, leads to the need for more strategic EIA tools. This need has become coupled with the increasing demands that sustainable development principles have placed on coastal planners and managers. One of the outcomes of this dual need has been strategic environmental assessment (SEA), defined by the Australian Commonwealth Environmental Protection Agency (Court *et al.*, 1994, p. I.3) as the:

> process of consideration of environmental impacts of policies, plans and programmes applied to higher levels of decision making with the object of attaining ecologically sustainable development.

SEA can be considered as an EIA for programmes, plans and policies, but can also be considered as a planning tool in its own right. It allows the consideration of environmental impacts over a larger geographic area and development time frame. In addition, SEA enables subsequent EIAs on a project-by-project basis to focus on the details of the project and possible alternatives. Wood and Dejeddour (1992) considered SEA as the first stage of a two-stage EIA system, with traditional project-by-project EIA as the second stage. There are currently two main approaches to SEA: the first the extension of traditional EIA principles, and legislative procedures and requirements; and the second, adopting a policy and planning rationale where environmental principles tend to be tailored in the formulation of policies and plans.

SEA assists in coastal planning and management in a number of ways. In decision making it enables managers to raise the importance of coastal concerns to the same level as the aspects of traditional development planning. SEA also facilitates consultation on a range of coastal issues between various organizations as well as the public. It can make EIA unnecessary, or reduce its importance for specific activities if they are considered sufficiently at the plan or programme level. Mitigation and compensation measures can be formulated for certain types of developments as a result of SEA as well as assisting in formulating or modifying codes of conduct.

A benefit of SEA is that it encourages the consideration of environmental objectives during policy, plan and programme-making activities within organizations which traditionally avoided incorporating environmental considerations in their decision making. Some project EIAs may also be redundant within an SEA. By reducing the need for or scope of EIAs, providing for a wider range of alternatives to be considered, and recommending suitable sites for projects, SEA allows managers to focus on specific aspects of EIA. Subsequent projects benefit when SEA facilitates the formulation of generic EIA best practices. SEA enables managers to investigate other areas of impact assessment such as cumulative, secondary, long-term and delayed; as well as the impact of specific policies (based on Wood and Dejeddour, 1992).

To summarize, the EIA process, like many other tools for managing the coast, continues to evolve to meet society's expectations and the needs of environmental managers, especially those in the coast. It is a well defined process which has been embraced and modified as a major environmental management tool in a number countries. Early in its history EIA focused on assessing developments on a case-by-case basis, which could consume considerable financial and human resources. However, as managers have come to understand EIA they have also explored how it can be modified and extended into a more cost-effective and efficient process. Cumulative impacts and SEA are two major outcomes of this exploration. Through SEA the role of EIA in planning and managing coasts is now varied and applied at operational and strategic levels.

4.3.2 Risk and hazard assessment and management

> Risk management, although a very useful process, is not a precise science, nor is it a particularly well developed art form.
>
> (Gerrard, 1995)

There is no such thing as a zero risk. Each day we take risks, from the small risks like whether we will get wet while taking a beach profile, to the large risks associated with undertaking a coastal development in an erosion-prone area. The larger risks can be financial, such as the financial viability of building multi-million-dollar marina developments on the coast; ecological, such as mitigating the risks posed to natural coastal eco-systems from heavy industries; or planning, associated with determining the proper location and design of structures to minimize the impact of natural hazards.

As the name suggests, risk and hazard assessment is concerned with assessing the probability that certain events will take place and asssessing the potential adverse impact on people, property or the environment that these events may have. Coastal examples include failures of a chemical refinery on the coast causing damage to the plant itself, and to surrounding residents and the environment through the release of toxic chemicals into nearshore waters. An example in the natural environment is analysing the potential impacts on a coastal region of severe storms. Not surprisingly, methods of managing risks once they have been assessed are called 'risk management' techniques.

This section considers all types of risks which influence coastal planning and management decisions. Before the details of risk assessment and management are described, some basic concepts of risk and hazard are introduced.

(a) Concepts of risk and hazard

> Difficulties often arise when scientists and technical people use jargon, especially when quite common words are used differently from everyday chat. Risk is such a word.
>
> (British Medical Association, 1988)

Specialists working on human-induced hazards, like chemical or nuclear plants, share a common goal with those working on natural hazards such as floods and earthquakes: to reduce the impact of hazards on people, property and the environment. The problem is that though specialists in the man-made and natural hazard fields know what they mean when talking of risk and hazard, the two groups often use different definitions for the words. The Royal Society of London (1983) gives a very British distinction between risk and hazard:

> Consider the existence of Nelson's Column as the hazard. It may be damaged by wind and lightning and, as a consequence, pieces may fall off and cause harm to people in Trafalgar Square. Risk would

measure the probability of specified damage or harm in a given period.

The difference between risk and hazard is interpreted differently by those studying natural hazards. For example, in studies of earthquakes Ambraseys (1983) states:

> Seismic hazard: i.e. the probability of occurrence of ground motions due to an earthquake. Earthquake risk may be defined as the probability of the loss of property or loss of function of engineering structures, life, utilities etc.

Thus some choose to include a component of the impact of a hazard, such as on life or utilities, in their definition of risk. A further complication is introduced by others who study natural hazards. For example, Jackson and Burton (1978), citing White (1974), stated:

> No natural hazard exists apart from human adjustment to it: the notion of risk or hazard automatically implies some human or social component.

The easiest way to avoid the issue of semantics is to choose working definitions at the outset of a project, an approach taken here, with the chosen definitions shown in Box 4.17. Note that there may be cases where different definitions will be used, especially in non-western societies where different cultural norms may exist (Gerrard, 1995). Nevertheless, the simple definitions shown in Box 4.17 provide a basis from which to start discussions.

Box 4.17

Simple working definitions of risk and hazard for use in coastal planning and management

- Hazard: An event or process with potential for harm to people, property and the environment.
- Risk: The occurrence probability of hazardous events.
- Hazard impact: The consequences of hazardous events to people, property and the environment.

Now that simple working definitions of risk and hazard have been established, some basic concepts of risk can be addressed. Perhaps the most fundamental of these is that risk management must extend the 'scientific' notion of risk shown in Box 4.17 to (O'Riordan, 1995, p. 296):

a culturally framed concept which acts as a metaphor for individual feelings about loss of control, powerlessness and the drift of social change away from what is good for the Earth towards what seems to be bad.

This may seem to be a very big step from, for example, working out the impacts of a storm surge on a coastal community. Nevertheless, the concept that risk, and therefore the management of risk, is deeply ingrained in how societies function is now widely accepted as the central tenet of risk management. The concept, which has taken many years to form, has mainly developed around the experiences of managing the risk of the nuclear industry or the impacts of extreme natural hazards such as tsunamis and cyclones (Box 4.18). Managers of hazardous industries have now realized that trying to convince a sceptical public that a particular industry is safe by bombarding them with high-powered science does not work, it can even be counterproductive. The focus today is on shared 'multiway' communication of risks (see Figure 4.14).

Box 4.18

Cyclone risk management in Bangladesh (Kausher et al., 1994, 1996)

The Bangladesh Cyclone Preparedness Program (CPP) provides an impressive example of the effectiveness of cyclone hazard mitigation measures. Cyclones of similar severity devastated the area in 1970 and 1991 (Box 2.11), but the loss of life in the later event was less than half of that in the former; and this despite significant population growth in the area during the intervening two decades. The difference in impact was generally attributed to the effectiveness of the CPP, set up in 1973 in response to the 1970 cyclone.

There are three components to the CPP: warnings; shelter construction and evacuation to shelters; and disaster relief. The system relies heavily on a grass-roots support system, based on 2089 units of 10 volunteers each. Volunteers have specific tasks such as cyclone warning, shelter management, rescue, first aid and relief.

The warning system, whereby volunteers supplement warnings over public radio by using megaphones mounted on rickshaws, is obviously a critical element of the CPP. Surveys undertaken after the 1991 cyclone revealed that 64% of people took some precautionary measure on hearing the cyclone warnings, such as moving to cyclone shelters or well built houses or embankments (Hossain et al., 1992). How this figure compares to the total response of coastal residents to the warning will never be known. Since it refers only to the survivors, it is tantalising to speculate about

continued . . .

whether those who died in the tragedy heeded the warning in the same proportion as those who survived.

Reinforced concrete buildings designed to save human life from cyclones are the second part of the CPP programme. A five year shelter construction programme was initiated in 1972, but was abandoned after only 234 shelters were built, or an estimated 10% of actual demand (Talukder and Ahmad, 1992). The programme was re-started after a cyclone in 1985, and a further 62 shelters were constructed, but few of the survivors of that event reported that shelters were available to them, stating among the reasons they had not sought shelter that the distance to the shelter was too great, they felt their household goods would be at risk if they left them unguarded, and lack of proper facilities in the shelters. The post-1985 shelters also suffered from a design defect which had serious cultural ramifications. Having only one large communal room, unlike the three-storey shelters built after the 1970 cyclone, men and women were crowded together, something which is considered to be a violation of purdah for women (Talukder *et al.*, 1992).

There are many lessons to be learned from the Bangladesh experience, including the fact that planning, even if in part flawed, can have outstandingly positive outcomes. But perhaps the most important lesson is the need to take a holistic approach to cyclone impact mitigation, embracing adequate (and culturally appropriate) shelters, evacuation routes, and improvements to the early warning system. Above all, however, an attempt has to be made to engender positive attitudes towards survival among those who, because of their poverty, lack of housing choice, and their historical close relationship to the coast, have a tendency to face nature's ferocities with resigned fatalism.

(b) The risk and hazard management process

Developing a risk management strategy involves a number of distinct stages (Table 4.11). These stages are very similar to the basic phases used in strategic planning, including coastal management plans, namely:

- scoping and investigation;
- analysis;
- implementation (mitigation); and
- monitoring.

The traditional way to carry out a risk management programme was to undertake each of the steps shown in Table 4.11 in turn, starting with hazard identification. This approach has now been found to be appropriate in only limited circumstances – mostly when there is little interaction with the public required, as in the case study shown in Box 4.19. In this example, a standard staged approach was undertaken,

Table 4.11 Definitions of the stages of the risk management cycle (adapted from Gerrard, 1995, and Soby *et al.*, 1993)

Step	Description
Hazard identification and prioritization	• Determines what can go wrong by identifying a set of circumstances and established priorities of the most urgent hazards requiring policy action.
Risk assessment and characterization	• Evaluates how likely is it that a set of hazardous circumstances will arise and estimates their consequences.
	• Determines subjectively the tolerable level of risk.
	• Considers how risks can best be avoided, reduced to tolerable levels and controlled
Policy decision and implementation	• Chooses and implements a particular policy option in the most inclusive and open manner possible
Risk evaluation	• Provides feedback about the net effects of risk mitigation

following the general phases described above. It demonstrates the 'technical' model of risk assessment and management – that is, with a focus on the technical analysis of risks coupled with technical risk mitigation measures.

Risk communication is not a large part of the risk management approach shown in Box 4.19, mainly because the management of risks in this case is essentially an internal management issue to do with the operational management of port facilities. However, when conveying the message of risks, and how best to manage them, to others – especially the general public – a different approach is required which places 'risk communication' at its heart (Figure 4.14).

The risk management cycle differs from the linear, scientific view of risk management in that it emphasizes the importance of feedback, so much so that the starting and finishing points for risk management become blurred (Gerrard, 1995). The key element in the risk management cycle is the central role of risk perception and risk communication, which goes beyond the traditional view of 'public' (non-expert) perceptions, and the one-way communication of risk information from experts to the public. Applying this principle, the management of risks in the coastal zone which affect the general public – for example, cyclone risk management strategies – requires the voice of the public to be heard at a very early stage in the

Box 4.19

The management risks for the transportation of hazardous goods in UK ports (adapted from Gavaghan, 1990)

Technical risk assessment was used in the United Kingdom to analyse the chances of a major accident occurring due to the transportation through ports and loading/unloading of hazardous cargoes such as flammable petroleum products, radioactive waste and potentially explosive fertilizers (Gavaghan, 1990). The study used the normal steps in undertaking a technical risk assessment (Figure 1), each of which is described briefly below.

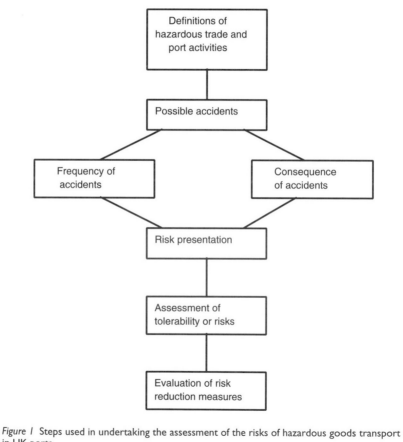

Figure 1 Steps used in undertaking the assessment of the risks of hazardous goods transport in UK ports.

continued . . .

The first step in the technical risk assessment is to study how a port is used in the movement of hazardous cargoes, by analysing the various loading and offloading actions, storage procedures and how the ships actually move around the port. This analysis is undertaken in such a way as to assist in describing possible accidents (step 2), and specifically the frequency and consequences of these accidents (step 3) (Figure 1). Possible accidents studied included:

- the collision of two moving vessels;
- a passing ship striking a berthed ship;
- a ship running aground;
- a shipboard fire spreading to the cargo;
- the splitting of cargo during loading or unloading;
- ship failure due to construction defects;
- a ship being struck by falling aircraft;
- failure of shore-based equipment (pipelines, etc), and
- the 'domino effect' of minor accidents combining to cause a major disaster.

The way in which each of these events is analysed is through 'event tree analysis', a standard method for such analyses in the cases of engineering process risks. One such event tree is shown in Figure 2.

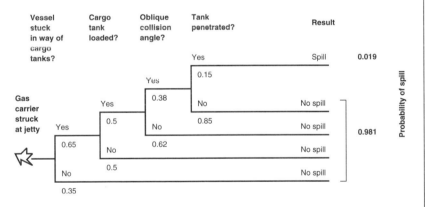

Figure 2 Events and probabilities which must take place for a damaged tanker to spill its load.

Figure 2 also shows how the probability or frequency of such accidents is analysed (step 3). At each branch in the event tree a probability is analysed for that particular event. For example, a ship struck at the jetty must be hit at an oblique collision angle for a spill to occur. The probability of this 'yes' case is 0.38, or 38% (Figure 2). Therefore, the probability the collision will not be oblique is $1 - 0.38 = 0.62$ (62%). The results of this analysis show the

continued . . .

overall probability that a gas carrier struck at a jetty will produce a gas spill is 0.019 or 1.9%.

The next step in the technical risk assessment process is to analyse the consequences of each 'risky' activity. In this case study, the consequences to human life were studied. The results showed that a frequency of one human death every 2.5 years is likely to occur in British ports as a result of the movement of hazardous cargoes.

The final two steps of this risk assessment process are when judgements are made on the tolerability of risks. In the UK, the government's Health and Safety Executive (HSE) divides risk tolerability into:

- negligible: acceptable to most people and does not require action;
- intermediate: industry must apply risk reduction measures, unless the cost is grossly disproportionate to the lives saved; and
- intolerable: requires risk reduction regardless of costs.

The results of this risk assessment are currently being implemented in the United Kingdom.

process. Furthermore, their opinions should be incorporated into the decision-making process: managers can no longer afford to dismiss these views as uninformed or irrational. In other words, present-day coastal risk management must be undertaken extremely sensitively and comprehensively (Gerrard, 1994). These factors should also be taken into account if the study and management of risk is to be incorporated into planning procedures, including the production of coastal plans (Chapter 5).

A key part of the risk management process is the evaluation stage, which comes between assessing what the levels of risk are, and deciding what, if anything, should be done about them. A number of methods can assist in the evaluation process, the most common of which is risk–benefit analysis. As the name suggests, this method is based on evaluating the economic efficiency of risk mitigation measures by adapting cost–benefit techniques (section 4.3.4). Other techniques are described in various specialized texts on risk management listed above.

A central concept in the risk evaluation process is that of 'tolerability' (O'Riordan et al., 1987). This is a carefully chosen word which says that people essentially 'put up with' (tolerate) risks; they do not necessarily accept them, as was previously thought by risk experts. This notion then flows into the decisions which are made to reduce risks, depending on whether they are assessed to be 'tolerable' both by regulatory agencies and by the general public who will be affected by risk-related decisions. For example, in the United Kingdom much of the government's effort in risk assessment and management is focused on the principle that efforts

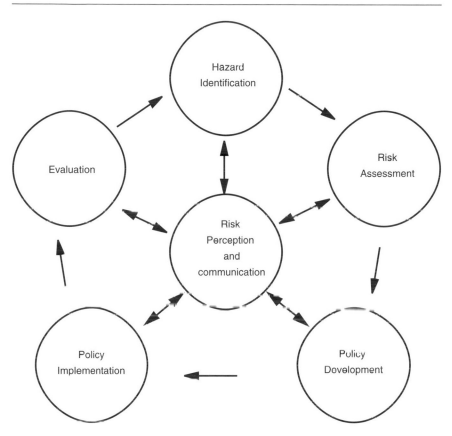

Figure 4.14 The risk management cycle (Soby et al., 1993).

should be made to reduce risks only when they are either entirely intolerable, or sharing the cost of reducing them further is still reasonable given the extra risk reduction gained (Gerrard, 1995). This is known as the 'as low as reasonably practicable' (ALARP) approach, which yields three risk management areas (Figure 4.15).

(c) Mitigating risks and hazards in coastal planning and management

Mitigation of hazards and risks is required where the risk management process has determined that risks are currently intolerable, as described in the previous section. The range of ways to mitigate risks is wide, and there are just as many ways to classify them, the simplest being to divide mitigation options into the following (not necessarily mutually exclusive) categories (Standards Australia and Standards New Zealand, 1995):

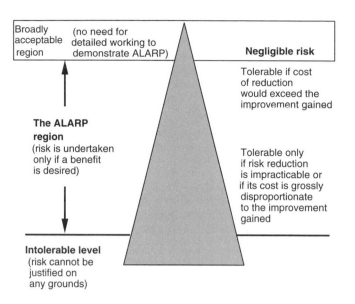

Figure 4.15 The 'as low as reasonably practicable' (ALARP) approach to risk management decision making (HMSO, 1988; Gerrard, 1995).

- avoid;
- reduce the likelihood of occurrence;
- reduce the consequences;
- transfer; or
- retain/accept.

The wording of each of the above choices gives an indication of how they are applied, 'avoid' meaning that risky activities are either not undertaken, or that hazardous areas are identified and avoided. It is important to note that these options are generic, and will vary according to the types of hazards being addressed and the decision-making context (Table 4.12).

Details of the background and application of the options shown in Table 4.12 are given in numerous studies of natural hazard impact management (e.g. White, 1945; Burton *et al.*, 1978; Ericksen, 1986; Platt *et al.*, 1992). Each of the options can be applied by governments through a range of measures, including legislation, or the application of policy and economic instruments such as tax concessions (Burby *et al.*, 1991).

The key issue here for coastal planning and management is that there are many ways to address risk and hazard issues. Risk and hazard management can be either undertaken as a stand-alone exercise, as described above, or included as part of a coastal management plan.

Table 4.12 Example options and measures for coastal erosion hazard management (adapted from Kay *et al.*, 1994)

Subject	Steps
• Event Protection	'Hard' engineering (e.g. sea-walls, groynes) 'Soft' engineering (e.g. beach nourishment, dune enhancement)
• Damage Prevention	Avoidance (e.g. prevent development) Mitigation (e.g. relocatable or flood-proofed buildings, building codes)
• Loss Distribution (transfer)	Individual measures (e.g. insurance) Community measures (e.g. insurance, cost pooling, disaster relief and rehabilitation)
• Risk Acceptance	Do nothing

Perhaps the principal advantages of the latter are: first, that if the planning process is undertaken in a consultative manner, the often difficult issues of risk and risk management can be openly discussed, and solutions found jointly by affected communities and decision makers; and second, the choices for risk and hazard management can be increased by inclusion into a management planning process. There may be good reasons for undertaking certain actions in the name of good coastal management practice, but which also contribute substantially to risk management. Management of coastal dunes in front of hazard-prone beach front property, for example, can be for the conservation of biological diversity, recreation and public access, and a way of providing a sand-buffer during erosive storms.

4.3.3 Landscape and visual resource analysis

One of the most commonly appreciated aspects of the coast is the scenery. Rugged cliffs, picturesque bays, idyllic beaches, rolling dunes, rocky headlands, marshes and tidal flats, forested hinterlands, breaking waves and expanses of water all help create scenery which is highly prized by people. The coast, as the interface between water and land, often has a richness and diversity of scenic features not found in inland areas. The value of this coastal scenery hardly requires substantiation. It is a vital component of people's enjoyment of the coast. It is the setting for all their activities and is a strong influence on their sense of well being and quality of life. Coastal scenery adds to property values and provides the settings, and often the attractions, for tourism. This economic value of scenery is increasingly being determined by environmental economists as part of the coastal planning process (section 4.3.4). Coastal scenery is now regarded

---

Box 4.20

Terminology in landscape and visual assessment

'Landscape' is used by many different people for a variety of purposes, making it a rather ambiguous term. It has three main usages: the first refers to a scene (as in a landscape painting); the second refers to an area which has a common pattern of bio-physical features (as in a landscape ecology); and the third usage refers to the interpretation and experience of the environment by people (the landscape as we know it) (Zube *et al.*, 1974; Lowenthal, 1978; Meinig, 1979). The landscape architectural, environmental psychology and related professions to a certain extent use all definitions but specialize in an understanding of the latter.

- 'Landscape values' are the values people derive from their interpretation and experience of the environment.
- 'Landscape assessment' usually refers to the assessment of landscape values (the resource), and put simply is a process of analysing and mapping environmental characteristics and, using known criteria, determining those which contribute most to the experience and enjoyment of people.
- 'Landscape impact assessment' is a process of determining how changes to the environment will affect landscape values.
- 'Evaluation' is the process where assessment results are examined and used to make decisions about alternative futures.
- 'Valuation' is providing a value based on professional judgement, public preference and economics.
- 'Qualitative judgements' normally express results using criteria which are not themselves readily reduced to simple or precise numerical values. Most landscape assessment requiring judgement is qualitative even if results are expressed numerically (Litton, 1979).

---

as a resource, partly because of this economic value and partly because it is an accepted component of resource assessment programmes (regardless of economic value).

Scenic values are only one component of what can be called coastal 'landscapes' (see the definition provided in Box 4.20). Other landscape values include aesthetic (based on the sensory perception of a place by the community), social (based on the association between community, including Aboriginal people, and place), and historic (based on a connection with a historic figure, event, phase, or activity) (Blair and Truscott, 1989). The relationship between visual, aesthetic and other landscape values is illustrated in Figure 4.16 (Cleary, 1997). Landscape values are also closely linked to recreational values, but whereas the former is focused on values related to perception, understanding and enjoyment of

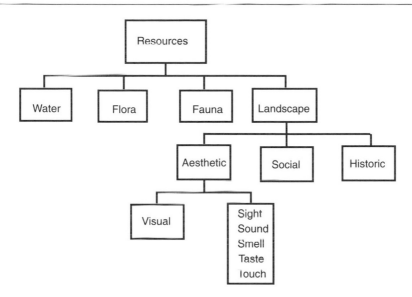

Figure 4.16 Relationship between landscape and visual values (Cleary, 1997).

the environment, the latter has been traditionally oriented towards the management of settings to provide for different recreational activities.

There is a great need to assess and manage landscape values in coastal areas. The increased popularity of the coast as a place to live and visit, and the consequent development, has put immense pressure on many natural resources, including landscape values. The will to protect landscape values generally stems from one of three areas:

- management agencies and individuals recognizing the long-term benefits of good resource management;
- lobbying by interest groups to protect the resource and to have their sentiments included in the decision-making process; and
- legislation and related legal cases requiring that landscape values be protected.

(a) Procedure development

Landscape assessment is a relatively new field and universally accepted methods of assessment do not exist. Formalized procedures for the assessment of scenic quality were first used in the 1960s and 1970s and developed rapidly as a response to legislation and related legal cases. In Britain and the United States, various Acts including the UK Countryside Act (1968) and the US National Environmental Policy Act (1969) provided the initial impetus which ensured that aesthetic values were addressed in land management. One of the problems that these early procedures faced was the view that scenic values are relatively 'intangible' given that assessment results are based on 'subjective' judgements of observers (Williamson, 1978). This comment was answered by pointing out that the standards for measuring the quality of any resource are based on subjective judgements, and that the issue is not subjectivity versus objectivity but rather the acceptance and consensus of standards and techniques used to measure quality (Brush, 1976; Daniel, 1976).

The trend in landscape assessment of including and integrating a wide range of aesthetic, social and historic values has occurred only in recent times. In the past, a large proportion of landscape assessments have been directed purely at visual values (Fabos and McGregor, 1979). A number of reasons have been put forward for this: there is a long history in many countries of people prizing the visual attractiveness of places (Johnson, 1974; Laurie, 1975; Fabos and McGregor, 1979); people's perception of any place is largely visual (USDA Forest Service, 1973); and there is a substantial body of legislation requiring management of visual values (Zube et al., 1974). Now, with the changing and broadening of definitions that has occurred in the landscape perception research field, it is recognized that the complex process of interaction between humans and the environment produces many outcomes other than measures of scenic quality (Zube et al., 1974).

KEY ASPECTS OF PROCEDURES

Excellent reviews of procedures by Fabos and McGregor (1979) and Ribe (1989) reveal a wide variety of procedures designed to assess a range of landscape values. The range of methods used shows variations in four key aspects: purpose, public input, paradigms, and techniques (Cleary, 1997). The choice or development of any assessment procedure should consider these aspects. Following is a brief discussion of these taken from Cleary (1997).

PURPOSES OF THE PROCEDURES

The purposes of the procedures can be grouped into four main categories. The first purpose is to identify landscape values. This is usually undertaken over a large study area (regional or local scales as defined in Chapter 5). The second category of purpose is to identify impacts, usually undertaken as part of an environmental impact assessment or management evaluation of development (section 4.3.1). The third purpose is to assist in the planning of development and can be specific to a development or apply to more general guidelines or policies. The fourth category of purpose is to identify public preferences. This can be done to identify attitudes towards specific projects or to provide general preferences.

PUBLIC INPUT

The second aspect of procedures, public input, varies from none to a heavy reliance on public preferences. Public interest has proven to be a key criterion in evaluating proposals in a statutory or legal context (McCloskey, 1979; Castledine and Herrick, 1995). Findings related to public interest have important implications for both assessing and protecting landscape values (Johnston, 1989) and are the basis for a number of criteria which indicate the public interest value of a landscape feature. Key criteria are:

- there is a tested public preference for the feature;
- the significance of the feature passes a threshold;
- people can experience the feature; and
- the number of people experiencing the feature passes a threshold which indicates its scale of significance (e.g. regional).

THEORETICAL PARADIGMS

The main theoretical paradigms for procedures are based on two quite different views of the nature of landscape values (see Zube *et al.*, 1974).

One view is that landscape qualities are inherent in physical characteristics of places and that assessment will reveal these qualities. This approach has resulted in a range of descriptive inventory or expert approaches, usually categorized as either a 'landscape approach' (McHarg, 1969) or a 'formal aesthetic approach'. The underlying assumption of these approaches is that experts are the best people to judge which places or landscape elements are of the greatest significance. This view is becoming less relevant as land use decisions relating to landscape values place greater emphasis on public interest and on landscape assessments based on researched public preferences.

A second view is that landscape qualities are linked to a person's experience of the environment. There are a number of approaches based on this view: the 'psycho-physical approach' endeavours to identify the relationships between environment characteristics and a person's response; the 'psychological approach' endeavours to identify the reasons, particularly the psychological processes, which make people respond in different ways to the environment; and the 'experiential approach' endeavours to gain an understanding of people's interaction with the landscape in terms of personal experience, feelings and meanings.

The psycho-physical approach, which involves eliciting, measuring and analysing public response to a series of environments, provides the most practical outcomes and is the most used (Zube *et al.*, 1974). Even so, as Ribe (1989) points out, findings about the relationships between environmental attributes and perceived beauty tells us little about the aesthetic magnitude of particular differences. They also fail to tell us when the emotional nature of perceptual change is sufficient to constitute a new state, as between beauty and ugliness, or approval and disapproval. This information would be particularly useful for landscape impact assessment.

The fourth aspect of procedures to consider, and the one that reflects all the other aspects, is techniques.

(b) Techniques

There is a wide range of techniques available for landscape assessment and typically used ones are illustrated in Figure 4.17, showing their general relationship to one another. These techniques are not specific to coasts and can be equally applied to inland areas. The understanding of the role of these techniques and the adequacy of their application in assessment studies and environmental impact statements is varied (Coles and Tarling, 1991). The main techniques are discussed below and key points relating to process and standards are highlighted.

RESOURCE ASSESSMENT

Resource assessment identifies existing landscape values (Box 4.20), providing the baseline knowledge for a variety of further planning and management processes. Although there is considerable variation in assessment methodologies, the intentions of most are directed at

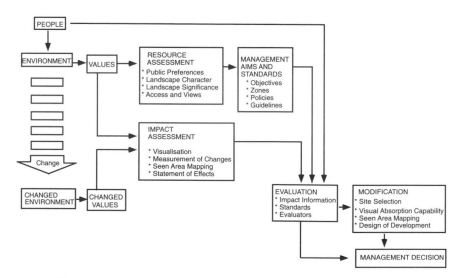

Figure 4.17 Main components of landscape management (Cleary, 1997).

Box 4.21

Main components of visual resource assessment (Cleary, 1997)

- Preference testing, which identifies the characteristics which are most important to the experience and enjoyment of people (see discussion on theoretical paradigms). Findings from specific studies are often applied in a general way to other studies. Holistic appraisals of places or developments should be restricted to evaluation rather than assessment and should be employed with caution (Castledine and Herrick, 1995).
- Inventory, which involves identifying and mapping of data on characteristics relevant for the assessment.
- Landscape character description and classification, which identifies and describes broad patterns of characteristics, allowing consideration of broad areas rather than individual sites. It provides an understanding of the diversity of places within a region, and the sense of identity of the whole region. It also gives a broad indication of appropriate land use and the basis for assessing significance, by providing an inventory of characteristics and by classifying their common patterns.
- Assessment of significance, which uses the preferences of the local community, or established criteria (aesthetic, social and historic) which have been determined by research elsewhere, to identify the quality of landscapes.
- Access and view assessment, which is a measure of how people experience the area. It identifies and classifies access routes, and views from these access routes, according to their actual or potential contribution to the experience of people who use them. This information is vital to establish the degree of public interest and can also be used to gauge the likely visibility of development and provide the basis for seen-area mapping when detailed impact assessment is required. It also identifies and provides some of the basis for managing scenic views in highly used tourist and recreation areas.

---

answering three simple questions: What is there? What is most valuable? How do people experience these? The processes employed to answer these questions generally have main components as described in Box 4.21.

The results from these components are sometimes combined to produce resource zones which, depending on landscape management and land use strategies, may form the basis for management zones.

These components, together with other considerations such as a sound theoretical basis, comprehensiveness and consistent application over large areas have formed methodologies which have proven to be most useful for planning purposes and the most defensible against public scrutiny and court challenges. Further criteria relating to the development of procedures is provided by Fabos and McGregor (1979), Ribe (1989), Zube *et al.* (1974) and Castledine and Herrick (1995) One of the most widely used procedures, particularly in the United States and Australia, is the Visual (Resource) Management System (Williamson and Calder, 1979) which has been adapted or developed by many others, an example being the Western Australian State Department of Conservation and Land Management (CALM) (1997).

IMPACT ASSESSMENT

Compared with resource assessment, impact assessment is a less developed, less understood and poorly applied process (Lange, 1994; Institute of Environmental Assessment (UK) and the Landscape Institute (UK), 1995). It is essentially a process of comparing the conditions before a development with the projected conditions after, and then determining the corresponding effect on landscape values. It is not the purpose of impact assessment to determine whether a development is acceptable or not, but rather to produce a statement of effects. Acceptability is decided in the evaluation of the project. Other typical failings in impact assessments are (Cleary, 1997):

- there is no comprehensive reference to existing landscape values;
- changes are not comprehensively identified or are not measured;
- there is no statement of effects (sometimes erroneously substituted by a judgement of acceptability) or the statement of effects does not refer to comparable variables (i.e. landscape character, significance and access and views); and
- visualizations are used to substitute rather than support all of the desirable components highlighted in the above points.

Many impact assessments are relatively simple and all should involve the following few steps (Cleary, 1997):

- Determine the degree of physical changes to the site. Changes to vegetation and built form are very common and can be measured (e.g. area, height).
- Determine the degree of visibility of the development. Like physical changes, visibility is also easily measured, and can be expressed in terms of magnitude (most easily determined by pixel counts on digital visualizations from locations chosen from the resource assessment), contrast (shape and colour), area affected (determined by seen area mapping), and duration (timing of the changes described above).
- Determine the impact the physical and visual changes will have on

the values of the area (i.e. landscape character, significance and access and views – which should have already been assessed – e.g. the development, in an otherwise natural area, clearly visible from major lookouts, will permanently remove the natural character; the development results in the permanent removal of a significant vegetation feature; the development will temporarily block views). Use this and relevant supporting material from the steps above to produce the effects statement.

AIMS AND STANDARDS

Management aims and standards set the desired future for landscape values. They often initially emerge in the parts of assessment reports dealing with management. Ideally these aims and standards will then be incorporated into plans, strategies or planning schemes which have statutory force as this will add to their defensibility (Castledine and Herrick, 1995). In some cases the standards might be so high that some form of land reservation is recommended in order to protect values. Regardless of the statutory force of the standards, they should be widely promoted in the community to ensure that they are included in the early stages of development planning.

Typical ways to express these aims and standards are:

- objectives, which detail the desired future of the resource in tangible terms;
- zones, which provide the geographic base for applying standards;
- policies, which are rules designed to fulfil management aims; and
- guidelines, which are usually recommended ways of achieving resource management aims.

EVALUATION

Evaluation is usually undertaken by the responsible planning or management authority in the lead-up to making a management decision (which usually incorporates many other considerations). It is essentially a process of comparing the development's effects with the landscape management and land use standards of the area to determine its acceptability in relation to these standards. The determination may be either approval or disapproval, with or without modifications recommended. The evaluation process is often weakened because of the lack of an adequate effects statement or inadequate management standards (Lange, 1994). Without this information, the evaluation will revert to personal judgement, which will vary depending on whether they are staff of the responsible management authority, or proponent, planner, community member or relevant expert. In some cases, evaluators have been reduced to making judgements on landscape impacts using photographs of the site and descriptions of the development.

MODIFICATION

Modifications to the development to reduce landscape impacts can be recommended as part of the evaluation process. These will generally focus on one of the following:

- Site selection, which usually offers by far the greatest opportunity to reduce impacts. Tools which can be employed for choosing sites with the least visual impact are: Visual Absorption Capability (Williamson and Calder, 1979), which combines maps of slope, proximity to ridges, vegetation (pattern, density, height) and soil (colour, erosion potential) to determine sites which visually absorb changes the best; and Seen Area Mapping, which identifies the area seen from nominated sites and consequently the least seen sites.
- Design, which can reduce impact through appropriate layout, form, colour, texture, scale and pattern of visible elements of the development.
- Screening, which uses ground modelling and planting to reduce the visibility of the development.

(c) Application of the techniques

Landscape assessment techniques have a number of strategic uses. They provide an understanding of the diversity within the region and the sense of identity across the region. Landscape character mapping provides a broad indication of appropriate land use (using landform, soil, vegetation, water and existing land use). Assessment of significance identifies areas of highest landscape value which might be best protected by land reservation or statutory controls. Mapping of various characteristics can help in recreation and tourism planning by identifying attractive features and key view points. It can also identify opportunities for landscape improvements such as vegetation corridors. Recommendations can be made as to the types of land uses appropriate for different areas based on the impact of various land uses on landscape values. Assessment also forms the basis for zones, policies and guidelines which can be applied over a broad area. It avoids ad hoc decision making, helps control incremental development, and allows planning for sustainable use of the resource.

At a development project level, assessment provides the basis for protection of landscape values and the creation of developments with sense of place. Some examples of application at this scale are as follows.

- Development can be kept away from areas with significant values, allowing these values to remain an attraction for people – including those who might use that development.
- Sites for development can be chosen which minimize the impact of the development on other users of an area while still retaining important features such as good views.
- Design principles can be developed from local characteristics and can reduce the visual impact of the development while creating interest and a local identity.
- Developments can be designed to offer a range of environmental experiences.

(d) Conclusions

Landscape values are an everyday part of our lives and are increasingly being considered in coastal management programmes. There are obvious benefits from landscape assessment such as better resource management, improved experiences for people using the coast, protection of tourism assets, less land use conflict, stronger local identity for developments, and greater support for management authorities. As with other resource areas, the best results are obtained by specialists who understand the role of the various assessment techniques and who recognize the opportunities that assessment reveals. This is not to exclude the important role that the public plays in determining values and lobbying for their protection, for it is the response to public interest which has been the driving force behind most visual and landscape assessment work.

4.3.4 Economic analysis

Traditional economic analysis of coastal management problems has not always provided appropriate solutions and has often confused coastal managers by not supplying realistic guidelines for their actions. Economists analyse issues by using a set of economic rules which can appear to distort, or even misrepresent, real-world coastal issues. Indeed, many economists would now recognize that such distortion has taken place in the past, using apparent economic rationalism to lead to poor and/or short-sighted decision making.

However, during the past 30 years or so problems with the way 'classic' economics views the environment have been tackled by the rapidly expanding field of 'environmental economics'. Environmental economics attempts to provide valuations of the 'non-market' goods and services provided by the environment. This is the value 'of' and not the value 'in' the environment.

Environmental economists would be the first to say that their work is still relatively new. As such, its application has only recently moved into the mainstream of coastal management decision making, and then only mainly in Europe and North America. Nevertheless, as free-market economies spread, together with the community's growing awareness of the importance of environmental quality and its value, economic considerations in coastal management have also grown.

In addition, in an era of increasing accountability in government decision making and reduced budgets, the questions of 'How much will this cost?', 'Are you sure there are not cheaper ways of doing this?' or even 'Is this coastal management programme cost effective' are never very far away. Coastal managers 'thinking economically' can help answer the 'how much?' question, which could relate to deciding between spending money on a coastal management programme and something completely different, such as building a new school or hospital. Or the question could relate to setting internal priorities within a coastal programme – for example, which coastal wetlands in a rehabilitation programme require urgent attention versus those which are of a lower priority. Thus, environmental economic analysis can assist in short-term decision making, such as the allocation of funds between government initiatives, and in the long-term decisions regarding the costs and benefits of specific developments along the coast.

Recently the values placed on the coastal resource have expanded to include the value of scenic views, beach access and other aesthetic qualities. Coastal land with ocean views and beach access creates high land values, with the beach becoming more popular and environmental quality expectations more acute. Thus, there is a strong demand for uncrowded and good quality (clean) beaches and natural vistas. People are now willing to pay more to enjoy the coast, and are also willing to pay for its protection. A key issue, then, is how these non-market values can be quantified and incorporated within government coastal management decision-making processes.

As is the case with many of the coastal management tools described in this chapter, economic analysis, if not properly applied, can cause far more harm than good. The often complex procedures of economic analysis mean that it is really quite easy to present misleading or false opinions to decision makers. Fortunately, there are rapidly growing numbers of environmental economists and some very good texts on the subject. These include books specifically on applying economic analysis to coastal management issues, such as the excellent texts produced by Edwards (1987), the United States National Oceanic and Atmospheric Administration (Lipton and Wellman, 1995) and the United Nations Environment Programme (Grigalunas and Congar, 1995). In addition there are more general introductory texts on environmental economics such as those by Turner *et al.* (1993), Pearce *et al.* (1989) and Dixon and Sherman (1990) and more advanced texts such as those of Costannza (1992) and Callan and Thomas (1996).

The increasing use of economic analysis tools in coastal planning and management has stimulated debate on how sustainability is factored into economically-based decision making. The debate is focussed on the choice of discount rates used to calculate net present values (Turner, 1993). Discount rates are used simply because people prefer money now, rather than the same amount of money in the future. The ethical dimension to this argument is that by 'discounting the future', there is an implicit assumption that the present is worth more than the future. Also, arguments about discounting relate to how long into the future the economic analysis is undertaken. In other words, how many future generations are considered when looking at issues of intergenerational equity in sustainable development. The point to stress, is that the choice to use particular discount rates in any economic analysis in coastal planning and management should be carefully made, either through reference to the texts listed above, or to an expert environmental economist.

Before the range of economic analysis techniques are described, some fundamental economic concepts, and how these are applied to coastal management issues, are introduced in the next section.

(a) Economic concepts

As Turner *et al.* (1995) summarized, economics is fundamentally concerned with the concept of scarcity and with the mitigation of scarcity-related problems. Viewed in this way, economic concepts have an important part to play in management decisions at the coast where the resolution of conflicts over sought-after space and resources is a fundamental element of coastal management and planning. Economics helps managers consider options for the most efficient allocation of resources, balanced by the needs of buyers (demand) and sellers (supply).

The classical economic concepts of supply and demand, and the cost of the opportunity foregone by society once a decision to allocate financial resources is implemented, played a central role in converting natural coastal areas into various 'developed' uses such as ports. The central economic rationalization in these developments was to increase the speed of economic growth as measured by, for example, a nation's gross domestic product (GDP) and in doing so excluded reference to environmental costs and social amenity.

Classical economic analysis placed little value on natural environments, including those on the coast. Such places were seen as a bottomless sink, and therefore were valued at close to zero. Not until it was realized that the very environment enjoyed by people was becoming degraded to such

Table 4.13 Examples of uses and environmental functions of mangroves (from Ruitenbeek, 1991, 1994)

Sustainable production functions	Regulatory or carrier functions	Information functions	Conversion uses
• Timber	• Erosion prevention (shoreline)	• Spiritual and religious information	• Industrial/urban land-use
• Firewood	• Erosion prevention (riverbanks)	• Cultural and artistic inspiration	• Aquaculture
• Woodchips	• Storage and recycling of human waste and pollutants	• Educational, historical and scientific information	• Salt ponds
• Charcoal	• Maintenance of biodiversity	• Potential information	• Rice fields
• Fish	• Provision of migration habitat		• Plantations
• Crustaceans	• Provision of nursery grounds		• Mining
• Shellfish	• Provision of breeding grounds		• Dam sites
• Tannins	• Nutrient supply		
• Nipa	• Nutrient regeneration		
• Medicine	• Habitat for indigenous people		
• Honey	• Recreation sites		
• Traditional hunting, fishing, gathering			
• Genetic resources			

an extent as to reduce amenity value, cause serious illnesses and deplete commercial fish stocks, did economists realize that the coastal environment had value; their classic economic models and solutions had failed to provide satisfactory answers (Smith, 1996). Economists term such things 'market failures'. Following the increases in environmental awareness since the 1960s, this concept has been extended to include the economic goods and services supplied by an environment in its natural state – extending the fundamental economic concept that 'nothing is free'. This principle can be illustrated by looking at the economic valuation of mangroves (Table 4.13) (Ruitenbeek, 1991, 1994; Tri *et al.*, 1996).

Perhaps the most important economic concept relates to 'economic value'. Its importance is summarized by Lipton and Wellman (1995, p. 10) as follows:

> A fundamental distinction between the way economics and other disciplines such as ecology use the term 'value' is the economic emphasis on human preferences. Thus the functionality of economic value is between one entity and a set of human preferences.

The characteristics of economic value are summarized in Box 4.22. Lipton and Wellman (1995) describe economic value through the different values placed on a polluted coastal area. In economic terms, this polluted area would only have less value than a pristine area because people prefer non-polluted areas to polluted ones. Clearly, an ecological view of the same issue would place different values on the pollution, especially if ecological damage had occurred – but how do you determine the monetary value of an ecosystem?

An important problem in valuing the uses, functions and amenities provided by coastal environments is that many of these are provided 'free'. That is, no market exists through which their true value can be revealed by the actions of buying and selling (Pearce *et al.*, 1989).

Box 4.22

Characteristics of economic value (from Lipton and Wellman, 1995)

- Products or services have value only if human beings value them, directly or indirectly.
- Value is measured in terms of trade-offs, and is therefore relative.
- Typically, money is used as a unit of account.
- To determine values for society as a whole, values are aggregated from individual values.

Table 4.14 Categories of economic value (from Grigalunas and Congar, 1995)

Economic value category	Description	Coastal example
Direct use value	Value obtained from direct, on-site use of a good	Recreational value of beaches
Indirect use value	Value obtained indirectly from a good, where you use another good that depends on the good in question	Coastal wetlands contributing to fish populations
Consumptive use	Good is consumed when used, such that the good is not available for others to use	Fish harvesting
Non-consumptive use	Good is not consumed through use, such that the good remains for others to use	Scenic coastal drive
Non-use value or passive value	Value obtained without the need to use the good	Knowledge of coastal biodiversity existence
Option value	Value obtained by maintaining the option to use the good at some time in the future	Conservation reserves

Environmental economists find many ways to account for these values, as shown below. The important point to stress here is that different categories of economic value can be derived depending on the use and/or service they provide (Table 4.14).

An emphasis on quantifying the economic value of coastal resources is the centrepiece of all the economic analysis tools, the basic guiding concept being the distinction between the valuation of goods or services which are traded in a market (market goods) and those which are not (non-market goods). An example is when a commercial fish catch can be traded in the market place, whereas the natural coastal habitats which support the fishery cannot. A mixture of these values occurs when private coastal residential land is traded in the market place: the value the purchaser places on the land will be strongly influenced by the amount they are willing to pay to experience the benefits of living at the coast.

Both market and non-market valuation techniques rely on the economic concepts of 'consumer surplus' and 'producer surplus' to estimate the net 'willingness to pay' (Table 4.15). Without going into the intricacies of these concepts, they are fundamentally to do with the laws of supply and demand; that is, the more of a good that is supplied above what is demanded, the more the price is lowered to attract more buyers.

Table 4.15 Economic valuation definitions (from Lipton et al., 1995)

Term	Definition
Consumer surplus	Excess of what consumers are willing to pay over what they actually do pay for the total quantity of a good purchased
Producer surplus	Excess of what producers earn over their production costs for the total quantity of a good sold

Some economists argue that the easiest way to consider economic valuation is to think about the difference between 'value' and 'price'. In many cases price does not correspond to value. Consider the situation in the United Kingdom where sewerage sludge was dumped in the North Sea for many years. In that case the waste assimilation function of the marine environment was priced at virtually zero, despite its immense value in dispersing waste (Bateman, 1995).

The most commonly used way to value resources traded in markets is the estimation of producer and consumer surplus using market price and quantity data. This is one of a range of well established methods which rely on the direct observation of market behaviour and value preferences. They are thoroughly described in most texts on resource valuation (e.g. Edwards, 1987).

In contrast, the valuation of goods and services not traded in a market requires some assumptions to be made about how human preference for an environmental good or service is expressed. Three major types of procedures are available to measure these (Lipton and Wellman, 1995): direct (including the Travel Cost Method and Random Utility Models); comparative (Hedonic methods); and experimental (methods which elicit preferences, either by using hypothetical settings, called contingent valuation, or by constructing a market where none exists).

Non-market valuation techniques such as these are essentially what separates environmental economic techniques from the classical approaches. However, in reflecting this difference, the application of non-market valuation techniques is fraught with difficulty. This difficulty arose from problems of estimating economic values of non-market goods, especially in view of the recent gain in momentum of advancing environmental economics into economic decision making in industry, government and the community. Refinement of these methods continues. Meanwhile, it is as well to note that the assumptions made in applying non-market valuation techniques can play a critical part in the results obtained.

Non-market valuation techniques and their underlying assumptions are too complex to be described here. The introductory texts of Lipton and

Wellman (1995) or Grigalunas and Congar (1995) provide sources of further information.

(b) Economic analysis tools

A number of general tools are available for economic analysis of coastal management issues (Lipton and Wellman, 1995), ranging from relatively simple studies of the most cost-effective means of achieving a clearly defined goal (cost-effectiveness analysis and economic impact analysis), to analysing the regional costs and benefits of a range of interacting environmental, social and economic impacts (benefit–cost analysis). Benefit–cost analysis is the most widely used economic analysis tool because of its flexibility and broad applicability. Benefit–cost analysis normalizes (usually to monetary values) the potential economic penalties and benefits of particular actions. It is frequently used in assisting coastal management decisions, most notably in Europe and North America, but is less applicable (as with all economic analysis tools) in places where detailed economic information is lacking (Kay *et al.*, 1996a). As a result, economic analysis used in day-to-day coastal management is generally restricted to the developed world. Neverthless, as Boxes 4.24 and 4.25 demonstrate, economic analysis is increasingly being used in the developing world, albeit on a project-by-project basis at present.

Box 4.23

Steps in benefit–cost analyses (adapted from Lipton, 1995)

Step	Description	Issues
Describe quantitatively inputs and outputs	Analyse the flows of goods and services	Choice of discount rates
Estimate the social costs and benefits	Assign economic values to input and output streams. Use appropriate economic valuation techniques – market and non-market	Choice of economic valuation technique(s)
Compare benefits and costs	Total estimated costs compared with total estimated benefits. If the net value (benefits minus costs) or a project or action is greater than zero, then it is considered to be economically efficient	Policy issues – is a net value greater than zero sufficient justification to go ahead with a project?

Box 4.24

Benefit–cost analysis applied to the management of Indonesian coral reefs (after Cesar, 1996)

In order to evaluate the financial viability of protecting coral reef resources in Indonesia, an economic analysis has been applied by Cesar (1996). Cesar's analysis highlights individual economic benefit versus the economic impacts to society of various human activities which impact coral reefs (see table on page 223). These activities are from direct impacts (blasting/poisoning/over-fishing and coral mining) and indirect impacts from increased sediment loads from land-based sources (urbanization and logging).

Although Cesar admits that his figures are first estimates only, reflected in many cases by the large range in his estimates, a number of coastal management issues are brought into focus. Perhaps the key issue is that all threats which could be quantified produced a total net loss to society greater than the net benefits to individuals; in some cases the net loss can be up to 50 times the net gain. The economic values presented in the table show the net benefits of conservation for a single threat per km^2 of reef. Given the large reef area (75 000 km^2) and the multiple threats to which many of the Indonesian reefs are exposed, the net benefits of reef conservation for the whole of Indonesia will probably run into the tens of billions. This guesstimate can be compared with the total 1994 Indonesian Gross Domestic Product of US$175 billion (World Bank, 1996).

Economic analysis was used to assess who was responsible for the various threats to coral reefs shown in the table, and the economics of their actions. This analysis concentrated on the size of the economic stakes per person and the location of the individuals causing the threat. The size of economic stakes concentrated on the benefits which accrue to the individuals, families, boats and companies involved, including consideration of the economics of management of the operations, such as 'side payments' rents. The location of the individuals causing the threat focused on whether they were local to an area (insiders) or came into an area (outsiders). Outsiders, for example, include large-scale fishing operations. The analysis enabled the division of the threats to coral reefs (see table) to be classified, and policy responses designed (see figure).

The result of this approach is that the flows of money vary considerably for the different threats to coral reefs. For example, coral mining was found to be essentially a local activity with small economic stakes, and as a result a local 'threat-based' management approach, such as community-based management, is recommended. In contrast, poison fishing is a 'big-outsider' management problem with large-scale operations threatening extensive areas of remote coral reef. Here a national threat-based approach is recommended, with enforcement as its key element.

continued . . .

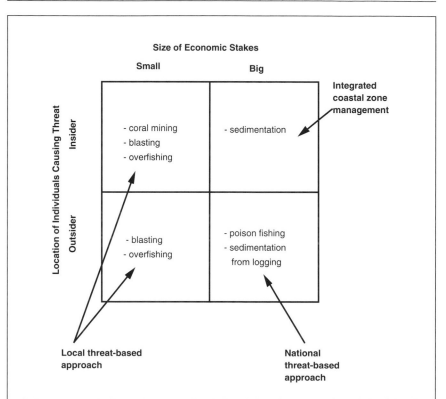

Policy responses to threats to coral reefs in Indonesia based on economic analysis of size of economic stake and stakeholder location (Cesar, 1996).

This approach has formed the basis of assessing the benefits and costs of the large Coral Reef Rehabilitation and Management Programme, funded by the World Bank/Global Environmental Facility. The values determined by this programme form the basis for the World Bank's decision whether or not to provide a loan to the Government of Indonesia for coral reef management. Only if the economic rate of return is high enough (above the so-called opportunity cost of capital) will the loan be forthcoming. This practice is standard in World Bank loans to traditional sectors (e.g. agriculture, infrastructure). However, the difficulty of quantifying environmental benefits has impeded the adoption of this approach to natural resource management investments. The coral reef analysis presented above is a good example of how the World Bank is extending the application of its economic analysis approach to coastal, marine and other natural resources management projects worldwide.

continued . . .

Total Net Benefits and Losses resulting from threats to Indonesian coral reefs (Cesar, 1996).

Threat	Function							
	Net Benefits to Individuals (1000$US)		Net Losses to Society (1000$US)					
	Total net benefits	Fishery	Coastal protection	Tourism	Food security	Biodiversity	Others[1]	Total net losses (quantifiable)
Poison fishing	33	40	0	3–436	n.q	n.q	n.q	43–476
Blast fishing	15	86	9–193	3–482	n.q	n.q	n.q	98–761
Mining	121	94	12–260	3–482	n.q	n.c	>67[2]	176–903
Sediment-logging	98	81	–	192	n.q	n.q	n.q	273
Overfishing	39	109	–	n.q	n.q	n.q	n.q	109

[1] 'Others' includes loss of food security and biodiversity values (not quantifiable)
[2] Estimated costs of logging for fuel in lime-burning kilns
n.q. Not quantifiable
Assumptions: Present value; 10% discount rate; 25 year time-span; all figures are per km^2 of coral reef; intrinsic values not included.

Box 4.25

Regional benefit–cost analysis applied to the management of an Indonesian mangrove system (after Ruitenbeek, 1991)

The government of Indonesia's plans to develop Bintuni Bay (Irian Jaya) focused on the removal of vast areas of mangroves to establish a port and related infrastructure. These facilities were to service resource development projects, such as mining and logging, in adjacent inland areas. In response to this proposal, options for managing the Bay's mangrove systems were examined using a benefit–cost economic analysis by Ruitenbeek (1991, 1994). This study used a case study area to investigate the usefulness of economic analysis techniques in modelling the potential impacts of the woodchip export industry on mangrove management.

Household surveys were undertaken among the estimated 3000 households which use the waters and coastline of Bintuni Bay for traditional mangrove cutting, fishery and agriculture. Data were collected on the value of such traditional uses of the mangrove. This information was linked to studies of the 'formal' market economy to assess the linkages between it and the subsistence economy.

A benefit–cost analysis was undertaken which modelled different scenarios of 'ecological linkage' between mangrove management practices offshore, fishery production, traditional uses, and the benefits of erosion control and biodiversity maintenance functions. The different scenarios were used both because of the intrinsic uncertainty in the linkage between mangroves and fishery production, and the added uncertainty in an area which has been little studied in the past. These five ecological linkage scenarios were:

- none;
- weak;
- moderate;
- strong; and
- very strong.

These scenarios were differentiated according to two parameters which assumed the relative intensity of linkage between mangrove clearing and fishery and other impacts and the time for that impact to occur. For example, the 'strong' linkage scenario assumed a 100% linkage, and five years for that linkage to take place, whereas the 'very strong' scenario assumed 100% linkage and no time delay.

The importance of the mangrove systems of Bintuni Bay to the local economy is shown when 'no linkages' are assumed. When the mangrove and fishery ecosystems are assumed to be independent, the total value of the resource totals Rp3000 billion ($US1.5 billion) in present net value (NPV) terms (Figure 1).

continued . . .

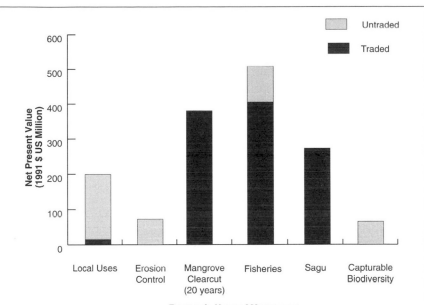

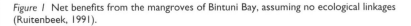

Figure 1 Net benefits from the mangroves of Bintuni Bay, assuming no ecological linkages (Ruitenbeek, 1991).

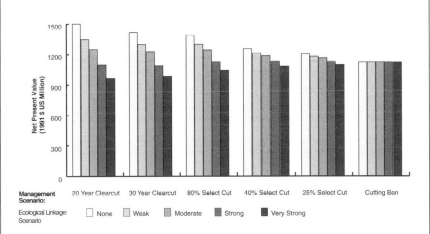

Figure 2 Summary of net benefits from the mangroves of Bintuni Bay under modelled ecological linkages and management regimes (Ruitenbeek, 1991).

The five ecological linkage scenarios were modelled assuming six mangrove management options ranging from 100% clear cutting to a complete cutting ban (Figure 2).

continued . . .

Figure 2 reveals the extent of the challenge facing the managers of the mangrove system. If a cutting ban is implemented (and enforced), the same NPV for all ecological linkage scenarios is predicted. This NPV is around US$1100 million. At the other end of the management spectrum, total clear cutting over 20 years is predicted to produce a wide range of economic impacts depending on the strength of ecological linkage. With no linkage, NPVs are around 35% higher than with cutting ban. However, with strong linkage NPV is around 15% lower. This great range of NPV decreases with the stringency of environmental management controls.

The strength, then, of Figure 2 is that it shows what managers often 'feel': that different management decisions carry with them different degrees of risk and return.

Based on the management options presented in the report, and their potential economic impacts, the Indonesian government has reconsidered the nature, scale and location of the proposed port and related infrastructure in the Bay. The Government is currently evaluating its options for attempting to balance development of this infrastructure with mangrove conservation.

Benefit–cost analysis compares the present value of all social benefits with the present value of opportunity costs in using resources (Field, 1994). In its essentials, a simple benefit–cost analysis requires inclusion of time series data (e.g. depreciation and discounting of future costs and benefits to present-day values) regarding natural resources, environmental quality, and social and economic values. All benefit–cost analyses have three basic steps once the project has been fully specified, including the management options to be analysed (Box 4.23) (Lipton and Wellman, 1995).

The three steps shown in Box 4.23 make actually doing a benefit–cost analysis look easy: it isn't! Results of such analyses rely heavily on the choice of economic valuation techniques, discount rates and a host of other factors described in a number of texts on environmental economics (e.g. Folmer *et al.*, 1995). However, as demonstrated by the case studies which apply benefit–cost analysis to coral reef management (Box 4.24), mangrove management (Box 4.25; Figure 4.18) and beach nourishment (Box 4.26), the technique can be a powerful weapon in the armoury of coastal managers.

Benefit–cost analysis can be carried out at a number of geographic scales, ranging from analysing the economic activities of a single activity to broad national or even international resource uses and environmental functions (Table 4.16). As shown in Box 4.25 for mangrove-related management issues, benefit–cost analysis can be used to assist coastal management plans undertaken from local site to national/international scales; it can therefore be used to assist in all the geographic scales of coastal management planning described in Chapter 5.

Figure 4.18 Mangrove destruction, Irian Jaya, Indonesia (credit: Reg Watson).

Box 4.26

Economic valuation of Adelaide's metropolitan beaches (after Evans and Burgen, 1992)

The State Government of South Australia spends in the order of US$1.4 million per annum artificially re-nourishing Adelaide beaches (R. Tucker, South Australian State Government, personal communication). To assist in the justification of this expenditure the State Government contracted local universities to calculate the economic value of Adelaide's beaches (Evans and Burgen, 1992). Three components were used to evaluate the economic value of beach amenity:

- property value effects;
- day visitor effects; and
- other factors including tourism, special events and public finance.

Property value effects related to the valuation of property with direct beach front access, within easy walking distance of the beach, with beach views and no beach views. Samples of properties at various distances from the beach and with different views were taken and analysed. The total impact on property values of proximity to the beach for the 27.4 km of Adelaide Metropolitan coast is shown in the table.

continued . . .

Summary of benefits of Adelaide beach resources

Benefit	US$ million per annum
Impact on property values	4.2
Day visitor effect	5.6–10.1
Public finance (council property tax only)	0.2
Total	**10.0–14.5**

Previous surveys of beach use in Adelaide showed that an estimated 6.3 million visits occurred in 1986 (an average of around six visits each year for all Adelaide residents). Various techniques were used to estimate the value of beaches expressed by day visitors. The results of previous travel cost studies were compared with the property-value effects of living close to the beach and the cost of other recreational amenities in the area, such as swimming pools.

Other economic effects analysed included the spending by visitors to the beach, the effect on local Council property taxes from higher property values, and the value of entertainment events held on the beach. However, the study concluded that only sufficient data existed to estimate the component of property tax levied by local Councils (see table).

The total amenity value of the entire Adelaide beach system was estimated at US$10.0–14.5 million for the 27.4 km of its length, or US$0.36–0.53 million per km (see table). It is interesting to note that these values are significantly lower than the estimated US$1.2–1.4 million per km (1992 dollars) estimated for the USA by Dean (1988). The difference in these values may reflect cultural differences, perhaps including the greater willingness by Americans to pay a premium for beach-front properties, or the relative ease of public access to the beach in the two countries.

The US$1.4 million per annum cost of the re-nourishment programme is significantly less than US$10.0–14.5 million. Hence the current re-nourishment programme can be seen to be an 'economically efficient' policy response to the beach erosion problem of the Adelaide beaches.

(c) Economic instruments

Economic instruments deliberately intervene in the free market as a means of aligning production costs, as defined by the market, with social costs (total cost of production, including environmental costs) (OECD, 1989a). This notion of production cost adjustment as applied to mangrove management is shown in Figure 4.19.

There are three main groups of economic instruments (OECD, 1989b):

- direct regulation;
- charges, taxes and subsidies;
- market creation.

Table 4.16 Scales of benefit–cost analysis (adapted from Ruitenbeek, 1991)

Scope	Planning scale	Role of benefit–cost analysis	Mangrove related example
Single operator	Site	Optimize production	Evaluation of forestry profitability under different forest management options
Key resource uses	Site/Local	Optimize joint production of two or more graded commodities	Evaluation of joint profitability of fisheries and forestry under different forest management options taking into account linkages between forest and fishery
Traditional production of local populations	Local/Regional	Valuation of production	Accounting of physical flows of hunting and gathering and valuation of these flows at local and shadow prices
All resource uses and environmental functions in region	Regional/National	Optimize value of all uses and functions in region	Evaluate joint value of fishery forestry, traditional uses and erosion control under different management options, taking into account linkages between forest, fishery and other ecosystem components
All resource uses and environmental functions (global scale)	Regional/International	Optimize value of all uses and functions	Evaluate joint value of fishery, forestry, traditional uses, erosion control and international benefits of biodiversity maintenance or climate control under different management options, taking into account linkages between forest, fishery and other ecosystem components

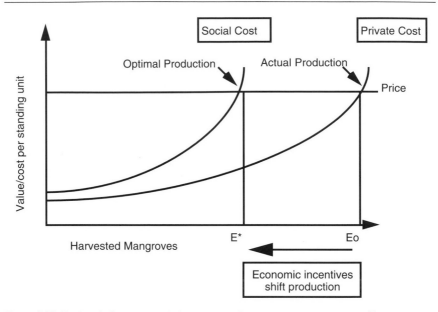

Figure 4.19 Rationale for economic instruments in mangrove management (from Ruitenbeek, 1991).

Table 4.17 The main groups of economic instruments and examples for mangrove management (adapted from Ruitenbeek, 1991, and Turner, 1995)

Economic instrument group	Explanation	Example in mangrove management
Direct regulation	Direct alteration of price or cost levels	• Greenbelt regulations + fines • Trawling regulations + fines
Charges, taxes and subsidies	Direct or indirect alterations of prices or costs via financial or fiscal means	• Royalties and licence fees • Land or input taxes • Production or export taxes
Market creation	Creation of a priceable market in the use of an environment	• Auction of leases • Tradeable permits • Property right reforms

Table 4.17 shows the three groups, together with a brief explanation of each and examples of their application to mangrove management.

Box 4.27

Example of using the control of subsidies in coastal zone management: the US Coastal Barrier Resource Act (CoBRA)

In 1982 the United States Federal Government enacted the Coastal Barrier Resource Act (CoBRA). The legislation designated various undeveloped coastal barrier islands, depicted by specific maps, for inclusion in the Coastal Barrier Resources System. The Act removed certain Federal Government subsidies that otherwise encouraged development in the specified environmentally sensitive coastal areas. The Coastal Barrier Resources System now covers 1 271 395 acres comprising 1211 miles of United States shoreline.

CoBRA's stated objectives (Beatley *et al.*, 1993) were to:

- reduce growth pressures on undeveloped barrier islands;
- reduce threats to people and property from storms and erosion and minimize the public expenditure in such cases; and
- reduce damage to fish, wildlife and other sensitive environment resources.

The Federal subsidies that were denied in CoBRA areas included:

- new flood insurance policies (which also cover coastal erosion);
- federal monies for new roads, bridges, utilities, erosion control etc; and
- non-emergency forms of disaster relief.

However, a number of funding measures not included in the Act were general 'revenue sharing' grants and a range of maintenance and repair grants, including those to existing public roads, jetties and infrastructure associated with energy resources (Godschalk *et al.*, 1989).

These Federal subsidies can be considerable, so their removal could be expected to have a marked effect on the level of development of barrier islands.

However, studies which analysed whether the Act had reduced development of undeveloped barrier islands showed it had not stopped the pressures, but had 'some effect' in discouraging new development (Beatley *et al.*, 1993). Evaluation studies also highlighted the complexities in using the modification of subsidies as a coastal management tool, and how they can be bypassed by using funds from other levels of government or the private sector. Also, the role of other federal subsidies, such as those offered by US Federal Government tax laws (such as tax breaks on second homes) and other subsidies specifically excluded from the original Act, were unclear in how they encouraged, or discouraged, development.

The range of application of economic instruments in coastal zone management is enormous – from 'reef taxes', or levies placed on tourist visitors to coral reefs, to charging for fishing licences. As shown in Table 4.17, these approaches vary from using fines to enforce regulations to highly market-oriented approaches. As Ruitenbeek (1991) points out for mangrove management in Indonesia (Box 4.25 and Table 4.13), each approach has its advantages and disadvantages: direct regulation is the most administratively feasible, yet can be economically inefficient; whereas market creation is the most economically efficient, but is often the most difficult to implement. Between the conflicting goals of economic efficiency and administrative efficiency is the use of charges, taxes and subsidies (Box 4.27).

Economic instruments are relatively blunt policy tools: their application is generally not directed at a specific geographic area, and can cover several issues (OECD, 1989b). Depending on how the instrument is developed by a government, its application can cover the whole coast or be focused on particular areas. But it is far less narrowly defined than, for example, a policy to require environmental impact assessments for new developments.

As well as those economic instruments that are deliberately developed by governments to improve coastal management, there are those 'inadvertent subsidies' which actively promote unsound coastal management practices – for example, in Indonesia, where trade and regulatory distortions create inadvertent subsidies to the over-exploitation of mangroves

Table 4.18 Recommended steps to developing a charging system to assist mangrove management in Indonesia (Ruitenbeek, 1991)

Step	Explanation
1 Identify key resource areas	• Economic and ecological importance
2 Identify resource use conflicts	• Forestry vs. offshore fisheries • Tambak vs. ecological functions • Offshore fisheries vs. traditional uses
3 Identify inadvertent subsidies	• Low or unenforced fines • Low fees, charges or royalties • Trade distortions • Land tax or fiscal distortions • Input subsidies
4 Identify and evaluate corrective measures	• Reforms to existing system • Introduction of new measures
5 Implement preferred measures	• Pilot basis in critical areas • Broad basis regionally or nationally

(Ruitenbeek, 1991). In order to correct these distortions, Ruitenbeek (1991) recommended that the Indonesian government consider developing an economic incentives system to improve mangrove management. He recommended that five steps be followed in order to develop a charging system (Table 4.18) which must fulfil two key functions:

- providing adequate incentives to private operators to develop the mangrove system's sustainably;
- providing adequate funds to local or regional authorities to monitor, and where necessary regulate, mangrove development.

4.4 Chapter summary

This chapter has provided an overview of a number of tools used by coastal managers. Some of these are commonplace, whereas others are used only in specific circumstances and only in some parts of the world, but we believe are likely to become more widespread in coming years.

We chose to present a balanced view of current management tools available – not just the technical ones. Administrative, social and cultural management techniques are equally part of the modern coastal manager's toolkit.

Each of the management tools can be used individually, or in combination with others presented in this chapter and any other tools applicable to a particular situation. However, the effectiveness of these tools is often optimized when used as part of a coastal management plan.

Chapter 5

Coastal management planning

Coastal management plans can be very powerful documents. They can chart out a course for the future development and management of a stretch of coast and/or assist in resolving current management problems. This dual benefit is the greatest strength of coastal management plans: they can have an eye to the future, but still be firmly based in the present.

Coastal management plans can also be used as part of any coastal programme aiming to bring together (integrate) the various strands of government, private sector and community activities on the coast. As such, coastal management plans have the potential to play a vital role in the successful integration of various coastal management initiatives.

Finally, coastal managers' use of coastal management plans can act as a kind of melting pot which helps blend together the various tools described in the previous chapter to deal with a range of issues. In doing so coastal management plans can assist in resolving conflicting uses and ensuring that management objectives are met. As will be shown below, this can enable coastal managers to tackle difficult and/or sensitive issues in a holistic, non-threatening way.

In order to present a structured discussion of the various types of coastal management plans the first section of this chapter presents a discussion of the different ways in which they can be classified. One of these classification types is then used to structure the description of coastal management plans – whether they are 'integrated' or 'subject' (non-integrated) plans. Last, the processes by which coastal management plans are produced is described with special attention paid to designing a planning process which engenders not only a sense of ownership of the plan with stakeholders, but also a commitment to its implementation.

5.1 Classifying coastal management plans

Plans used in the management of the coast can be classified according to a number of criteria which form the basis of the terminology used to describe plan types in this chapter. The most common are the classification methods shown in Table 5.1.

Table 5.1 Coastal management plan classification methods and plan types

Classification method	Plan types			
Geographic coverage	International	Whole-of-jurisdiction	Regional	Local	Site
Focus	Operational	Strategic			
Degree of integration	Subject	Integrated			
Statutory basis	Statutory	Non-statutory			
Reason for plan production	Required for funding	Required to clear statutory works conditions	Legislation which requires management plans	Direct response to management problem	

Some of the classification methods in Table 5.1 are mutually exclusive but most are not; indeed most coastal management plans produced today can be described according to one or more of the criteria shown in the table. Often a classification is required to accurately describe a coastal management plan by including information about its scale, focus and/ or degree of integration. For example, a plan may be required in order to obtain funding, be integrated and strategic in nature, and cover a particular geographic region.

Any one of the five methods shown in Table 5.1 could be used as the basis for structuring this chapter. Each has advantages and disadvantages. Choosing one classification method over any other could create an impression that one style of plan is more important than another; however, for purposes of clarity we have chosen the simplest classification method – by the degree of integration – to form the basic divisions in this chapter. Subject plans which have little or no degree of integration are described first, then integrated plans which attempt some form of integration are outlined. Within the discussion of subject and integrated plans the geographic coverage of plans is used as a way of structuring their analysis. However, before subject and integrated plans are discussed, it worth discussing the other plan classification methods (Table 5.1) in more detail.

5.1.1 Coastal management plan focus

Coastal management plans can also be examined according to their focus on either strategic or operational issues (Figure 5.1). Strategic planning issues are concerned with the long-term future development of the coast, such as siting of ports or the location of future coastal urban developments. As described in Chapter 3, operational management issues are concerned with the day-to-day management of the coast, such as the issuing of permits, or on-the-ground management works, such as rehabilitation. Plans assisting in operational issues are usually called

Management		Planning	
		Strategic Planning	Operational Planning
	Strategic management	• Strategic plan • Strategic management plan	• Strategic operational plan
	Operational management		• Operational plan • Management plan

Figure 5.1 Coastal management plan types according to strategic or operational focus.

'operational plans' or simply 'management plans'. The same terminology can be applied to plans which result from strategic management decisions (Figure 5.1), being termed 'strategic management plans'. There is also linkage between strategic management and operational planning. Strategic management decisions can set the framework for management planning in specific areas. For example, strategic decisions on the siting, design and operations of tourist pontoons in coral reef areas will influence the day-to-day planning of those areas.

(a) Strategic planning

> A strategy must be realistic, action oriented, and understood through all spheres of management. A strategy must be more than a cluster of ideas in the minds of a few decision makers, rather the concepts must be disseminated and understood by all managers.
>
> (Thorman, 1995)

Strategic coastal planning attempts to set broad, long-term objectives, and defines the structures and approaches required to achieve them. It is an ongoing process so that changing needs and perspectives of society can be accommodated, and as a consequence is often multi-dimensional and multi-objective. Strategic planning does not attempt to give detailed objectives, nor give a step-by-step description of all actions required to achieve the objectives. Strategic planning is the highest order of planning; it attempts to provide a context within which more detailed plans are designed to set and achieve specific objectives as well as the development of government policy.

Strategic planning is a process in which the major elements determining the form, structure and development of an area are considered together and viewed in a long-term and broad perspective. The key functions of strategic planning are (AMCORD, 1995):

- providing a long-term 'vision';
- planning, prioritizing and coordinating; and
- providing broad regulation.

Strategic planning is an important part of management because it provides guidance in managing development within a longer-term framework than operational planning. Strategic planning is often on 5- to 25-year time frames, while operational planning is undertaken on an annual to triannual basis. Although strategic planning has long-term time frames, it is still an ongoing process so that changing needs and perspectives of society can be reviewed, generally at 5-year intervals. Strategic planning is also important because it is one of only a few frameworks which are multi-dimensional and multi-objective. Strategic plans can simultaneously focus on time and space while examining a range of competing issues and objectives. The Shark Bay Regional Strategy is a good example of strategic planning applied to coastal areas (Western Australian Planning Commission, 1996b). It uses a horizon of 5 to 10 years over a large spatial area and seeks to broadly manage a range of issues from World Heritage Values to rural development (see Box 5.23).

The long-term, broad geographic focus of strategic planning and its position as the highest order of planning, setting specific short-term objectives as well as the development of government policy, influences the use of other strategies within the planning hierarchy. It might seem from this that strategic planning is only appropriate at national, state and regional levels. However, while most strategic planning does occur at these levels, it does not preclude its application at the local or site level. Strategic planning is also relevant at these lower levels because local or site plans can incorporate a broad range of objectives such as sustainable development, improving access to the coast, and the sustainable use of particular resources. To achieve these objectives a long-term view is needed to produce fundamental changes in the local society's view of how areas or resources should be managed at all planning scales.

The long-term and broad perspectives taken in strategic planning facilitate a number of activities necessary for sound management (AMCORD, 1995), which are also relevant on the coast. Strategic planning provides a channel for communication with the community and other stakeholders (e.g. steering committees, workshops). It enables managers and stakeholders to anticipate change in a well defined framework and to define a vision of how this change could be accommodated (e.g. tourism). In doing so, long-term objectives can be set and a long-term framework for a range of initiatives such as environmental quality can be established. Strategic planning provides a framework for other long-term or short-term strategies and policies for specific issues (e.g. fishing or tourism). Strategic planning through its long-term and multi-objective framework helps to identify action areas, establish priorities for action (e.g. structure plans or tourism development projects) and mechanisms to coordinate these actions. Along with prioritizing, the resources needed to effect these actions can be identified.

Strategic plans generally deal with broad categories of management such as the appropriate uses of specified areas such as marine waters; particular resources such as fishes; development – economic, social and infrastructure; and environmental management. Again the multiple objective nature of strategic planning is highlighted, and to accommodate these objectives in a planning framework a strategy can be based on a number of mechansims such as broad planning statements, policies, recommendations for exisiting and future programmes or initiatives, a zoning scheme, or a combination of the above. Most of these mechanisms are discussed in detail in Chapter 4.

Like all planning initiatives, stakeholder participation is a fundamental component of strategic planning. Meeting the needs of all stakeholders through the multiple objective nature of strategic planning is difficult and there may not be agreement by all parties. Nevertheless there usually needs to be consensus on a shared vision and agreement on actions to realize that vision. This can only be accomplished through meaningful public participation as discussed in section 5.5.1b.

Strategic plans and resulting action programmes can and should incorporate monitoring and evaluation to ensure that the strategy is working and that management can respond to changes in societal values and expectations.

(b) Operational planning

At the operational level, goals specific to the area's physical and socio-economic conditions are formulated, and form the basis of the area's coastal zone management plan.

Goals or aims at the operational level will be guided by broad international, national or regional strategies, and stakeholder participation, but in ways specific to local conditions. Area-specific goals may be to improve the livelihood of coastal residents through appropriate species and habitat management, or to maintain traditional-use opportunities.

Operational planning is concerned with how on-the-ground and on-the-water management actions will be realized. At the broader planning scale level this generally involves the allocation of financial and human resources, where necessary the formulation of statutory mechanisms, and the establishment or coordination of other organizations to undertake the activities required to give effect to the plan. Operational plans at the local or site level define the financial, infrastructure and human resource requirements needed to meet specific management objectives. This is usually done in the medium term (three to five years) to provide the time needed to budget for major capital works and projects, and the short term (annual) which enables agencies to implement the plan. The scope of these operational plans will vary with the available resources, administrative arrangements, and budgeting requirements for the agency responsible for managing the area.

5.1.2 Statutory basis of coastal management plans

The formal power of a coastal management plan as defined by its statutory basis has a large degree of influence on both plan contents and the approach to its formulation. Some management plans, most commonly those associated with formalized land or water-use planning systems, have the full force of law in their implementation. In contrast, other coastal plans may have been undertaken without such statutory force. These two groups of plans are generally called 'statutory' and 'non-statutory', respectively.

Statutory plans usually contain provisions regarding the use and management actions for particular areas of land or water. The most common of these are zoning provisions in statutory urban planning documents such as town planning schemes, and marine management zones related to marine protected area planning (section 4.1.3).

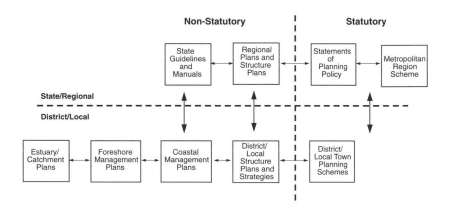

(a) **Western Australia**

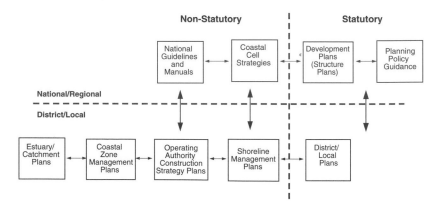

(b) **England and Wales**

Figure 5.2 Comparison of statutory and non-statutory plans influencing coastal management in Western Australia and the United Kingdom (from Kay *et al.*, 1995).

Planning legislation aimed at the control of urban development is a common legislative requirement for the production of integrated plans. These plans are usually focused on land-use planning, and at present rarely cover both land and water. Nevertheless, integrated land and water use plans are beginning to emerge from this essentially land-oriented process (some examples of these are shown later in this chapter).

Examples from Western Australia and the United Kingdom (Figure 5.2) illustrate the divisions between statutory and non-statutory coastal plans which influence coastal management. In some cases the division between statutory and non-statutory coastal management plans is blurred by legislation forming the framework within which they can be developed; in other cases the division is specified by legislation which does not make plan preparation a legal requirement, but specifies plan contents. An example of this approach is the United States where the preparation of State integrated coastal plans is voluntary, but if the States choose to do so there are requirements specified in Federal law (Chapter 3). These requirements are imposed to ensure that Federal coastal management objectives are met.

5.1.3 The requirements of coastal management plans

The word 'requirements' for coastal plans is used here to refer to the reasons why a plan is produced. This may seem rather obvious, in that coastal plans are produced to assist in addressing coastal management issues and problems (Chapter 2). However, this reason may be the direct cause of the production of coastal plans in some circumstances only. The direct cause and effect relationship (ie. a problem produces a plan) can often be influenced by legislative requirements, influenced by inter-governmental relations, or be in response to community or political pressures. Coastal management plans may be encouraged, or sometimes a prerequisite, for obtaining funding for coastal management activities. The most frequently cited example of such a system is in the United States, where States must produce a Coastal Zone Management Plan in order to obtain Federal Government funding for various coastal management activities in their State (see Box 3.8).

Other requirements for the production of coastal management plans include statutory provisions, such as those linked to Environmental Impact Assessment requirements or planning approvals (see section 4.3.1). For example, in Western Australia management plans for foreshore reserves (site level plans) are usually required for planning approval for some types of coastal urban developments. The requirement for such plans may also be linked to permit, licensing and other related statutory provisions (see section 4.1.4a). In some cases coastal plans may not be a legislative requirement for the granting of permits or licences, but may be encouraged by the authorizing government departments in order to provide a context for individual decision-making actions on the coast.

Finally, there may be direct legislative imperatives that require management plans to be produced in areas potentially subject to the impacts of coastal erosion and flooding, or for conservation areas such as national parks. Legislation which proclaims marine protected areas may require management plans to be produced ahead of proclamation, as is the case in Western Australian marine protected areas. In Indonesia a marine park can be declared without a management plan, but management actions cannot be initiated without such a plan. However, all Indonesian national parks (marine or terrestrial) require a management plan once declared. These approaches attempt to avoid the 'paper park'

Box 5.1

Consultation requirements for zoning plans in the Great Barrier Reef Marine Park, Australia

The Great Barrier Reef Marine Park Act specifies that zoning plans will be prepared for Sections of the park and to meet the following objectives:

- conservation of the GBR;
- regulation of the use of the park so as to protect the GBR while allowing reasonable use;
- regulation of activities that exploit the resource of the GBR Region so as to minimize their effect;
- reservation of some areas for appreciation and enjoyment by the public; and
- preservation of some areas in their natural state undisturbed by man except for the purposes of scientific research (Government of Australia, 1975).

The Act also specifies that the public are invited to make representation on two occasions: the first when it is decided to prepare a zoning plan, and once a zoning plan has been drafted. The GBRMP Authority is required to consider any representation made and if it thinks fit, alter the plan accordingly (Government of Australia, 1975). The draft plan is forwarded to the Minister responsible for the GBRMP who either accepts it or returns it to the Authority with comments for reconsideration.

Once accepted, the plan is laid before Australia's two houses of parliament for 15 sitting days. If neither house passes a resolution to disallow the plan, it is passed and comes into operation on a date specified by the Minister. If the plan is disallowed a new plan must be prepared, and the process begins again.

syndrome of declaring marine protected areas without providing a framework of resources to manage the area for its conservation values (Alder *et al.*, 1995b). Simlar management planning requirements may be specified through legislation for terrestrial reserves protected for conservation purposes.

A key issue with coastal management plans which have some external requirements – be it funding, legislation or other reasons – is that these requirements place constraints on some aspect of the plan. Such constraints could include the contents of the plan, information needs, how the plan should be formatted, who should be consulted, the timeframe for plan finalization, or the steps that must be taken to obtain approval (Box 5.1). The formulation of zoning plans for the Great Barrier Reef Marine Park is one example of how legislation directs the planning process (Box 5.1).

Requirements for plan production can have a profound effect on the overall shape of coastal plans. Clearly, plans must be produced to satisfy those constraints, such as being formatted correctly in order to obtain funds. If the constraints adequately reflect the practical issues of coastal management planning within a nation's administrative and political framework, this should not detract from management outcomes. However, where this is not the case, there is clearly a risk that satisfying the constraints imposed on the production of a plan can impede or even override sound coastal management practices.

An often overlooked requirement for coastal plans is community expectation. This is, after all, a major reason for undertaking coastal plans – that the community expects the best management of coastal resources. If the local community or stakeholder group is not satisfied with the outcomes of a plan, they can actively work against it through lobbying, or by simply boycotting its implementation actions. The most commonly used method for avoiding this problem is a consensus-based model for producing the management plan, described in section 5.5.1.

5.1.4 Degree of plan integration

Perhaps the main division in coastal management planning is between plans which attempt to assist in the management of issues through their integration with others, usually through the use of spatial management techniques, or managing issues through sector-by-sector prescriptions.

Plans which cover one particular aspect or sector of coastal management are termed 'sector' or 'subject' plans (Gubbay, 1989). These include, for example, some natural resource management plans, such as a fishery management plan, coastal engineering, nature conservation plans and various industry-sector plans, such as a tourism strategy. Plans concerned with particular coastal management tools also fall into this category, such as the plans and strategies associated with the various coastal management techniques described in Chapter 4.

In contrast, plans that focus on the bringing together of various government sectors or management approaches, or attempt to address conflicts and the multiple use of a geographically defined area, are usually labelled as 'integrated coastal management plans'. The use of the term 'integrated' follows the sense described in Chapter 3 of generically joining together and does not imply the degree to which this joining occurs. Other words – for example coordination or harmonization – could equally be used to describe such plans.

Integrated plans can also be called 'area plans' to denote their coverage of a specific area of coast. Area plans only equate with integrated plans where there is some element of integration attempted in the planning exercise. Without attempts at integration, area plans simply become subject plans which cover a particular area. An example of the differentiation between subject and area plans for the United Kingdom has been developed by Gubbay (1994) (Figure 5.3).

Nevertheless, subject plans can be included or accommodated in integrated plans at similar spatial scales. For example, in the Shark Bay area of Western Australia (Box 5.2), tourism planning and integrated coastal planning have been joined at a number of spatial scales (Figure 5.4). The current Shark Bay Plan was drafted in 1996, at the same time as a tourism plan for the Gascoyne (including Shark Bay) was drafted. The region plan which highlights the need to manage the World Heritage values also recognizes the Gascoyne plan and recommends that many of the action statements of the tourism development plan specific to Shark Bay should be initiated.

The broader planning perspective in the Shark Bay area also demonstrates the evolution of subject and integrated planning over time. Figure 5.5 shows how the integrated planning cycle both incorporates a number of subject plans and results in the production of others. The initial Shark Bay plan identified a number of subsequent subject plans which needed to be developed, and number were formulated and implemented. The outcomes of these initiatives provided input into the second region plan in 1996, which in turn has identified further subject planning.

In some cases coastal management issues can be managed simply through a series of policy statements and initiatives, examples of which were described in Chapter 4. In these situations the level of integration is generally low, but agency coordination and cooperation is usually still required.

Clearly integration is the best planning option for many coastal management cases for number of reasons: it has a holistic approach to solving issues, it is effective and efficient in its use of resources and easily handles multiple objectives. Another important feature of integration is its

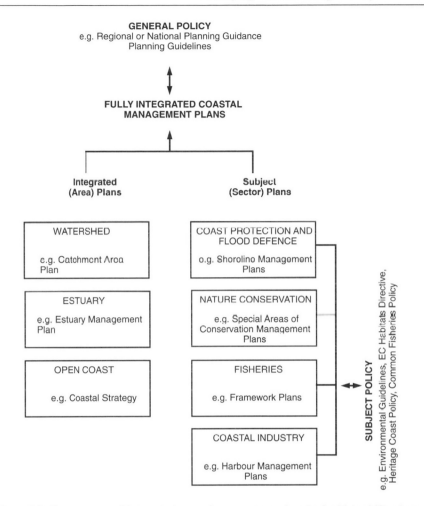

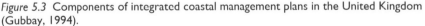

Figure 5.3 Components of integrated coastal management plans in the United Kingdom (Gubbay, 1994).

independence of spatial scales; that is, integration can be used at various planning scales. Nevertheless, there are numerous cases where a subject-by-subject approach is preferable. These cases are described in the next section.

(a) Coastal management subject plans

Subject plans are those developed to address a single issue, subject or sector and, as a result, may be deliberately non-integrative, or may be developed as a consequence of a recommendation of an integrated coastal

Box 5.2

Tourism planning and its relationship with integrated coastal planning in Shark Bay, Western Australia

The Shark Bay Region Plan from Western Australia is a good example of how a number of planning initiatives can be integrated (Figure 5.5). It is also a good example of how subject plans – tourism plans in particular – can be integrated into a broad planning framework.

Nature-based tourism is a growth industry in Shark Bay. Fishing, interactions with dolphins, four-wheel driving and diving are the major attractions. Two approaches to planning for tourism have been taken at Shark Bay. The World Heritage Plan makes several provisions for a range of nature-based tourism opportunities and their management within protected areas. Specifically the World Heritage Plan (Dowling and Alder, 1996) aims to:

- protect the dolphin population and their habitat;
- enhance visitor experiences with dolphins and increase visitor awareness of the conservation values of the region's marine and arid environments;
- maintain conservation values while providing and encouraging recreation and tourism activities; and
- promote and undertake scientific studies and monitoring of the Reserve's biophysical and social values.

In addition, the Gascoyne Regional Ecotourism Strategy, which includes the Shark Bay Region, guides the development of a sustainable nature-based tourism industry. The strategy clearly recognizes the importance of maintaining the World Heritage values and to achieve this it recommends the assets base of the region be extended as follows to reduce the impact of growth on existing product:

- the geographic area on which tourism depends be extended;
- preparations begin for the infrastructure needed to accommodate growth;
- the assets base be managed in a coordinated way;
- marketing to target groups who are empathetic with the objectives of environmental preservation;
- educating visitors and potential visitors about the environmental values of the region;
- regularly addressing competing interests and evaluating likely outcomes and cost/benefits;
- providing for the ongoing monitoring of the environment and local cultures;

continued . . .

- optimizing the use of limited management resources through cooperation and coordination; and
- addressing the needs of local communities and fostering their participation in the industry (Gascoyne Development Commission 1996).

The key to effective management of the tourism sector and its impact on the fragile resources of the Shark Bay region has been the use of integrated regional planning. The first Shark Bay Region Plan was completed in 1988 and revised in 1996 after extensive public consultation and environmental sensitivity analysis (Figure 5.5). All stakeholders in the region, including tourist operators, agreed that sustainable economic development in the area is only possible if the environment is carefully managed. This consensus is reflected in the 1996 plan.

plan (see Figure 5.5, for example). Subject plans can cover a range of topics – in fact any issue facing the management of the coastal zone described in Chapter 2. For example, they commonly include resource management plans (e.g. fisheries management plans) and industry sector plans (e.g. a transport or tourism strategy).

Subject plans are used for coastal management in a number of circumstances. Perhaps the most common of these are when they are used as a contribution to a broader approach to either an integrated coastal management plan or coastal management programme. For example, in England subject plans are viewed by government as an important part of

Figure 5.4 Range of plans for the management of Shark Bay, Western Australia.

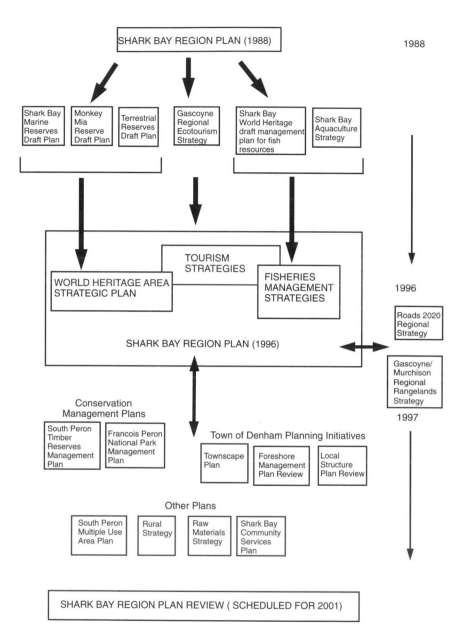

Figure 5.5 Integrated regional coastal planning and subject planning in Shark Bay, Western Australia (Dowling and Alder, 1996).

that nation's coastal zone management efforts (Gubbay, 1994; Kay *et al.*, 1995) (Figure 5.3). The United Kingdom government recognizes that the effectiveness of these plans is maximized through their inclusion in a broader integrated coastal management programme (Figure 5.3).

Integrated plans are described in more detail in section 5.3, and subject plans in section 5.4.

5.2 Designing a coastal planning framework

Before describing subject and integrated coastal plans in the next two sections it is worth reflecting on how an overall framework for coastal planning can influence the approach and style of individual coastal plans.

A simple way of examining this issue is by considering the management of a typical coastal problem, such as the degradation of a coastal dune due to recreational pressure. There are a number of ways the problem could be addressed through direct management actions, but there are effectively only three approaches which involve the use of coastal plans (Figure 5.6). The first approach is to undertake immediate management actions, such as revegetation, access management, etc., without first producing a plan. In a situation where issues are few, or management actions simple and/or unlikely to cause conflict between different coastal user-groups, such direct action is the most appropriate approach.

The second approach is to write a coastal plan to guide management actions, then undertake those actions. This course of action may be the most appropriate where there are conflicting issues and/or users, or complex management issues.

The third option is to develop a coastal planning framework which considers the various types of plans available to address the particular management action and how the plans would interact with other issues and overall management objectives to assist in achieving desired management outcomes. Subsequently, a plan is produced and implemented by undertaking management works. Which option is taken again depends on available resources, legislative basis, social and cultural factors, and political priorities and acceptability.

It is important that coastal managers be able to distinguish between the different types of plans described in the previous section, and between different geographic scales of integrated plans outlined below. This way managers can make an informed choice regarding the need for a planning framework and which plans, or combination of plans, are the most appropriate for their circumstances.

The development of a coastal planning framework usually occurs when there is the need to resolve more than one issue or to formulate more than one plan. Thus, in the many and varied circumstances where management

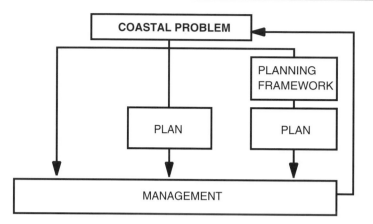

Figure 5.6 Options for coastal planning frameworks.

needs are greater than what can be addressed through a single ad hoc plan, a coastal planning framework is often designed. Such a planning framework is often part of, or closely linked to, an overall coastal management programme. The form of coastal programmes is usually dictated by the administrative, political, economic and social circumstances of particular coastal nations, as described in Chapter 3.

Assuming a coastal planning framework is required, the issues which require consideration in its design can be broadly grouped into four main areas (Figure 5.7):

- relationship with an overall coastal management programme (including the type, number and intensity of management issues and problems) and other government policies, strategies and plans;
- choice of plan types and production styles;
- linkages between plan types; and
- scales and coverage of plans.

The most important factors influencing a coastal planning framework are the type, number and intensity of management issues and problems. This has a direct bearing on the choice of particular styles of coastal plan and the tailoring of plans to fit particular objectives. These factors also have an indirect bearing on framework design through their influence on the shape and nature of an overall coastal management system. As discussed in Chapter 3, coastal management programmes are constructed to reflect the management issues being addressed and the particular cultural, social, economic, political and administrative issues within individual coastal nations (Figure 5.7). Well designed coastal management programmes emphasize the central role of coastal planning; therefore a

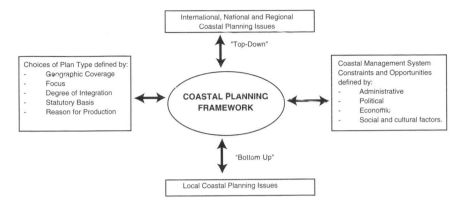

Figure 5.7 Major factors influencing coastal planning frameworks.

coastal planning programme as such often does not have a separate identity from an overall coastal management system.

A coastal planning framework helps to choose between the wide variety of coastal plan types described in the previous section. The choice of coastal plans available to a coastal manager in any coastal nation will be constrained to a large degree by its systems of governance and, in turn, any overall coastal management system. This issue is particularly relevant to the statutory basis of coastal plans, the reason for their production and geographic coverage (Table 5.1). The latter is often constrained by the relative distribution of power, human and financial resources between levels of government and how these levels of government interact. For example, local-level planning may be constrained in countries where local government has small staffs and/or budgets. Similarly, the statutory planning systems in coastal nations, and how these powers are shared between levels of government, will largely constrain the choice of statutory or non-statutory coastal plans. A comparison of the coastal management plan types in the United Kingdom and Western Australia, shown in Figure 5.2, illustrates the point. Legislative requirements may also dictate the approach to coastal plan production by defining, among other things, those who should be consulted. Where there are no such constraints coastal planners are free to produce plans using the various techniques described in section 5.4.

Fitting together coastal plans that have been designed to have different scales, foci, degrees of integration, etc. can be compared to putting together a complex jigsaw puzzle. To take this analogy further, the task is made even more difficult by having a poorly defined picture to guide the assembly, with no well defined edges to the jigsaw. Pieces of the coastal planning jigsaw include how plans at one scale relate to those at another scale and how different styles of plans interrelate with each other in time,

Table 5.2 Example coverages of different scales of coastal plans

Level of plan	Plan coverage (km of coast)	Number of plans required for 1000 km of coast[1]
Whole-of-jurisdiction	1000	1
Regional	100	10
Local	10	100
Site	1	1000

[1] For an example national coastline 1000 km long

space and in the coverage of management issues. A nation's 1000 km of coast could be covered by one overall national scale coastal management plan, 10 regional-scale plans each covering 100 km of coast, and so on (Table 5.2); however, attempting to undertake 1000 separate site-level plans covering 1km each (Table 5.2) would clearly be a daunting exercise, even for the best resourced government.

However, attempting to cover long lengths of coastline with detailed management plans could in most cases be counterproductive unless undertaken in an extremely well structured, organized process over a long time period. The obvious danger in embarking on a large number of detailed plans is that the overall context of those plans is lost. There is also the danger of each plan attempting to produce similar outcomes for the coast; such as, for example, uniform types of coastal access which do not reflect site specific characteristics – the very purpose of site-level coastal planning.

The opposite of attempting to cover a coast with a plethora of detailed plans is attempting to achieve detailed management outcomes with international, national or regional plans. In this case, the higher-level purposes of such plans, including identifying areas which require more detailed coastal management plans, becomes lost in an attempt to fix all management problems. This can also be counterproductive if there are different levels of government involved at the various management planning scales. For example, a national government may become embroiled in site-specific problems more effectively addressed by local governments or community groups, and vice versa. The solution to the competing pressures for site-specific (operational) coastal management planning and higher-level strategic plans is to develop a structured programme which identifies management priorities at regional, local and site level.

A hypothetical case of such a structured integrated coastal planning programme is shown in Box 5.3 for a generic coastal nation with a 1000 km

Box 5.3

Integrated coastal planning programme of a hypothetical coastal nation

Imagine a coastal nation with 1000 km of coastline embarking on a coastal planning programme. The various stakeholders in the management of the coast have decided that a multi-level integrated coastal planning approach is needed. They decided to develop national, regional, local and site-level coastal management plans which aim to assist in resolving issues of critical environmental degradation, conflicts over the current use of coastal resources and future sustainable use of the coast. The decision makers consider that a long-term approach with priority areas tackled in just five years is the best course of action. After that, programme priorities will be reviewed and the overall success of the approach evaluated.

The identification of priorities results in the development of a national-level plan, four regional, eight local and 12 site plans (see figure). In some cases coastal problems are so acute and complex that a full 'cascade' of management plans from national to site levels will be developed (Location A in the figure). In other areas, such as Location B in the figure, site-specific plans are warranted, but not a regional-level plan. Other areas, such as Location C, required local-level planning, but not regional- or site-level plans.

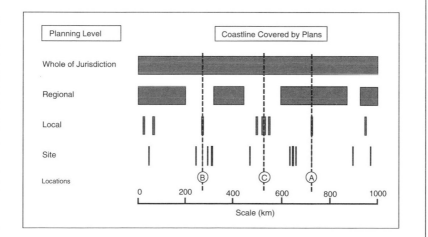

Coverage of integrated coastal management plans of an imaginary nation.

In this example, only the national-level plan covers 100% of the coast, with progressively lower percentages covered by regional, local and site-level plans. This was judged by the designers of the coastal planning

continued . . .

programme to be the most efficient mix of the various geographic coverages of plans within budgetary and human resource constraints.

It is interesting to speculate how this imaginary coastal management planning programme would evolve after its first five years. Assuming that an evaluation has taken place, it may have been found, for example, that a particular level of plan was particularly effective, or that cascading plans had been found to be too complex in practice. This may have been especially so in relation to which levels of plan should be completed first. Other issues are likely to be whether the whole coast is covered with regional, local or site plans or whether existing areas are re-planned if the first round of plans did not meet their objectives.

coast. In this example, whole-of-jurisdiction, regional, local and site-level plans are developed for priority areas under a five-year planning programme.

Subject plans, as well as integrated plans, may seek to address a particular issue at a range of spatial scales. For example, a nation's fisheries management planning system may contain national, regional and local-level plans. The recommended management actions of such plans may be included in integrated coastal plans at the equivalent spatial scale (assuming that the integrated plans cover both coastal land and water). Of course, the opposite situation may occur with the outcomes of integrated plans being included in subject plans.

Which of these cases occurs is simply down to timing: the plan produced first will influence the second plan, the third plan will be influenced by the first and second plan, and so on for subsequent plans. This simple sequence assumes that coastal management issues have not changed over time – usually the exception, and hence the sequence of plans is likely to be affected by evolving circumstances, including the incorporation of previously unforeseen issues. Also, previous planning exercises may have uncovered new issues or problems which may have been considered unimportant, or were not considered at all. The result could be that plans are seen to exacerbate or even create coastal management problems, although in reality the plan merely brought the problems to the attention of planners and managers.

Of course, this sequencing effect will depend on the time elapsed between plans. If this time is long (say, over five years), then previous plans may be out of date and of little relevance to subsequent planning initiatives. Plan sequencing is also determined by the statutory basis of any coastal planning framework. There may be statutory requirements to formally adopt the outcomes of previous plans.

In countries where some form of coastal planning has already taken place, the issue of which type of plans to produce first, in which areas, and to address which problems, may have already been addressed. In this case, it may be assumed that plan sequencing reflects coastal management priorities. Of course this may not be the case, with the order in which plans were produced reflecting other priorities, such as political imperatives or the need to satisfy funding requirements.

In contrast, in coastal nations where little or no coastal planning has previously existed, it is worth considering what may be the optimum sequence for plan production. A rather generalized answer to this question is that the most effective sequence will depend on the opportunities and constraints inherent in the governance of a coastal nation. The result would be that a suitable sequence falls out from an analysis of governance issues, which in turn reflect the social, economic and cultural circumstances.

These sequencing issues also determine to a large extent the overall design of a coastal management programme, of which coastal planning initiatives may be a part (see Chapter 3). The nation-by-nation approach to the sequencing issue is supported to some degree by analyses of the various national approaches to coastal management and planning listed in the bibliography. This literature is supported by the various international guidelines for coastal management programmes, including those of the World Bank, IUCN and OECD, which stress a case-by-case approach to the design of coastal management and planning programmes. Though this conclusion is the best available at present, it remains rather unsatisfying in that there is little general guidance through comparative analyses of coastal planning programmes. Consequently, there are no definitive answers to the most effective overall design of coastal planning programmes in general, and to the plan sequencing issue in particular. Clearly, this is an area worthy of future study.

5.3 Integrated coastal management plans described by geographic coverage

As the name suggests, integrated coastal management plans aim to bring together environmental, social and economic considerations which influence the use of coastal resources into a plan or plans which provide a coordinated direction for coastal managers. When integrated plans are formulated using these three considerations the framework is often set for effective decision making in the coast. Historically, decision making has been made independent of these considerations, contributing to inappropriate or conflicting decisions about how the coast and its resources are managed.

Integrated coastal plans are now widely used as a mechanism to draw together disparate and uncoordinated decision-making processes of coastal resource management (Chapter 3). They can be developed in response to a number of coastal management issues, but the most common is simply conflict between different uses which cannot be adequately addressed by a number of single subject plans. These conflicts are often due to differing social, economic and environmental values held by coastal resource users. They can be managed in a number of ways, such as using legislation, policies, zoning provisions, and the many other techniques described in Chapter 4. However, in many cases the most effective and efficient management option is the formulation of an integrated coastal plan. Integrated plans are currently the most widely used approach to addressing multiple and/or conflicting issues by providing a framework for focusing the efforts of those charged with managing the coast. This focuses managers towards a common goal, and in doing so assists in coordinating and integrating their actions.

As described in Chapter 3, integration is not a tangible management outcome, but rather a way of thinking about the designing of planning processes which use communication, negotiation and coordination skills to help stakeholders reach informed decisions about how the coast and its resources will be used. These methods are used to bring stakeholders together to open up and maintain dialogue, and to develop mutual understanding and commitment. Once established such an integrated planning framework can then focus stakeholders on discussing, analysing and prioritizing coastal issues. Management prescriptions can then be agreed to, and a commitment made by the plan's authors to its implementation, ideally through a coordinated implementation system.

The different levels of understanding and awareness of often disparate coastal issues can be addressed through integrated coastal planning designed to accommodate differing needs. Training, capacity building and information exchange (section 4.2.3) can strengthen integration mechanisms such as collaborative and community-based management, cooperation and coordination, as discussed in Chapter 4.

Integration can occur vertically (between administrative levels), horizontally (between sectors) or combinations of both. Whatever level of integration is used, integrated plans are rarely developed without some form of overall national direction. This direction can be supplied through a range of measures as described in Chapter 3, but most commonly through either legislation or the development of national coastal management guiding statements. Guiding statements are usually part of a national coastal zone management strategy, which itself can be considered as a form of an integrated coastal plan – a view we support, with national coastal management strategies becoming the 'whole-of-jurisdiction' scale of integrated coastal management planning (Table 5.3).

Integration in coastal management planning can be between levels of government, coastal users and the community, or between different sectors of one level of government. It can therefore provide an important

Table 5.3 Scales of coastal management plans

Level of plan (scale)	Key role
International	• Transboundary issues • Creating a common purpose
Whole-of-Jurisdiction[1]	• Administrative arrangements • Setting national objectives and principles • Focus on priorities
Regional	• Translating international and national goals and objectives to local outcomes • Aggregate local needs and issues to formulate national and international priorities and programs
Local	• Community involvement in setting management options
Site	• Managing well defined problems • Tangible results of all planning levels can be seen

[1] National jurisdiction used here as an example

mechanism for coordination between one or more sectors and levels of government.

Planning scale refers to the geographic coverage of plans; or, more literally, the scale of any maps produced as part of the plan. For example, a coastal zone management plan covering 1000 km of coast would include small-scale maps depicting the study area, whereas a much more localized plan covering 1 km of coast would have much larger-scale maps.

It is important to note at the outset of this section that coastal management plans which operate at various scales (Table 5.3) are very different from each other. As will be shown, such plans can range from broad statements of intention by international organizations, to detailed site design plans developed by a community group. Nevertheless, all these plans, at whatever scale they operate, share the fundamental elements of planning: they define a future direction, and describe steps in order to achieve that direction. At each scale of planning, the purpose and scope of planning differs. Which level of planning to undertake is determined by the issues and level of future planning and management of the study area; it is also strongly influenced by its location within a planning hierarchy (Table 5.3).

Coastal nations often choose whether to develop their coastal management planning approach with the geographic hierarchy shown in Table 5.3. Planning at the international, whole-of-jurisdiction and regional levels is generally strategic in keeping with broad guidelines and policies.

Table 5.4 The hierarchy and characteristics of Western Australian coastal management plans (adapted from Alder et al., in preparation)

	Regional plans and strategies	Local coastal plans	Foreshore management plans	Site design plans
Scale	1:100,000 to 1:25,000	1:25,000 to 1:5,000	1:5,000 to 1:1,000	1:1,000 to 1:200
Coverage	100–1000 km of coast	10–100 km of coast	1–10 km of coast	100 m–1 km of coast
Scope of plan	Includes large areas of the coast incorporating several local governments	Usually covers a single local government and designates development setbacks, coastal reserves and other areas of coastal utilisation	Encompasses a coastal area within a local government and designates rehabilitation areas, amenity sites and access ways	Detailed planning of infrastructure, landscaping and rehabilitation works within a foreshore reserve
Responsibilities for plan production	Usually State government in association with local government	Local governments in consultation with State government	Local governments or developer acting under the approval of State and local governments	Local government, developer or community groups
Common issues addressed in the plan	• transport networks • urban development • tourism development • port and industrial development • conservation estate planning	• development setbacks • assessing land capability • protecting structures from erosion	• reserve frontage adjacent to private land • determining foreshore widths • access to environmentally sensitive areas	• resolving conflicting uses • vehicle–pedestrian access • appropriate species for revegetation

continued . . . .

Table 5.4 continued

	Regional plans and strategies	Local coastal plans	Foreshore management plans	Site design plans
Nature and contents	• broad scale patterns of land use • designation of development zones • transport networks • coastal reserves • recreation nodes and major conservation (park) areas	• detailed designation of development zones • detailed transport networks • provision of local amenities • designation of areas for particular activities – recreation, conservation, wildlife habitats, and scenic, historical or scientific value	• detailed information on coastal reserves • detailed location of development and conservation and recreation areas, amenities and access • detailed allocation of areas for recreational amenities and facilities, conservation purposes, and scenic, historical or scientific value • designation of areas for access ways	• location and design of infrastructure and recreation facilities • detailed provision of access • detailed designs of amenities and facilities, including landscaping • detailed provisions for the protection of significant vegetation communities, wildlife or wildlife habitats
Implementation responsibilities	State agencies and local governments	Local governments	Local governments, developer or community groups	Local governments, developer or community groups
Review period	Ten years	Five years	As determined by local government	Upon completion of works, and as per local government requirements

Local and site-specific plans provide more prescriptive guidelines and management actions for specific activities, development, infrastructure and use of marine and coastal resources. Whether to use such a 'cascade' of integrated coastal plans (Environment Committee, 1992) will depend on the particular coastal issues being addressed, and the legislative basis of coastal management.

Integrated plans at different scales can be linked in a number of ways: through the deliberate flow of recommendations from higher level plans, or through the linking of common guiding statements. Another method is to encourage linkage through grant-in-aid schemes. The most widely cited of these is the United States, where state coastal management plans are encouraged by Federal Government grants. There are also common linkages between integrated coastal planning at the international and national level through the encouragement by donor agencies of national coastal strategies which form part of international initiatives, such as the United Nations' Regional Seas programme.

Higher levels of integrated plans can also actively encourage the production of similar styles of localized plans. This can be through recommended actions of the higher order plans, as a means to address the localized problems that higher levels of plans cannot address specifically, because of their more strategic nature. High-level plans are often called special area management plans and are in common use throughout the world, Sri Lanka and Western Australia being examples.

A critical issue when considering scales of integrated coastal planning is how plans are to fit together. Ten local area coastal management plans each covering 20 km of coast do not achieve the same things as one higher order plan which covers the entire 200 km. There is a danger in thinking that a plethora of local area coastal plans will achieve the same objectives as higher-order plans; they usually cannot. A related issue is that because local area integrated plans are often focused on areas experiencing major localized problems, the 'local fix' approach can become endemic in a nation's approach to coastal management planning. The result can be that higher levels of coastal planning become neglected as coastal planning moves from crisis area to crisis area with the symptoms of coastal problems being addressed, but not their cure (Donaldson et al., 1995).

An example of a hierarchy of coastal plans used in Western Australia and the nature, contents and common issues addressed by each are shown in Table 5.4.

A hierarchy of coastal plans exists in Australia, involving three spheres of government. Various mechanisms are used to link these plans which vary not only in scale but also jurisdiction. National–state government initiatives are often linked by memoranda of understanding (MOUs) and agreements, whereas links between state and local government are forged through regional planning.

There is no doubt that integrated planning is one option for managing the coast at any planning scale. Integration is not an easy concept to describe, let alone translate into planning actions. There is no single recipe for effecting integration since, as described below, it varies with planning scales, and social and economic conditions.

5.3.1 International integrated plans

Coastal management planning at international scales is highly strategic and focuses on developing broad strategies and actions plans to ensure common efforts between coastal nations. International-scale initiatives include global programmes and those developed between groups of countries. National groupings can be dictated according to the regional boundaries drawn by international organizations, such as the United Nations, or by economic or political groupings, such as the Association of South East Asian Nations (ASEAN).

At a global scale, international organizations can play an active role in the development of international initiatives focused on particular issues, or promote and coordinate the development of a particular coastal management tool or approach. Such international organizations include governmental institutions like the International Maritime Organization (IMO), the United Nations Environment Programmeme (UNEP) and the Organization for Economic Cooperation and Development (OECD); and non-governmental organizations such as the International Union for the Conservation of Nature (IUCN) and World Wide Fund for Nature (WWF). An example is the global action by a number of organizations which have banded together to promote an international initiative for the study and management of coral reefs (Box 5.4).

Global-scale initiatives can be effected through voluntary programmes, such as the Coral Reef Initiative (Box 5.4), or through formal mechanisms such as memoranda of understanding, agreements or action plans. Formal mechanisms can have various levels of statutory force: some are legally binding conventions, such as the Convention for the Protection of the World Cultural and Natural Heritage (1972), whereas others are non-statutory agreements. International environmental law is a relatively new activity, and the mechanisms by which international agreements are implemented within individual nations is a poorly understood area of law. However, in some cases the situation is clear when governments implement national laws which clearly define the requirements for meeting international agreements, for example with MARPOL requirements (Box 5.5).

The single most important global-scale plan of action which influences coastal planning and management is Agenda 21 – the outcomes of the United Nations Conference on Environment and Development (UNCED,

Box 5.4

International Coral Reef Initiative

The International Coral Reef Initiative (ICRI) (Drake, 1996) was launched in 1995 in response to serious concern about the increasing and widespread degradation of coral reefs and their related ecosystems, including mangrove forests, seagrass beds and beaches. ICRI is a global partnership of governments, international and regional organizations, non-governmental organizations (NGOs), multilateral development banks and private sector groups. This partnership aims to increase the capacity of countries and local groups to effectively conserve and sustainably use coral reefs and related ecosystems. The key to ICRI's success will be global cooperation, effective use of existing resources and identifying effective mechanisms for implementation.

This global cooperation commenced in the early 1990s with subsequent agreement to develop an agreed approach and framework for action. The goal of the International Coral Reef Initiative is to:

- raise global commitments to conserve, restore and sustainably use coral reefs and associated environments; and
- use and better coordinate the efforts of governments and regional organizations as well as catalyse and facilitate the development of new activities to ensure the conservation, sustainable use and management of coral reefs.

The Framework for Action is based on the following principles.

- The full participation and commitment of governments, local communities, donors, NGOs, the private sector, resource users and scientists is required to achieve ICRI's purpose.
- The overriding priority is to support actions that will have tangible, positive and measurable effects on coral reefs and related ecosystems and on the well being of the communities which depend on these ecosystems.
- Human activities are the major cause of coral reef degradation; therefore managing coral reefs means managing those human activities.
- The diversity of cultures, traditions and governance within nations and regions should be recognized and built upon in all the ICRI activities.
- Integrated Coastal Management, with special emphasis on community participation, provides a framework and process for the conservation and sustainable use of coral reefs and related ecosystems.
- A long-term commitment is required to develop national capacity to conserve and sustainably use coral reefs and related ecosystems, and the continued improvement of coral reef management requires a permanent commitment to an adaptive approach.

continued . . .

- Strategic research and monitoring programmes must be an integral part of the ICRI because management of coral reefs and related ecosystems should be based on adequate scientific information.
- Actions promoted under the Framework for Action should take account of, and fully use, the international agreements and organizations that address issues related to coral reefs and related ecosystems.

Because ICRI is a non-statutory 'partnership' approach, the implementation of ICRI initiatives is at the discretion of participating countries. For example, Indonesia has chosen to incorporate ICRI into its COREMAP programme (Box 5.32), while Australia is currently incorporating various initiatives into the management of the Great Barrier and other reef systems.

Box 5.5

MARPOL requirements for the management of garbage generated at sea

Box 2.8 highlighted the issue of port development and operations. In the Port of Victoria in the Seychelles, land reclamation and waste disposal are the major issues. The approach suggested to address the issue of waste management has been to apply the MARPOL 73/78 Convention. The Seychelles acceded in 1990 to the International Convention for the Prevention of Pollution from ships. Changes to the convention in 1973 and 1978 (generally referred to as MARPOL 73/78) also apply to the Seychelles.

The convention deals with pollution through five annexes, with Annexes I, IV and V having direct application to the port. Annex I deals with oily wastes from ships with wastes generated by spent lubricants, bilges, ballast, and fuel processing. In addition, there are sources of land-generated oily wastes such as oils from transport systems, industry and solvents. Under MARPOL, oily waste treatment facilities must be provided at the port before vessels are required to dispose of their wastes in an environmentally acceptable manner. To facilitate disposal, an oily water treatment unit and incinerator must be constructed.

Annex IV deals with sewage. Sewage generated by ships in the port is not an issue at the moment, but it should not be ignored since it has the potential to be a significant issue if the cruise industry increases. It is recommended that vessel owners, especially fishing-boat owners, be encouraged to install holding tanks and that where feasible sewage should be collected by tanker trucks and discharged into Victoria's sewage system or sewage treatment plant. However, financial constraints are slowing progress on this issue.

continued . . .

Annex V is focused on garbage. It includes a policy of discharging brine solutions from seiners during passage in the open sea so that the ship's propellor will assist in diluting the concentrate. By-products of the tuna cannery and other fish processing works should be used to produce fishmeal. It is recommended that other garbage be disinfected and fumigated before transport to landfill sites.

1992). Agenda 21 is essentially a global plan of action for sustainable environmental management and economic development. Of the 40 chapters of Agenda 21, the chapter addressing coastal and ocean management issues (Chapter 17 – Protection of Oceans and Seas) is the longest and most detailed. As described in Chapter 3, Agenda 21 effectively laid out a new paradigm for the planning and management of coastal areas, based on the principles of sustainable development. Although a non-legally binding document the global consensus reached in its adoption has pervaded the coastal programmes of many nations through the adoption of sustainable development principles into statements of coastal programme goals and objectives (Chapter 3).

Global-scale initiatives are complemented by international efforts between groups of countries, or bilateral agreements between two countries. An example of the latter is agreements regarding the conservation of migratory species, such as the Agreement for the Protection of Migratory Birds in Danger of Extinction and their Environment, between the Governments of Australia and Japan (1974). The combined effect of global-scale, bilateral and regional-scale international initiatives on national and more detailed coastal management planning can be significant. For example, in 1995 the Australian Federal Government listed 28 international treaties with significance for Australian coastal management (Commonwealth of Australia, 1995).

The Regional Seas Programme is a good example of a global initiative of the United Nations which is implemented by groups of nations (Box 5.6). The application of this regional approach to coastal planning in the East African region is shown in Box 5.7.

International initiatives are also important for building the capacity of coastal nations to implement coastal planning and management programmes. Training, professional development, scientific research and data management (Chapter 4) are undertaken by a host of international organizations to assist coastal nations which may lack such facilities. An example of this is the United Nations Environment Programme's Network for Tertiary Training in the Asia Pacific, which contains a Coastal Zone module (see Box 4.11).

Box 5.6

The Regional Seas Programme

The United Nations Environment Programme (UNEP) Regional Seas Programme is an international initiative to control marine pollution, and manage marine and coastal resources between neighbouring nations. The programme was initiated in 1974, and by 1993 a total of 140 coastal States and Territories were participating (Schröder, 1993).

The Regional Seas Programme concentrates on the development of broad-scale international coastal management plans, called Action Plans, which cover groups of countries. Currently there are 12 programmes and one under development (UNEP OCA/PAC, 1996). (Each programme is detailed in a publication produced by the United Nations. In addition, there are numerous background and technical reports on Regional Seas and these can be obtained from the Oceans and Coastal Areas Programme Activity Centre, UNEP, PO Box 30552, Nairobi, Kenya.)

- Existing:

  - Kuwait (1979)
  - Mediterranean (1980)
  - Caribbean (1981)
  - West and Central Africa (1981)
  - South-East Pacific (1981)
  - Gulf of Aden and Red Sea (1981)
  - East Asian Seas (1981)
  - South Pacific (1982)
  - Eastern African (1985) (see Box 5.7)
  - The Framework Action Plan for the Black Sea (1993)
  - Northwest Pacific (1994)
  - South Asian Seas (1994)

- Under development:

  - Southwest Atlantic

Each Action Plan is written according to the perceived needs of the governments concerned (Schröder, 1993) to:

- link assessment of the quality of the marine and coastal environment and the causes of its deterioration with activities for its management and development and the rational use of its resources; and
- promote the parallel development of regional legal agreements and of action-oriented programme activities.

continued . . .

A review of action plans highlights the common concerns managers have in managing marine and coastal areas – coastal developments, habitat loss, eutrophication, and increased health hazards associated with seafood, and fouling of beaches by tar and litter. The Regional Seas Programme recognizes that each region is unique and that there is no one model which can apply to every region. Nevertheless a common suite of management prescriptions is used to recommend a range of initiatives to address these issues. Where possible these initiatives aim to be at the regional level, and each government should coordinate its actions. Additional programmes specific to the region may be recommended. For example, the Gulf of Aden and Red Sea Action Plan recommends the formulation of national contingency plans for combating oil pollution (UNEP OCA/PAC, 1986) while the East African Region Action Plan recommends regional co-operation in tourism (UNEP OCA/PAC, 1982)

All Regional Seas Programmes have their own actions plans and financial mechanisms (trust funds), but only nine have associated Conventions; South Asia, East Asia and the Northwest Pacific have no legal instrument.

All Regional Seas Programmes were initiated with the support and guidance of UNEP. UNEP serves as the Secretariat to the Mediterranean, Caribbean, East Africa, West and Central Africa, Black Sea, Northwest Pacific and East Asia programmes. In all the other cases, autonomous inter-governmental bodies provide this function.

In addition, UNEP collaborates with a wide range of international organizations. UNEP's lack of implementation ability has been criticized as one of the problems with the Regional Seas Programme (Hinrichsen, 1994). However, in programme areas where countries have worked under the Action Plan framework and committed major funds, such as in the Mediterranean Action Plan, there has been more success (Hinrichsen, 1996).

Box 5.7

The East Africa Regional Seas Programme

> The coastline of the Eastern African region is an area rich in natural marine resources and breathtaking scenic beauty ... yet this is being seriously threatened by marine pollution, habitat destruction and the pressure of growing populations, urbanisation and industry.
>
> (Iqbal, 1992, p. 1)

East Africa is one of the 13 Regional Seas Programmes, which aims to provide a framework (described in Box 5.6) for regional cooperation, to conserve and develop the natural marine resources, and to combat coastal and marine pollution problems in the region. The East African Regional

Seas Programme covers the countries of Comoros, France (Réunion), Kenya, Madagascar, Mauritius, Mozambique, Seychelles, Somalia and Tanzania.

The Action Plan was initiated in 1981–1982 with the production of various baseline reports on the status of the region's coastal zones, summarized in UNEP (1982). The results of this work were to develop a draft regional action plan, recommend a number of priority actions in the region (basic and baseline study and environmental monitoring capability; environmental assessment programmes; training and assistance; institutional changes; and specific programmes such as improved use of fuelwood to reduce deforestation, and regional cooperation on tourism) and recommend that the draft action plan and two regional protocols (for cooperation in combating pollution in case of emergency; and for specially protected areas and endangered species) be endorsed by the member governments. These entered into force on 30 May 1996.

The first meeting of the contracting parties occurred in March 1997 and an outline set of operating procedures, including financial arrangements, was agreed to. The programme is currently focused on capacity building and public awareness raising on the integrated management of marine and coastal areas (R. Congar, personal communication, 1997).

A central component of any Regional Seas Programme is the contribution of governments into a Trust Fund to implement their own decisions. In the case of East Africa, the trust fund is also being contributed to by donor agencies.

The recent ratification of the legal instruments underlying the programme has helped to revitalize coastal management in the region. The Government of the Seychelles and UNEP are establishing the Regional Coordinating Unit of the East African Region on Sainte-Anne Island. In addition, a Regional Centre for Coastal Areas Management is planned to be established with donor and government support in Mozambique. In addition, there are a number of ad hoc expert groups established under the Regional Seas umbrella. Also there are a number of coastal management initiatives being undertaken by individual nations in the region (e.g. Russ and Alcala, 1994; Intercoast, 1995) which are aimed to be supported by the Regional Seas Programme.

Despite the long lead time required to establish the programme, the East African Regional Seas Programme is now beginning to gain significant momentum to focus and prioritize regional coastal management funding and action. It is becoming clear that this is particularly important for directing the attention of donor agencies and international development assistance towards management problems which are seen as priorities by governments in the region. Also, the programme is the regional focus for the implementation of global initiatives, such as the International Coral Reef Initiative (Box 5.4) and the Global Programme of Action for the Protection of the Marine Environment from Land Based Activities.

In summary, international initiatives, be they global, regional or bilateral, represent the broadest scale coastal planning initiatives. Therefore, they represent the top of the top-down view of coastal plan development.

5.3.2 Whole-of-jurisdiction integrated plans

The term 'whole jurisdictions' is used here to describe entire sovereign nations and those sub-national governments with significant legislative and/or budgetary powers. The most common type of such sub-national governments are State and Provincial governments within federal systems. The defining issue is the ability of governments to choose between legislating to develop a whole-of-jurisdiction coastal management approach or using an approach without the enactment of specific new laws.

Primary coastal planning foci at the whole-of-jurisdiction scale are on administrative arrangements for developing coastal planning frameworks, and articulating statements of goals, principles and objectives. Through the joint development of effective coastal planning frameworks and clear statements of what plans are attempting to achieve, more detailed coastal plans at regional, local and site levels are provided with an unambiguous 'space' in which to develop.

The combined effect of developing administrative arrangements and guiding statements of direction for coastal planning commonly results in broad strategic whole-of-jurisdiction coastal plans and policies. Depending on administrative, political, economic and cultural circumstances such plans can establish requirements for the development of coastal plans in subsidiary jurisdictions, such as local or state governments. In some cases, these requirements may be prescribed within national legislation or policy – for example, in the United States (Box 3.8) and New Zealand (Box 5.8). In other jurisdictions, the sharing of role and responsibilities between levels of government may mean that national-scale coastal plans provide a framework to encourage, through non-statutory means, other levels of government to adopt national approaches (Box 5.8).

The general approach of combining administrative arrangements with the formalization of guiding statements in the development of whole-of-jurisdiction coastal plans has been undertaken in numerous coastal nations. Here we focus on the nations used as case studies throughout the book in order to illustrate variations in this approach. The New Zealand example is used to show a legislative-based approach which specifies national requirements through a national statement of policy, backed by the national Resource Management Act (Box 5.8). This approach is contrasted to Australia, whose federal system of government dictates a different approach through the definition of national principles and

Box 5.8

National coastal planning in New Zealand

In the 1980s New Zealand embarked on a major process of legislative reform of its resource management legislation. This culminated in the passing of the Resource Management Act (1991) which is now the governing legislation for the management of New Zealand's land, air and water. The Resource Management Act rationalized more than 50 Acts governing the coastal environment (Rennie, 1993). The purpose of the RMA is the 'promotion of sustainable management of natural and physical resources' (RMA, section 5).

The Resource Management Act established a national framework for coastal planning. The Act authorized National Policy Statements which can address any issue covered by the Act. Importantly, all subsequent planning instruments cannot be inconsistent with them. The New Zealand Coastal Policy Statement (1994), which was prepared by the Minister of Conservation, is the only mandatory national policy statement required by the Resource Management Act. Therefore, the management of the coastal environment received special attention in the Act.

Draft New Zealand Coastal Policy Statements were released in 1990 for public comments. These were analysed, and it was decided that the level of comment required the release of a second draft in 1992 for additional public comment. These comments were formally reviewed by a Board of Enquiry, which published its findings and recommended changes to the Draft in 1994. Subsequently, a final Coastal Policy Statement was released in May 1994 by the Minister for Conservation which very closely resembled that recommended to him by the Board of Enquiry. The Coastal Policy Statement has a series of specific coastal policies, examples of which are shown in Box 4.2.

Importantly, the Resource Management Act also contains provisions which allow both the Minister of Conservation and Minister for the Environment to intervene in decisions when issues of national interest arise.

The Resource Management Act also required that each of the 16 regions in New Zealand must develop a Regional Policy Statement including a Regional Coastal Plan. Thus a formal hierarchy of coastal management planning in New Zealand was established, and is described further in Box 5.14. There are no specific guidelines for the production of such plans, but the strict requirements of the Resource Management Act has ensured that the plans produced are relatively similar in both their content and the way in which they were produced, by using draft plans and extensive consultation and public hearings.

Box 5.9

Indonesia National MREP

As part of Indonesia's commitment to manage its marine and coastal areas it has initiated a national programme – the Marine Resources Evaluation and Planning Project (MREP) (Ministry of Home Affairs, 1996). The project objectives are to improve marine and coastal management capabilities in 10 provinces, and to develop and strengthen the existing marine and coastal data information systems.

The major issues that the MREP programme addresses are the conflicts in planning and managing between national, provincial and local governments as well as the private sector and local communities. These conflicts have led to resource degradation which threatens the sustainability of Indonesia's marine and coastal areas. The situation is exacerbated by local governments' lack of jurisdiction in the coast, limited law enforcement and minimal human resource development in coast management.

The MREP project commenced in 1994 with funding support from the Asian Development Bank and has two major components:

- strengthening marine and coastal planning and management; and
- strengthening marine coastal information systems.

The two project components reflect the agreed national priorities for effective management of the nation's coastal waters. Although management of the marine resources below high water mark is under national jurisdiction, management of marine and coastal resources is undertaken at the provincial level. Therefore provincial agency participation is a fundamental part of the project.

Much of the mapping and Geographic Information System (GIS) studies has been done nationally as part of the development of marine coastal information systems. At the provincial level, however, training and capacity building has been conducted so that Marine Data Centres at the provincial level can update and maintain the relevant information. Overall GIS development and management is being coordinated nationally. Resource information gathering for ecological system processes and offshore mining has also been undertaken in the provinces and is used to support planning initiatives and update GIS databases. Once the Marine Data Centres are operational they will coordinate further resource assessment in the provinces based on national guidelines.

Within MREP there are 10 Marine and Coastal Management Areas (MCMA) which correspond to 10 provinces and three Special Management Areas (SMA) which can span more than one province. These areas will be selected to demonstrate the processes used to formulate and implement the coastal zone planning programme as part of the Strengthening of Planning

continued . . .

and Management component. The ultimate aim is for all 27 Indonesian Provinces with a coastline to have a coastal area plan implemented. The overall hierarchical framework for these case studies is discussed in Chapter 3 (Figure 3.7). The range of issues to be considered by these plans is shown by those being faced in the Province of Sulawesi Selatan (Box 5.13).

In each MCMA the tiered approach to developing a coastal management programme will be used as described in Chapter 3. Each province will develop its own strategic planning based on a vision and goals to reflect provincial priorities. The goals and objectives will be translated into policy, and policies will be implemented in the MCMAs using a number of tools, including zoning plans as described in Chapter 4. Policy will also guide the formulation of zoning plans for large areas within the province. In addition, where necessary, site or subject plans can be used to address issues or problems outside of the zoning plans and provincial policies. Implementation of zoning plans and other specific planning will be achieved using many of the tools described in Chapter 4.

Local management plans are also being developed in key areas and for critical issues (Box 5.18).

Box 5.10

National coastal planning in Sri Lanka

The Sri Lankan Revised Coastal Management Plan (1996–2000) is founded on six national strategies (Olsen *et al.*, 1992; Coast Conservation Department, 1996).

1. The coastal management programme will proceed simultaneously at the national, provincial and local levels with the collaboration required to achieve effective and participatory resource management by governmental and non-governmental agencies.
2. Implement a programme to monitor the condition and use of coastal environmental systems and the outcomes of selected development and resource management projects through the collaboration of national agencies.
3. Implement a research programme to provide a better understanding of ecological processes and social and cultural information.
4. Implement a programme to strengthen institutional and human capacity to manage coastal ecosystems.
5. Update and extend the scope of the master plan for coastal erosion management.

continued . . .

6. Implement a programme to create awareness, both by national and provincial government personnel and NGOs, of the strategies for coastal resource management and the issues they address.

Each national strategy (above) is accompanied by an explanation of why the strategy is important and a list of implementation actions.

An important addition to earlier coastal planning initiatives in Sri Lanka is an emphasis on coastal planning at regional and local scales in addition to the national level. The different topics and activities to be covered by each level of coastal plan are listed in the table.

Topics and activities addressed by Sri Lankan national, provincial and local coastal plans (Olsen et al., 1992).

National	Provincial	Local (Special Area Management)
• habitat monitoring and management • implementation of guidelines for resource use • access and resource use conflicts • a change in the practical and legal definition of the 'coastal zone' • guidance, incentives, regulations and procedures for provincial coastal plans • decentralisation of permit procedures • formulation of procedures for local-level plans and their implementation	• assessment of trends, condition and use of coastal resources and land use • identification of major regional coastal resource management issues, opportunities and constraints • mapping of areas of concern which may require local-level coastal planning • identification of areas suitable for conservation for development (e.g. roads, harbours, tourist resorts) • designation of green belts along the coast within which construction is prohibited or restricted	• range of local management problems, issues and concerns addressed through participatory plan production processes as described in Box 5.16

Implementation of the revised national Coastal Management Plan is being staged across national and local levels. Although the regional level (provincial) coastal planning is suggested in the revised plan, this mid-level planning is still in its infancy and will not be a focus for the next 5–10 years. Rather, the plan will focus on local-level plans (Special Area Management Plans) which are considered at present to be flexible enough to accommodate the major local coastal management issues in Sri Lanka. One of the two Special Area Management Plans produced so far in Sri Lanka, at Hikkaduwa, is described in Boxes 5.16 and 5.33.

objectives and legally non-binding memoranda of understanding with state governments. The Australian national coastal plan provides a contrast to the description and analysis of Indonesia's national marine resource planning initiative (Box 5.9). Finally, the Sri Lankan national approach to coastal management planning, which is closely tied to the overall structure of its coastal management programme, is described (Box 5.10). These two examples also demonstrate how whole-of-jurisdiction planning can set the framework and guide lower order planning, as described in the next three sections.

5.3.3 Regional-scale integrated plans

> Regional-level planning and analysis confers a number of advantages that are absent from local- and national-level planning. At the regional level, it is possible to address and resolve problems faced by entire ecosystems. Very often these issues cross a number of jurisdictions and can only be effectively addressed with a regional geographic focus.
>
> (Jones and Westmacott, 1993, p. 127)

Regional plans and strategies are used to address issues and problems which span a wide geographic range, generally covering more than a single local government authority. Typical lengths of coast covered by such plans are between 100 and 1000 km. Some regions are defined in legislation, other regions are defined according to the issues being addressed. Integrated regional coastal plans establish a regional framework for on-the-ground or on-the-water coastal management, implement policy developed at the whole-of-jurisdiction level, and can provide the stimulus for the formulation of local- and site-level coastal plans.

The key focus of regional-scale coastal plans is to provide a bridge between whole-of-jurisdiction plans and policies, and local- and site-level initiatives. The regional level of coastal plans is often the first planning level which is sufficiently detailed to become spatially oriented. International or whole-of-jurisdiction plans generally cover too much coast to translate broad economic, social and ecological considerations into tangible management recommendations or provide practical guidance on matters such as locations and/or mechanisms to spatially separate conflicting uses of the coast. Regional coastal plans can address issues of urban and infrastructure development, resource allocation, transport, tourism, access and conservation.

The form of regional integrated coastal plans can closely reflect whole-of-jurisdiction plans in that regional goals and objectives, and in some cases regional planning principles, can be developed. Depending on the

Box 5.11

Sample contents of regional coastal plans (adapted from Alder et *al.*, in preparation)

Section 1: Introduction

Describes or outlines:

- the history behind the preparation of the plan;
- justification for preparing the plan;
- the purpose of the plan;
- the scope and nature; and
- the planning process used including details on public consultation.

Section 2: Regional Characteristics

Describes the natural environment, and the social and economic profiles of the region. Depending on the major issues and purpose of the plan, existing land or resource use, as well as infrastructure or structure profiles, may also be introduced.

Section 3: Policy Framework

Outlines the legal basis of the plan, and how it relates to statutory and/or non-statutory policies and procedures. May also provide links to plans and/or policies at other planning scales.

Section 4: Guiding Principles

Outlines the key guiding principles which form the basis for developing objectives, actions and proposals. May be included in the policy framework section, depending on the statutory basis of the plan.

Section 5: Objectives

Provides a description of the objectives for responding to a particular activity or issue which may be supported by proposed actions to meet these objectives (these actions may be addressed in Section 6). The objectives generally relate to the values and issues of that particular area or activity and are guided by the plan's principles. Objectives can generally be grouped into the following categories:

- conservation and recreation/natural environment;
- settlement/urban and infrastructure development;
- agriculture/rural development;

continued . . .

- resource development (water, mining, agriculture, fisheries and forestry);
- tourism; and
- social development.

Section 6: Land and Water Use Plan and/or Management Actions

The land and water use plan allocates broad use categories to specific locations. The plan could consist of a map and a description of the preferred uses within a land or water use classification. These could be specific zoning provisions. Specific management actions can be recommended, or described on a sector-by-sector basis in Section 7.

Section 7: Planning Units

Provides the detailed description of the Land and Water Use Plan by geographic precincts which can be defined according to a surface water catchment framework or by specific aspects of the plan. In either case the detailed description usually consists of:

- a definition and description of the area;
- an outline of the issues, opportunities and constraints;
- list of preferred land and water uses; and
- description of the planning and management guidelines.

Section 8: Implementation

Outlines the actions which need to be taken to implement the plan, depending on the Land and Water Use Plan and other strategies described in the previous chapters. The range of mechanisms available to the various organizations and agencies involved in implementing these plans depends on its statutory basis, reason for production and focus.

Section 9: Monitoring and Review

Provides some guidance on monitoring issues and problems, and ways to address ongoing problems. The section also recommends the time frame for reviewing the plan.

links between whole-of-jurisdiction and regional plans (determined by legislative, funding or other factors described above) there may be potential to develop a specific regional identity. Even if the constraints on the contents and form of regional coastal plans are restrictive, there is usually scope for the inclusion of regional planning and needs and issues.

The content of regional strategies and plans will vary according to the issues addressed, the needs of the region and the approach used to formulate the plan. Example contents of typical regional coastal strategies and plans are shown in Box 5.11.

Regional coastal plans can often be the most difficult scale of coastal plan to develop. The primary reason for this difficulty is the 'bridging' role of such plans, situated as they are between tangible local issues and more strategic initiatives at national or international levels. The challenge, then, is to develop regional coastal plans which are tangible enough to provide clear guidance to local and site planning, and at the same time sufficiently strategic to assist in the implementation of national and international objectives.

There are several ways in which regional plan implementation can be achieved, including changes to town planning schemes, formulation of policy, or the drafting of specific detailed or sectoral plans. Implementation is usually a staged process which is managed through a forum to ensure that the process is consistent and ongoing. Ideally members of the steering committee charged with formulating a plan also participate in its implementation.

Three examples of regional coastal planning are presented below. Each illustrates how regional planning focuses on broad issues while providing guidance for local planning. The Central Coast Regional Strategy (Box 5.12) is a good example of integrated planning; it also highlights how land and marine use planning can be integrated. The Sulawesi Seletan case study takes a much broader view but provides a well defined framework for subsequent planning (Box 5.13). The New Zealand example shows how a structured approach to a coastal management plan is constrained by national policies and legislation (Box 5.14).

5.3.4 Local area integrated plans

Local integrated coastal plans will vary according to the particular issues addressed as well as the level of sophistication of the approach. Local plans tend to cover lengths of coast in the order of 10 to 100 km and generally involve only one local government. Planning at the local level is often a response to a particular set of issues requiring immediate attention or to facilitate current and future use of particular areas. Typical issues are dune stabilisation, demands for recreational facilities and access to coastal areas for development. In dealing with issues at the local scale it is important to differentiate between issues which are best managed at the regional level and those which can be managed at the local level.

The contents of a plan are usually determined by the process of its production. If a public consultation process is used to identify the important issues, much of the plan's content is defined. On the other hand

Box 5.12

Central Coast Regional Strategy (Western Australia) – planning context

The strategy is an example of a regional-scale integrated coastal management plan covering the 250 km of coast immediately north of Perth, Western Australia (Western Australian Planning Commission, 1996a). The issues promoting the strategy are shown in Box 2.4 and the principles used in its development are shown in Box 3.13.

The strategy is a non-statutory 'bridge' between state-wide policies and local plans, both of which can be statutory and non-statutory. The document is strategic in nature. The strategy is focused primarily on coastal land use, although it does address marine planning issues through the development of non-statutory marine 'precincts'. The committee charged

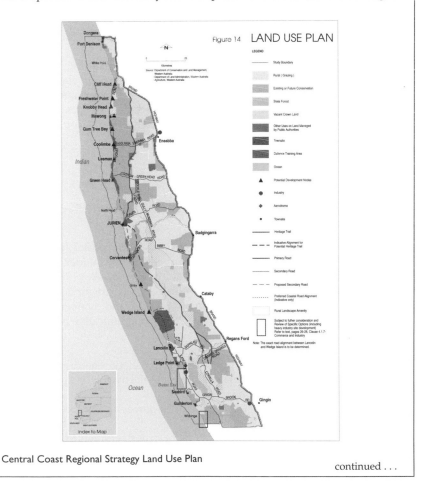

Central Coast Regional Strategy Land Use Plan

continued . . .

with implementing the plan is currently seeking funds under the Australian (Federal) National Coastal Action Plan to undertake a Central Coast Marine Planning Strategy to increase emphasis on marine planning and management.

The process for producing the Strategy was based on a steering committee made up of State and local government officials, local councillors and community representatives. Consultation was extensive, with community workshops, the production of a background information 'profile' report and a draft strategy for public comment.

The resulting coastal land use plan recommended a hierarchy of settlements along the coast, separated by existing or expanded conservation areas. Future tourist, industrial and residential development nodes were also identified (see figure). Thus, the regional strategy provided broad guidance as to which land and marine uses were appropriate in which locations. Local coastal planning could then concentrate on helping to manage well defined problems or use conflicts. For example, the rehabilitation of dune areas badly degraded by illegal 'shacks' can be transformed into tourist nodes.

In addition, the planning strategy has provided a mechanism to focus the activities of State government in the region. Both through the process of producing the plan, and the mechanism for its implementation, the emphasis has been on developing an integrated view of coastal land and marine use planning to ensure the region's sustainable development.

Box 5.13

Regional coastal planning – Sulawesi Selatan Province, Indonesia

The development of the Sulawesi Selatan provincial coastal management programme is following the tiered approach as proposed under the national Marine Resources Evaluation and Planning Project (Box 5.9). A draft provincial coastal strategy has been formulated which attempts to deal with a number of issues through the setting of a vision and management goals (see Box 3.12). To achieve these goals the following indicative policies have been recommended:

- All coastal planning and policy making will be coordinated at the provincial level and regional level subject to the oversight of a Provincial Coastal Steering Committee.
- Raise public awareness of the value of resources and processes so as to encourage responsible resource use.
- All planning efforts will contribute to the orderly implementation of the Sulawesi Selatan Coastal Planning System.

continued . . .

- Use the best available information when making decisions, and improve the information base for decision making in relation to coastal resource management whenever possible.
- Priority attention will be given in planning efforts to identify strategies for poverty alleviation in coastal villages.
- There shall be no further net loss of mangrove in Sulawesi Selatan and where possible efforts will be made to rehabilitate existing forests and replant forests in suitable areas.
- All coral reefs in Sulawesi Selatan waters will be protected from unsustainable exploitation and damage due to human activity.
- Marine and coastal tourism development will be actively encouraged and promoted, provided that such development is undertaken with due regard to the ecological and social carrying capacity of the development site.
- New coastal aquaculture operations shall only be permitted where they can be proven to have no adverse environmental impacts.

It is proposed that these policies be implemented using a range of tools – local integrated planning, site plans, community group consultation, formal and informal education within the Spermonde Marine Coastal Management Area. These approaches include areas designated for case study testing of local-area coastal management plans (Box 5.18). In addition, an oil spill contingency plan for the Makassar Strait and a mangrove rehabilitation plan are being developed.

Box 5.14

Regional coastal planning in New Zealand

New Zealand's Resource Management Act (1991) required each of New Zealand's 16 regions to develop a Regional Policy Statement including a Regional Coastal Plan. An interesting twist in the requirement of Regional Councils to develop Regional Coastal Plans is that they are are only compulsory under the RMA for the 'coastal marine area'. This is the area seaward of the mean high water mark at spring tide to the limit of territorial waters (see figure), or the 'wet' component of the coast (Rennie, 1993). However, where it is considered appropriate in order to promote the integrated management of a coastal marine area and any related part of the coastal environment (the 'dry' component of the coast) the RMA allows a regional coastal plan to form part of a broader regional plan. The result is that many regional councils chose to prepare a Regional Coastal Environment Plan in order to break down the artificial division between 'wet' and 'dry' components of the coastal environment.

continued . . .

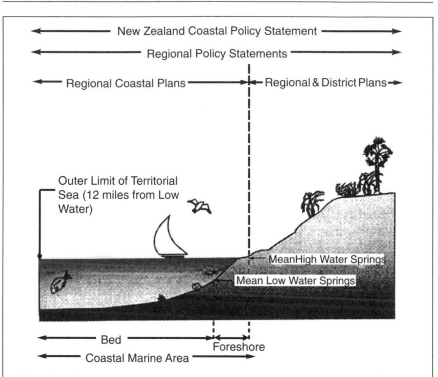

Hierarchy of policies and plans for guiding national, regional and local governments in managing the New Zealand coastal environment.

A Regional Coastal Plan is required to be consistent with the Resource Management Act and cannot be inconsistent with the New Zealand Coastal Policy Statement (Box 5.8). In addition, district councils must prepare District Plans which may not be inconsistent with any national or regional plan or policy.

Initial evaluation of the success of the regional coastal planning process has been that there have been some significant advantages and disadvantages of the legislative requirements controlling their production. Statutory timeframes were placed on the finalization of Regional Coastal Plans by 1 July 1994 to meet public notification requirements. This timeframe was unanimously condemned as unrealistic (especially since the New Zealand Coastal Policy Statement which controls the production of regional plans – as described in Box 5.8 – was not finalized until May 1994). Councils have had to decide between releasing proposed planning documents which they do not consider to be of a good standard or ignoring the statutory requirements. The public consultation process that has followed the release of discussion documents and subsequent proposed plans has been extensive and has yet to be completed for some plans. It is also clear that the

continued . . .

outcomes of revisions made as a result of the public process will be challenged in an already overloaded Environment Court. New Zealand is unlikely to see regional plans effectively in place for some time as a result (H. Rennie, personal communication, June 1997).

Despite these problems with a statutory regional planning process, the restrictions ensure a mimimum level of coastal planning by each Regional Council. The strategic nature of these docments has also undoubtedly enabled decision makers to assist in obtaining a long-term view of the use of coastal resources.

Box 5.15

Typical contents of local integrated coastal management plans (adapted from Alder et al., in preparation)

Section 1: Introduction

This gives the background to the plan: why the plan was needed, steps in its development and where it fits into the context of the region. Context setting may make reference to planning that has taken place at either a regional or site level and which has a large degree of influence on the plan.

It also gives a brief background to the study area; generally only included if the intended audience for the plan is outside of the study area – which is usually only the case if the plan is to be used to obtain funds for its implementation, and where the funding agency is outside the region.

Section 2: Description of the Natural Environment and its Resources

Overviews natural environment and resources, generally from the perspective of how they will influence management through the opportunities and constraints they present. Factors usually considered include:

- climate (including potential future climate change and sea-level rise);
- oceanography (including coastal processes);
- geology and geomorphology;
- hydrology;
- vegetation and wildlife (terrestrial and marine);
- visual resources;
- economic resources (e.g. minerals).

continued . . .

Section 3: Description of Social, Cultural and Economic Aspects and/or Issues

Describes the social, cultural and economic issues relevant to the study area. The main issues usually addressed are:

- land tenure;
- settlement history;
- settlement patterns;
- economic base and commercial activities;
- recreation and tourism.

Section 4: Formulation of Goals and Objectives

States the goals and objectives tackled by the plan. Depending on the process chosen to develop the plan, the goals and objectives can be derived through public workshops, or they may be selected from the legislation, policy, or higher level planning documents.

Section 5: Analysis of Planning Alternatives – by Issue

Alternatives for planning and management are discussed within the context of the natural, social, cultural and economic environments described in the previous sections. The section presents and discusses each issue of importance in the study area, possibly ranging from the management of local recreation to locations for heavy industry on the coast.

This section should aim to segment description of issues wherever possible; on-the-ground integration of these issues should be left to the following section.

Section 6: Description of Coastal Management Poposals – by Sector or Zone

The division of the coast into sectors or zones can be described at the beginning of this section, or in the introduction. It is usually easier to divide the coast into sections to simplify the description of on-the-ground or on-the-water management actions.

The management options for each section/zone are analysed, focusing on what choices were available and why particular ones were made. The analysis can draw on the issue-by-issue description given in the previous section.

The section is supported by plans and maps, which may be of varying levels of sophistication depending on the issue and the levels of accuracy required.

continued . . .

Section 7: Implementation

Explains how the sector-by-sector recommendations of the plan should be implemented, and by whom.

Section 8: Monitoring and Review

Sets out monitoring and evaluation criteria and procedures. Can also provide a timeframe for plan review.

---

if local plans have been recommended by broader regional or whole-of-jurisdiction planning, then there may be issues and/or approaches already identified. Typical contents of local coastal management plans for either case are shown in Box 5.15.

The goals and objectives in local coastal plans will be tangible, and action and development oriented especially on the foreshore. Often the coast is divided into areas and specific objectives are formulated for each area: one area may be developed for recreation and associated commercial facilities, another for recreations without commercial facilities. Because the goals and objectives are tangible they are easier to implement, and facilitate the identification of a set of criteria to evaluate the plan's performance. Information collected for planning is focused at the local scale. Requirements include information on biophysical features such as prevailing weather, seas and coastal conditions, and particulars of existing facilities and current and future access needs.

The number and scope of options available for policy and planning responses at the local level is often determined by community opinion. If sectors of the community are divided on the preferred outcomes of a plan, a number of options with a diversity of actions and recommendations will be needed. If the issues are broad, then most options are 'best-fit' outcomes which attempt to reach a consensus amongst the community. Other factors such as funding, expertise and community opinions will influence the range of options (Chapter 4), which can range from administrative (e.g. changing town planning schemes) to engineering works (e.g. reforming a dune).

Criteria should, if possible, be given against which the success of the plan can be measured. Given that the plans are intended to cover all relevant local coastal management issues, their recommendations may be expected to cover a wide range of subject areas. Recognizing this, the major recommendations of each plan are often divided into groups such as environmental, planning, administrative and miscellaneous/sociocultural.

Three case studies (Boxes 5.16 to 5.18) illustrate the range of approaches used in formulating a local coastal plan. All three include many, if not all, of the above features.

Box 5.16

Sri Lanka Special Area Management Plan

Hikkaduwa is a densely developed coastal tourist centre approximately 100 km south of Sri Lanka's capital Colombo and 150 km from the nation's International Airport (White *et al.*, 1997). Hikkaduwa is a popular international tourist destination, with over 300 000 tourist guest-nights in Hikkaduwa during 1992. However, tourism development is poorly planned and is causing significant impacts on the natural environment, including on Sri Lanka's first marine sanctuary which abuts the town (White and Samarakoon, 1994). Without the development of a planned approach to tourist management, and other local problems (most notably coral mining), there is the potential for a gradual reduction in the natural coastal assets which draw international (and local) tourists to the area. This phenomenon is well documented from other areas in Asia (Chapter 4).

The SAM process was chosen for this site because of the broad number of local user groups required to take ownership of management issues in order to improve environmental quality, and ensure a sustainable fishery and tourism industry. Additional emphasis was placed on the implementation of the plan being self-supporting locally, thereby reducing the need for continuing national or international financial contributions (the tourist industry at Hikkaduwa is mainly locally owned).

The processes for producing a SAM focus on consensual planning using locally based full-time facilitators to bring the plan together with participation from the broadest range of stakeholders possible. The strategies produced through the planning process and expected results are shown in the figure.

Each strategy is implemented through a set of defined actions co-ordinated through an implementation committee supported by national coastal resource management agencies.

The SAM links back to the higher level national plan through the recommendation that an urban growth management strategy be developed which is incorporated in the regional-scale Southern Area Development Plan. Thus, the SAM process is linked both to detailed site- and regional-scale planning.

The main lessons learned from the SAM process in Sri Lanka were summarized by White and Samarakoon (1994):

- The SAM process must be open, participatory and work towards consensus.
- Decisions must be clear and well documented.

continued ...

- National government agencies must understand and accept the process.
- Stakeholder groups must be equally represented in the management process.
- Implementation results should be apparent within three years.
- Monitoring and feedback results make the programme tangible.
- In Sri Lanka, collaborative management is a more appropriate concept than community-based management for coastal resources.
- Community groups can make the difference in success or failure.

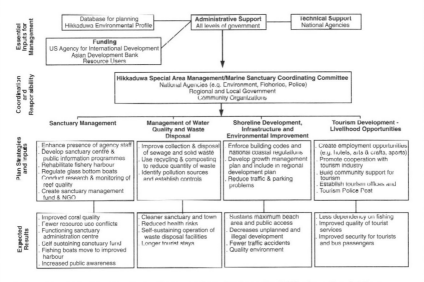

Management inputs, strategies and expected results from the Hikkaduwa Special Area Management planning process (Hikkaduwa Special Area Management and Marine Sanctuary Coordination Committee, 1996).

At present, the Hikkaduwa SAM is being implemented and its success monitored and evaluated in comparison with the second SAM undertaken in Sri Lanka. It is too early to determine the overall applicability of SAMs for other areas of the country, although early indications are promising because there are significant improvements in the management and use of local coral reef, lagoon, beach and mangrove resources in the two SAM sites of Hikkaduwa and Rekawa Lagoon. Of importance in Sri Lanka is that the local communities have played and are playing a major role in the planning and implementation process for the SAM plans. This process promises to provide a greater degree of sustainability than achieved through the National CZM Plan of 1990. Nevertheless, the National CZM Plan set the stage to promote and support the SAM efforts which are now incorporated formally in the revised National CZM Plan of 1997 (Box 5.10).

Box 5.17

Local planning on Christmas Island (Australia)

Christmas Island is located in the Indian Ocean and is a Territory of Australia. The island is a small mid-oceanic volcanic atoll; it is characterized by small sandy coves and bays between steep rocky headlands. Much of the coast is fringed by a narrow strip of coral reefs, and then drops steeply down to 200–4500 m. On the north side of the island, Flying Fish Cove provides an excellent anchorage and port (Figure 5.12). As a consequence, most islanders live close to the cove.

Historically the island's economy has been based on phosphate mining, but it will soon be mined out. Until now, the island's coast and resources have not been developed extensively. But, as the Christmas Island community turns to other sources of economic development, the potential for coastal and marine resources is high for a number of industries – fishing and tourism in particular. To ensure that development of these resouces is sustainable, an integrated management programme for the conservation of marine wildlife was drafted.

The Australian Parks and Wildlife Act was used as the legislative basis for the production and implementation of the coastal management plan. The Act provides for a 'management programme for the conservation of wildlife', but the definition of 'wildlife' is broad, and can be used to develop management on an ecosystem basis. The status of Christmas Island as an Australian offshore territory constrained the use of State planning legislation, used as an important coastal planning and management tool on the mainland.

The main issues addressed in the plan include:

- limited information base for decision making;
- accessible beaches are crowded during peak times;
- ad hoc marine tourism;
- over-exploitation of specific marine resources by commercial and recreational fishers and collectors;
- foreign fishing vessels operating in the management area;
- a range of user conflicts in Flying Fish Cove;
- waste management; and
- ballast water discharge.

The planning process was particularly challenging due to the dominant cultural groups on the island. Although Christmas Island is an Australian Territory, many of the island's 2100 residents are either from Cocos Malay or Chinese communities and do not necessarily share the same values, perceptions and beliefs as mainland Australians. From the onset of the planning programme the main priority was to produce a plan which all

continued . . .

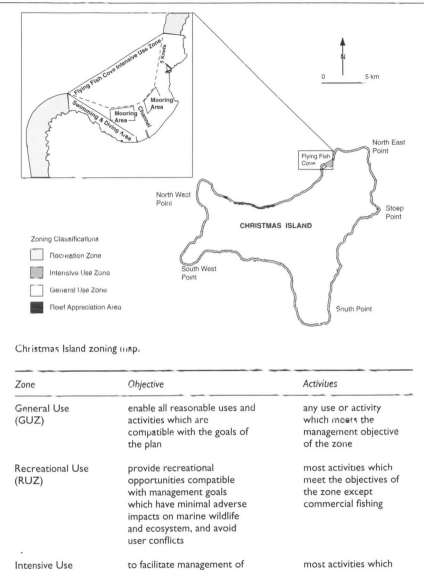

Christmas Island zoning map.

Zone	Objective	Activities
General Use (GUZ)	enable all reasonable uses and activities which are compatible with the goals of the plan	any use or activity which meets the management objective of the zone
Recreational Use (RUZ)	provide recreational opportunities compatible with management goals which have minimal adverse impacts on marine wildlife and ecosystem, and avoid user conflicts	most activities which meet the objectives of the zone except commercial fishing
Intensive Use (IUZ)	to facilitate management of activities that are consistent with the public and commercial uses of Flying Fish Cove and do not have adverse impacts on the marine ecosystem and avoid conflicts among users	most activities which meet the objectives of the zone except commercial fishing, collecting, and deballasting outside of the Australian Voluntary Ballast Water Code

continued . . .

community groups would accept and implement. Consequently an extensive community programme was initiated and a number of discussions with community groups, management agencies, commercial enterprises, users and fishers were held throughout the planning process.

Eventually a plan was formulated which included various provisions applying throughout the management area, and a zoning plan which detailed the site planning in Flying Fish Cove (Box 5.22). Provisions include: the banning of poisons, powerheads, explosive and SCUBA for fishing; catch limits for specificed species of fish; harvesting of gravid female lobsters; codes of practice; and constraints on the export of fish caught for recreation purposes. The plan was accepted by the community, and is now being implemented.

The Christmas Island coast is adjacent to a national park in several areas. Where possible, zoning complemented these park areas. For most of the close-inshore areas and the coast, Recreation Use Zone (see table) is sufficient for management of these areas. Two reef-appreciation areas within the RUZ are used to provide for 'look but don't take' areas. The Flying Fish Cove area is zoned (see figure) as intensive use and site-specific planning has been undertaken to address a number of coastal issues (Box 5.22). Offshore marine areas are zoned GUZ (see table) which allows for a wide range of uses.

The Christmas Island example illustrates how the views of ethnically distinct community groups can be accommodated into a planning framework at the local level. How this is translated into site-level planning is shown in Box 5.22.

Box 5.18

Spermonde Archipelago, Indonesia, local integrated coastal planning

The Indonesian national coastal and marine management project (MREP) is using Sulawesi Selatan as a case study in regional-scale coastal policy development (Box 5.13). The Spermonde Archipelago offshore of the province of Sulawesi Selatan is one of the Marine Coastal Management Areas (MCMA) under the MREP project.

Within the MCMA the focus is on more detailed planning through the formulation of zoning plans for specific areas. These zoning plans will support the provincial vision, goals and policies while addressing issues within the areas nominated for zoning. There are also provisions for local area planning. In the Spermonde Archipelago, Pulau Kapoposang and Papandangan are used as a case study for this level of planning. Pulau Kapoposang and Papandangan are located on the western boundary of the

continued . . .

Archipelago and close to the Makassar Strait. Kapoposang was selected because: it had an established community which expressed a desire to be involved in sustainable management of the island and adjacent reef; the potential for island and reef-based tourism; and an active local reef fishery around the island and adjacent reefs. Tourism and improved marine resource use offer the most promise for economic development on the island. Papandangan was included because the adjacent reef has tourism potential if destructive fishing ceases. The community on this island uses Kapoposang's resources and is a source of issues such as destructive fishing. This potential for management for the area was recognized by the Governor of Sulawesi Selatan who has recommended that the area be declared a marine tourism park.

A study of the islands and reef has been undertaken which addresses the following major issues (Salam *et al.*, 1996).

- Destructive fishing and collecting – cyanide and explosives are used in the commercial and subsistence fisheries. These methods are not sustainable and will ultimately lead to coral degradation.
- Low socio-economic conditions – the residents of both islands are poorly educated, health and education facilities are limited and access to financing is difficult. Many residents are caught in a cycle of borrowing money at above market interest rates to finance the next season's fishing while committing their future catch to the lender.
- Over-exploitation – commercial and subsistence fisheries are heavily exploited and showing signs of overfishing; catch rates are declining despite increasing effort.
- Lack of information – information of the distribution and abundance of resources in the management area is lacking. This lack of information constrains the identification of high conservation areas, intensive use areas, and areas for future development.

The study recommends that a zoning scheme should form the basis for the management plan; the primary objectives are the improvement of the socio-economic conditions for island residents, protection of endangered species, rehabilitation of degraded areas and the development of community-based resource management programmes. General provisions in the management plan include the banning of commercial fishing within the management area, prohibiting the mining of coral, substituting liquid gas for mangroves as a source of cooking fuel, and passive rehabilitation of habitats by prohibiting destructive fishing practices. Access to freshwater is a major constraint to island development although all islands have a brackish lens. To maintain or improve this water source it is proposed that mangroves be replanted and a ban on further harvesting on mangroves be imposed.

continued . . .

The development of tourism in the area will be facilitated by improved tourist facilities and transport to the area. In addition, tourism operator awareness programmes are proposed to ensure that operators are aware of the potential impacts their operations can have on coral reefs. Better education facilities and programmes are proposed to improve the islands' social situation. A study of the mariculture potential of the area is also proposed. The problem of a lack of information will be addressed by encouraging researchers to include the area in their study, especially for the management of fisheries resources.

Five zones are proposed:

- conservation for protection, research and regeneration of resources with access restricted to researchers only;
- traditional use for sustainable use of resources by traditional residents, commercial exploitation is banned;
- replenishment for protection for a specified time from exploitation to allow resources to recover for a maximum of five years;
- intensive use for protection of reef resources while allowing for general use; and
- buffer to provide a transition area between intensive and conservation zones.

Looking back at the provincial coastal policy developed through the MREP programme (Box 5.13), the Kapoposang study incorporates many elements of these policies. Furthermore, the assignment of zones to particular areas will be done as part of the programme to develop a community-based management programme in the area. The community-based management programme is part of the COREMAP programme which is discussed in Box 5.32.

5.3.5 Site-level integrated plans

Site planning is the art and science of arranging the uses of portions of land. Site planners designate these uses in detail by selecting and analysing sites, forming land use plans, organising vehicular and pedestrian circulation, developing visual form and materials concepts, readjusting the existing landforms by design grading, providing proper drainage, and finally developing the construction details necessary to carry out their projects.

(Rubenstein, 1987)

Site-level integrated coastal management plans are detailed strategies for the use, protection, development and management of a small coastal area. Site plans are prepared:

- to provide a local context for detailed land- or water-use decisions;
- as a condition of an approval for a planning application (e.g. subdivision, development, re-zoning, or as a requirement of an environmental impact assessment);
- to implement a local or regional coastal management plan;
- to review a previous plan; or
- to meet community demands.

Site plans may be used to consider the following issues: local coastal conditions, environmental values, pedestrian and vehicular access, parking, signs, conservation and rehabilitation works, buildings and other structures, recreation facilities, the maintenance of recreation areas and support facilities, and future vesting and management of coastal land and water (Box 5.19).

Box 5.19

Typical issues and problems addressed by site-level integrated plans

Site planning focuses on localized issues and problems. Site plans can as a result be extremely varied, given the potential range of coastal management issues (Chapter 2). Typical among such issues are:

- identification of degraded or coastal resources requiring rehabilitation or protection;
- identification of environmental, landscape, recreation and coastal resource values;
- threatened fragile coastal ecosystems, landforms or rare/threatened flora and fauna;
- natural hazard management (erosion, flooding, cliff-falls, potential sea-level rise);
- level of access to coastal areas and resources (multiple use paths – walking, cycling, skating; pedestrian ways; roads and car parks);
- demand for type, location and access to recreation facilities;
- need to upgrade existing recreation facilities;
- types and levels of environmental assessment; and
- clarification of ongoing operational coastal management responsibilities (who pays for what and where).

Site plans may be based on information contained in regional and local plans in the area and on consultations with government agencies and local authorities. Studies, however, may be needed to obtain details about specific sites conditions or issues (Box 5.19).

Box 5.20

Typical contents of integrated coastal management site plans (adapted from Alder et al., in preparation)

Section 1: Introduction

A short report outlining the reasons for undertaking the project, who is involved and the planning context (e.g. statutory basis, focus).

Section 2: Site Description

Describes the:

- biophysical aspects of the area and the current state of the resources;
- existing and future use of the area;
- existing facilities and access;
- existing management of the area (if any); and
- management issues relating to the project.

Section 3: Objectives

Lists and briefly discusses the major objectives for developing the site plans.

Section 4: Proposed Works

For each component of the design plan gives details of the location, material, species lists and construction methods. Where appropriate, maps, photographs, plans and diagrams are provided for further detail.

Section 5: Implementation

Provides advice on the priority, timing and approximate budget of the various components of the design plan. Timing is especially important for rehabilitation plans since the success of the revegetation depends on the time of the year the plants are established.

Also includes maintenance schedules for rehabilitation measures (e.g. plantings), structures and facilities proposed in the plan.

Section 6: Review and Monitor

Recommends when the plan and its implementation should be reviewed, and suggests which variables should be monitored to assess whether the site design is meeting its objectives. Some of the requirements for monitoring may be met through the maintenance schedules mentioned above. Other factors may need a specific monitoring programme. Which variables are monitored is determined by the objectives for the site design, and available expertise, techniques and resources.

The content of site plans will vary according to the particular issues and conditions of the area. For most site design plans, a common content is used which includes the sections shown in Box 5.20.

Site-level coastal planning commonly uses approaches developed by landscape architects and planners (e.g. Rubenstein, 1987; Thompson and Steiner, 1997), integrated with civil engineering skills (for earthworks, etc.) (Figure 5.8).

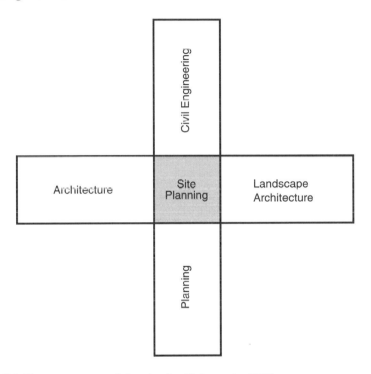

Figure 5.8 The components of site planning (Rubenstein, 1987).

The type of information needed in planning at this level will depend on the objectives for managing the site. If the main objectives for site planning are to stabilize coastal dunes, information collected should include those factors which assist in identifying the causes of degradation. If the plan is to provide low-key recreational facilities, such as a car park, lookouts and picnic area, a different range of information has to be obtained, including geomorphology, soil type, runoff characteristics and vegetation types, and user data such as potential levels of demand.

Site planning can often follow a relatively simple planning process, as conflicting uses or demands, which may require a consultative style of management planning, have been addressed by local or regional plans.

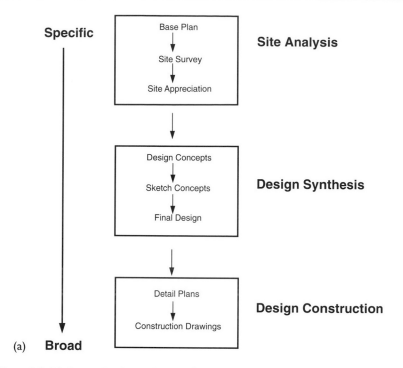

Figure 5.9 (a) Generalized site planning design process and *(b)* concept evolution (adapted from Schmidt, 1996, after Rutledge, 1971).

Consequently, a linear step-by-step approach can be used which follows the general design process shown in Figure 5.9 which begins with ideas, develops and then refines those ideas in relation to site opportunities and constraints. This process can provide the framework for working design concepts to be developed which can evolve into sketch designs (Figure 5.9) which articulate the relationship between the various elements of a site (Rutledge, 1971).

Site plans often divide the coast into sectors or precincts based on natural environmental features, beach/shore characteristics and current and proposed use or development. Within each sector, proposals for managing the foreshore are presented which may include provisions for access and parking, recreation facilities, conservation and rehabilitation, landscaping, marine facilities, protection works and possible commercial uses.

Plans usually contain an implementation programme, time frame and costings as well as an outline of who is responsible for funding, the undertaking and staging of proposed works, and responsibilities for interim and long-term maintenance. A schedule of work and estimated

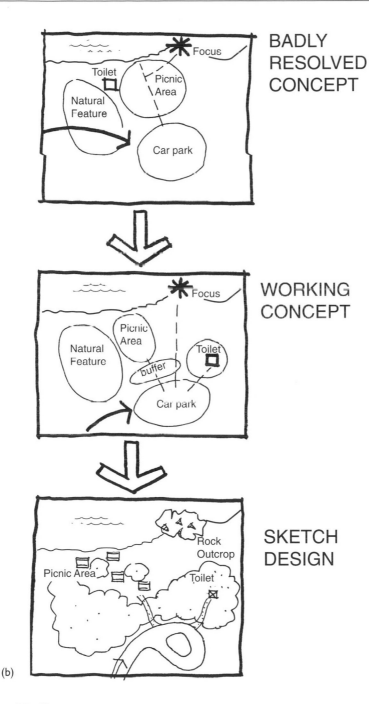

BADLY
RESOLVED
CONCEPT

WORKING
CONCEPT

SKETCH
DESIGN

(b)

Figure 5.9 (b).

Box 5.21

Dune rehabilitation planning, Warnbro, Western Australia

The coastal dunes at Warnbro in Perth, Western Australia, were badly degraded due to uncontrolled access causing vegetation loss and blow-outs (Figure 5.10). The area was used for many years by Perth residents for a range of destructive activities, including sandboarding, walking at random through the dunes, trail-bikes and four-wheel-drive vehicles.

The urban expansion of Perth extended to the Warnbro area in the 1960s, but did not significantly impact on the dunes until the early 1990s. The broad-scale planning of this expansion, including location of major roadways, and the overall boundaries of the subdivision and extent of the foreshore reserve, was undertaken through regional-scale planning under the statutory Perth Metropolitan Region Scheme (MRS) (Hedgcock and Yiftachel, 1992).

Under Western Australia's planning legislation conditions may be placed on developers to write and implement special area management plans as part of approval to subdivide land. It has been common practice in the State that management plans for public foreshores be undertaken as a condition of subdivision. These 'foreshore management plans' are site-level plans which specify the detail of environmental rehabilitation, managed access ways and the location of car parks and recreational facilities (Box 5.20). Subsequent to these works being undertaken the foreshore reserves themselves are transferred from private landowners to be managed by a government authority (State or Local) as part of subdivision approval.

Planning processes were supported in the Warnbro area by the State Government Department of Agriculture placing a Soil Conservation Notice over the proposed development area. The Notice required that the land was stable before development could take place (S. Clegg, Ministry for Planning, personal communication, June 1997). These actions were supported by a local environmental community group, the Warnbro Land Conservation District Committee.

Perth's dry and hot summers, linked with strong summer sea-breezes, requires sensitive and well planned dune rehabilitation. The foreshore management plan for the site detailed rehabilitation techniques, staging and costing (Quilty Environmental Consulting, 1991). Dune rehabilitation techniques used were:

- earthworks;
- brush and mulch for temporary stabilization;
- wind fences; and
- permanent revegetation (various species for the berm, frontal and hind dunes).

continued . . .

Individual site rehabilitation prescriptions were guided by a detailed site management plan, an extract from which is shown in the figure. The foreshore management plan also shows the potential location of access paths between the subdivision (and its car parks for visitors) and beach together with a coast-parallel dual-use path. ('Dual-use path' was originally used to describe wider paths for cycling and walking. However, with the increase in in-line skating and skateboarding, these are now 'multiple-use paths'. All new subdivisions are required to link with previous dual-use paths to form a continuous coastal recreational pathway.)

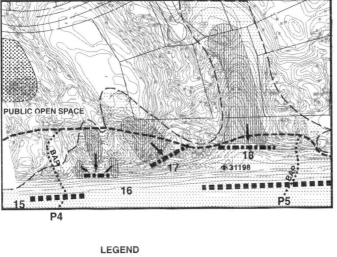

LEGEND

ZONE FOR STABILISATION TREATMENT	`[- - - - - -]`
FRONTAL DUNE CONSTRUCTION	■ ■ ■ ■ ■ ■
DUAL USE PATH (2 Tentative Options)	– – – – – –
BEACH ACCESS PATH (Tentative)	· · · · · · · · · · · · · ·
SITE NUMBERS FOR REHABILITATION PRESCRIPTIONS	**16, 17, etc**
SECTION OF PUBLIC OPEN SPACE TO BE RETAINED UNDER NATIVE VEGETATION	
FORDUNE DEVELOPMENT	■ ■ ■ ■ ■ ■ ■
SAND SOURCE AREAS FOR DUNE CONSTRUCTION	||||||||||||||||||

Extract from the Warnbro Dunes rehabilitation site plan (Quilty Environmental Consulting, 1991).

The result of the dune management work is shown in the aerial photograph taken to advertise the sale of the land-lots (Figure 5.11).

continued . . .

Rehabilitation of the Warnbro dunes was undertaken by the developer between 1991 and 1993, with follow-on maintenance for a further two years until the rehabilitation had successfully established a self-sustaining plant community (Figure 5.10c). A total of US$600 000 was spent on the rehabilitation programme. The result was that fencing, beach access pathways and a north–south dual-use path allowed for an increasing intensity of public use without the damage that past uncontrolled use had caused.

The foreshore land was transferred from the developers to the local government (the City of Rockingham) in 1995. Council's financial and human resource constraints, combined with a resurgence of sandboarding in the area and extreme winter storms, has resulted in marked deterioration of the condition of the dunes. These problems are being addressed by Council through the establishment of a foreshore advisory committee of community and Council representatives to advise on problems along the city foreshores and avenues for improving their management.

Box 5.22

Integrated site planning in Flying Fish Cove, Christmas Island

There were a number of issues in the Christmas Island coastal management plan (Box 5.17) which focused on Flying Fish Cove, including:

- access to boat launching and storage facilities;
- transport barges moving through recreation areas;
- access to moorings;
- powered boats being used in swimming areas;
- resource exploitation in areas popular with tourists.

These conflicts and issues were addressed through spatial separation through creating zones (see figure in Box 5.17). These zones reflect the desired use of the resources in Flying Fish Cove as identified by the Christmas Island community through the coastal management planning process.

---

costs may be required depending on the type and intensity of proposed works. Two examples of site planning are presented in Boxes 5.21 and 5.22. Warnbro is a good example of site planning with a local planning framework: the works and outcomes are illustrated in Box 5.21 and Figures 5.10 and 5.11. The Flying Fish Cove example (Box 5.22, Figure 5.12) illustrates how conflicting uses were resolved in a community with diverse coastal values and needs.

Figure 5.10 Warnboro Dune (Western Australia) rehabilitation sequence (credit: Quilty Environmental Consulting).

(a) May 1993

(b) June 1993

(c) August 1993

Figure 5.11 Warnboro Dunes and neighbouring subdivision at completion of dune rehabilitation (credit: Australian Housing & Land).

Figure 5.12 Flying Fish Cove, Christmas Island (credit: Greg Pobar).

A programme to monitor and review a plan's effectiveness is often developed to ensure that the works are meeting management objectives. Once the works are completed they should be reviewed to ensure that they are meeting their objectives: if rehabilitation works were undertaken, are they working; if recreation plans were implemented, are user expectations satisfied; did unpredicted hazards arise; and are maintenance costs within the allocated budget? Specific aspects of the plan may need to be monitored (e.g. dune stabilization) and changes made where appropriate.

5.4 Subject plans in coastal management

Subject plans are written to address one, or a limited number, of subjects or issues. Subject plans can be developed at a range of spatial scales and can have different foci and statutory bases, depending on the subject being addressed, the mode of implementation of the plan, and who is writing the plan. Subject plans can be written by individual government agencies, private companies or non-government organizations, depending on their involvement in particular coastal issues. The various coastal issues, opportunities and problems described in Chapter 2 can individually, or in combination, require the production of one or more subject plans (see Table 5.5).

Subject planning for most resources differs from spatial planning in a number of areas. At the site-specific level, subject planning appears as spatial; but at higher scales, planning is not spatially based. National programmes are too broad to have a geographic focus, especially if there

Box 5.23

Aquaculture and integrated planning in Shark Bay, Western Australia

Some forms of aquaculture, such as pearl oysters, rainbow trout and some edible oysters, have been farmed in Australia for many years. The past 10 years has seen a rapid expansion of the industry for human consumption, especially Atlantic salmon, rainbow trout, Pacific and native oysters, mussels, scallops and abalone (Anutha and O'Sullivan, 1994; Anutha and Johnson, 1996). The industry is currently focused on the cooler waters of southern Australia, but with Australia's enormous and varied coastline there is considerable potential to expand national, regional and local economies, to generate export revenues and to provide employment opportunities. Australia has the advantage of being able to learn from other countries' experiences in developing this industry, and this is reflected in the approach by the Australian government at national, state and local level.

The overall goal and direction for developing aquaculture is guided by the National Aquaculture Strategy. The strategy has a number of goals listed, including the need to develop an ecologically sustainable industry. One mechanism recommends the development of Industry Codes of Practice, including an environmental code. The National Aquaculture Strategy complements the National Strategy for Ecologically Sustainable Development (ESD) (Commonwealth of Australia, 1992) which, when linked with the National Coastal Action Plan (Commonwealth of Australia, 1995) and other national and Federal initiatives, sets an overall framework for the aquaculture industry. In fact a number of Australian states have amended their legislation which covers aquaculture development to incorporate ESD principles (see Box 5.25).

The ESD Strategy is implemented in a number of ways including measures which can be implemented at the State level along with industry, and training and research institutes. For example, Western Australia has developed a State Planning Strategy which recognizes the ESD Strategy (Western Australian Planning Commission, 1996c). In the State Planning Strategy a number of recommendations are made to facilitate development of the aquaculture industry including the recognition of aquaculture as a potential land use and its potential environmental and social consequences. The State Planning Strategy emphasizes sound management practices be used by the industry. It also views regional planning as a mechanism for identifying resources for aquaculture development and ensures that the needs of aquaculture are considered in regional plans.

In parallel with the State Planning Strategy the State government also produced an Aquaculture Development Strategy recommending that regional planning be used to meet the goal of development of an ecologically sustainable aquaculture industry. The State's aquaculture strategy also views regional planning as a mechanism to provide guidelines for

continued . . .

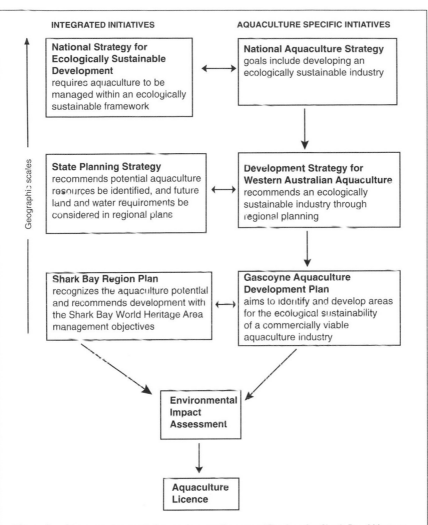

INTEGRATED INITIATIVES · AQUACULTURE SPECIFIC INTIATIVES

National Strategy for Ecologically Sustainable Development
requires aquaculture to be managed within an ecologically sustainable framework

National Aquaculture Strategy
goals include developing an ecologically sustainable industry

State Planning Strategy
recommends potential aquaculture resources be identified, and future land and water requirements be considered in regional plans

Development Strategy for Western Australian Aquaculture
recommends an ecologically sustainable industry through regional planning

Shark Bay Region Plan
recognizes the aquaculture potential and recommends development with the Shark Bay World Heritage Area management objectives

Gascoyne Aquaculture Development Plan
aims to identify and develop areas for the ecological sustainability of a commercially viable aquaculture industry

Environmental Impact Assessment

Aquaculture Licence

Geographic scales

Hierarchy of integrated coastal plans and aquaculture-specific plans for Shark Bay, Western Australia.

dealing with aquaculture by identifying potential sites and species, and guidelines for the assessment and approval of aquaculture applications.

An example of the use of regional planning to assist in managing aquaculture, and how the cascade of national and State integrated and subject plans interact, is shown in the Shark Bay area of Western Australia (see figure). In Shark Bay, the Shark Bay Region Plan (Western Australian Planning Commission, 1996b) interacts with the Gascoyne Aquaculture Development Plan (Gascoyne Development Commission, 1996) which aims

continued . . .

to develop aquaculture in the Gascoyne as a commercially viable industry. Sustainability of the industry in this environmentally sensitive area is possible only through sound ecologically sustainable development principles by ensuring that any aquaculture operation must be compatible with World Heritage Area management objectives. Therefore, aquaculture development proposals are accordingly assessed through EIA requirements. The Shark Bay Region Plan recognizes the potential for aquaculture and that it can be developed in a region dominated by World Heritage Area values. The plan recommends implementation of the provisions of the Gascoyne Aquaculture Development Plan.

is a wide range of ecosystems in the country. Subject planning is often sector specific, while spatial planning tends to go across sectors and integrates a number of sector management activities. Information needs are well defined, narrow and focused on that particular sector rather than a range of sectors. This sector-focused approach often limits the scope of community or stakeholder involvement to those who are directly involved in the particular sector. In sector planning, modelling is more prevalent and uses 'what if' scenarios extensively. There have been attempts to use models in spatial planning but their use is not as well developed since the number of variables such as representing various interest groups is much larger and hence more complicated.

While there are considerable differences between the two forms of planning, similarities between spatial and integrated plans can be found. Both forms of planning are issues driven, use similar planning principles and often have a legislative basis. The same approach to planning as described in the previous section is often used, and community involvement is integral to either form, with the planning process commonly being as important as the outcome.

There are literally thousands of subject plans pertaining to the coast and its resources which could be used to illustrate subject planning; however, there are few examples of subject planning integrated with high-order planning. The history of planning within the Shark Bay area is an example of how subject planning can be integrated into regional planning as well as providing input into formulation of future regional plans (Box 5.23).

Subject plans have a wide range of applications in a coastal planning framework (Table 5.5). They can be used when there is no apparent conflict between coastal uses, a circumstance which usually arises when such conflicts have been previously resolved, allowing single-issue plans to be developed and implemented. Subject plans can follow from the outcomes of integrated coastal management plans, often forming the action statements of such plans, assisting in their implementation. This is

Table 5.5 Example subject plans used in coastal management

Subject plan groups	Example subject plans
Resource exploitation	Capture fisheries Aquaculture Oil and gas
Natural resource management	Key ecosystems (e.g. mangrove, coral reef, saltmarsh) Water quality
Infrastructure management	Coastal engineering (flood and coastal defence) Ports Recreational boating (marinas, moorings, boat ramps) Sewerage Solid waste disposal Offshore tourist pontoons

especially common at the broader coastal planning levels, mainly from the international to regional scales. Sections of coast with one owner or manager can also lead to single-issue plans being developed, again due to a lack of conflict.

Subject plans can also be used as the forerunners of integrated plans. In such circumstances there may not be a willingness to develop a full integrated strategy without first undertaking some single-subject plans. This approach has the danger that a subject-by-subject planning approach further deepens divisions between sectors, possibly leading to a long-term increase in conflict between coastal uses.

However, in many coastal planning frameworks a temporal separation of integrated and subject plans does not occur, with integrated and subject plans being developed and implemented at the same time.

The development of tourism plans with integration between planning scales is illustrated in Box 5.24. The planning system for the development of aquaculture in Shark Bay is an example of how subject planning can be integrated with similar subject plans but at different scales, and how subject plans can be integrated with broad planning initiatives (Box 5.23).

5.5 Coastal management plan production processes

There are a number of ways to produce coastal management plans, depending on the type of plan and the issues to be addressed by it. The processes used to produce a regional-scale single-subject coastal plan are

Box 5.24

Levels of tourism plans for coastal management

Regional

Major recreation opportunities may be identified, based on an assessment of current and future recreation demand and a general appraisal of site characteristics. Appropriate recreation sites are generally those which are: likely to be heavily used, now and in the future; capable of sustaining that use for a range of popular coastal activities; appealing to several user groups; and readily accessible, close to urban centres or on major travel routes. It is also important to ensure that a region provides a variety of sites as suggested by the Recreation Opportunity Spectrum (Box 4.12).

Local

As at the regional level, appropriate planning steps included assessment of recreation demand and application of regional-scale planning and consideration of site capacity and possible environmental and social impacts. At a local level, basic facilities (e.g. parking, boat launching ramps, commercial enterprises, ablution facilities) can be assigned to specific localities, and their general standard determined. Strategies for altering sites can be devised; for example, by managing access, adding more facilities, upgrading existing ones, or developing new sites.

The general level of access to various segments of the coast should be decided on, as should any requirements to exclude existing or potential recreation activities in order to reduce conflicts between user groups or to maintain a site's desired position on the Recreation Opportunity Spectrum (Box 4.12).

Site

Recreation planning for specific locations should ensure that facilities, pathways, signs, commercial enterprises, etc. meet the needs of particular user groups and are suitable for the recreation activities occurring at that site.

At specific recreational locations users' recreation needs should be addressed, such as: determining exact design of parking, ablution blocks, barbecues, tables, pathways and other facilities. Recreation needs are also relevant to landscape plans, which should provide shelter, space for children's play equipment and/or space for setting up recreational equipment such as sailboards.

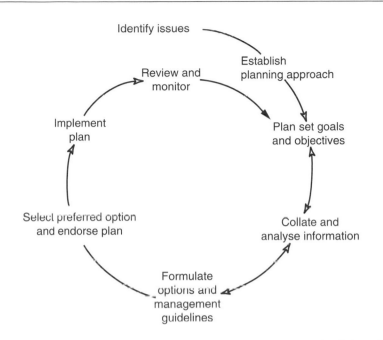

Identify issues

Establish
planning approach

Review and
monitor

Plan set goals
and objectives

Implement
plan

Select preferred option
and endorse plan

Collate and
analyse information

Formulate
options and
management
guidelines

Figure 5.13 General steps in the formulation of a coastal management plan (adapted from Alder *et al.*, in preparation).

likely to be very different to a site-level integrated coastal plan. Of course, this is to be expected, given the diversity of coastal management plans. Nevertheless there are some generic steps which are shared by the great majority of coastal plans, be they integrated or subject plans (Figure 5.13).

The general steps outlined in Figure 5.13 can, in some circumstances, be followed in a simple linear fashion; that is, the planning cycle is started at the top of the the figure and followed to its conclusion in a step wise manner. Circumstances where this is most commonly appropriate are when the need for consultation is limited, such as in the production of site-level plans, or for some subject plans.

Two common themes underlying the preparation of successful coastal management plans are, first, the division of the planning process into the gaining of specialist and/or technical information about the coastal environment, including the pressures placed by people on that environment; and second, gaining the views of local people and users on the best use and management of the coast. This division between 'technical' and 'user/social' information is obviously blurred by the fact that users often have considerable technical knowledge. However, the division is useful in that the two categories of information are usually acquired in different ways. Information from users is gained through methods including the staging of public workshops, questionnaires and surveys, focused public

---

Box 5.25

Subject plan production process – Queensland Fisheries Management Plan

The Queensland Fisheries Act (1994) provides an updated framework for fisheries management in the State. The Act shifted the focus from production to use of the resources while maximizing community, economic and other benefits and emphasizing fair access to fisheries resources for all community sectors.

The Queensland Fisheries Management Authority (QFMA) implements the Act. The primary implementation mechanism used by the Authority is a statutory fisheries management plan. The QFMA recognizes that it must ensure that the plans are coordinated and integrated across other aspects of natural resource management to be effective – such as catchment management, habitat protection and marine park management plans.

The QFMA also views community consultation as fundamental to effective management planning. To facilitate this consultation a system of Marine Advisory Councils have been established. The Councils are based on the State's major fisheries – currently there are six Councils: trawl, reef, freshwater, mudcrab, tropical fin fish, and subtropical fin fish. Their role is to provide advice on appropriate management of the fishery, resource use, development of the fishery and protection of fish resources. Each Council also prioritizes research, monitoring, surveillance and enforcement needs. Management options available to the Councils include restricting fishing gear, seasonal closure, spatial closures, a combination of seasonal and spatial closures, quotas as well as specific habitat management options.

To deal with local issues, Zoning Advisory Committees are also formed, and they have a broader role in some respects. These councils are regionally based and deal with local issues and impacts. Various community interests are represented on the council, not just the fishing groups. The council advises QFMA on local issues and management matters. They are also important because Zoning Advisory Committees facilitate getting information on fisheries management needs and responses to the broader community.

This approach to fisheries resource management in Queensland is gaining wider acceptance in Australia, with other States amending their fisheries legislation to form consultative groups.

---

forums, the production of a draft plan and the receipt of public submissions on the draft plan. In contrast, technical information is gained by using consultants, academics and other specialists.

Both technical and social information is used in the formulation of Queensland Fisheries Management Plans, where the Management Advisory Committees and Regional Advisory Committes, composed of interested parties from industry, government and the community, are invaluable sources of input into the planning process. This approach is described in Box 5.25.

The situation shown in Figure 5.13 is more complex when community consultation is involved. In these cases experience has shown that a linear step-by-step approach is generally less efficient than a more flexible community-driven approach. This is not to say that there aren't well defined processes for guiding, albeit gently, consensual styles of coastal management plans, as described below.

Consensual processes have evolved out of general everyday experience in undertaking planning exercises. Special planning studies which attempt to address complex issues and which involve a high degree of vested interest and public scrutiny have also contributed. Plan production processes for consensual-style coastal management plans are outlined in the next section.

5.5.1 Consensual-style coastal plan production processes

The consensual-style coastal plan is now the most commonly used plan production technique in integrated coastal planning exercises which attempt meaningful public consultation. There are three main reasons for this:

- If the people, organizations and government agencies affected by a coastal plan are included in its production, the plan's recommendations are likely to reflect their opinions, reducing conflict and making the plan easier to implement.
- The community is more likely to participate in the implementation of a plan if they have been involved in its production, creating the potential for future cost-savings.
- The involvement of relevant government agencies increases the likelihood that they will support the plan's implementation with either cash donations or the provision of staff time.

Coastal management plans produced through consensus building require a special approach. Those responsible for the production of the plan usually need to spend more time 'planning the plan' than when more traditional techniques are used. This is because a detailed planning framework has to be constructed which is rigid enough to ensure that a plan is actually produced on time (and within budget), but flexible enough to allow the consensus-building approach to work effectively. This delicate balance requires that plans:

- are issue driven;
- use steering committees with a wide range of representation;

- • are based on a rigorous public participation programme; and
- • are focused on goals and objectives which can be implemented.

Three major processes drive the consensual style of plan production: an administrative process, a public participation process, and the process of writing the plan itself (Figure 5.14). These three processes are discussed in separate sections below.

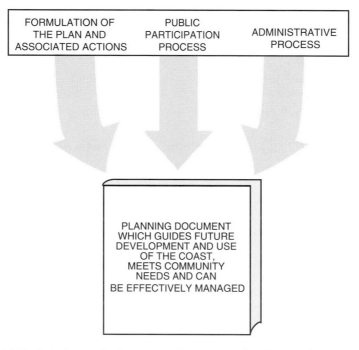

| FORMULATION OF THE PLAN AND ASSOCIATED ACTIONS | PUBLIC PARTICIPATION PROCESS | ADMINISTRATIVE PROCESS |

PLANNING DOCUMENT WHICH GUIDES FUTURE DEVELOPMENT AND USE OF THE COAST, MEETS COMMUNITY NEEDS AND CAN BE EFFECTIVELY MANAGED

Figure 5.14 Typical plan production process for consensual-style coastal management plans (adapted from Alder *et al.*, in preparation).

Unlike the linear step-by-step plan production process shown in Figure 5.13, the three processes shown in Figure 5.14 are generally run in parallel; however, despite this paralleling effect, the basic steps shown in Figure 5.13 still have to be carried out, the difference being that some are carried out at the same time or iteratively. As the planning process progresses through the various steps there are a number of activities and considerations to be undertaken, as detailed in Box 5.26.

An example of the processes used in developing a consensual-style integrated coastal management plan is shown by that used in the development of the Thames Estuary Management Plan (Box 5.27).

Box 5.26

The general steps in the formulation of a consensual-style integrated coastal management plan

- Identify issues
  Recognition that there are issues/problems which need to be addressed.

- Establish the planning approach
  Who will be responsible for the planning process? This is guided by decisions made in the administration process. The public participation programme and the planning framework should complement each other so that the community is well represented and has the opportunity to effectively participate in the planning process.

- Set broad goals and objectives
  Stakeholders must participate in formulating goals and objectives. Regard should be given to how the success of the plan in meeting its objectives will be evaluated. Tangible objectives such as maintaining ecosystems or reducing impacts to dune systems are easy to measure, but intangible objectives such as increasing stakeholders' enjoyment of a particular site are difficult to measure and monitor. As more information is obtained, goals and objectives need to be revised accordingly

- Collate and analyse information
  This includes all types of relevant information: biophysical, social and economic, and information obtained through public participation programmes. There may be areas where information is lacking. If funding or staffing is available, studies to obtain further information should be initiated.

- Formulate options and management guidelines
  These will address issues identified through the consultative process and technical studies, and make provision for future use and development of the study area in consultation with stakeholders.

- Select preferred options
  The planning team selects a final option after evaluating the range of options formulated. The option selected will depend on a number of factors such as available resources, community attitudes, existing plans and policies, and feasibility of implementation. The ease of implementing the selected option is critical to the long-term success of the planning project. The planning team also ensures that recommended management actions and guidelines are consistent, integrated and coordinated with other strategies and management plans; otherwise it

continued . . .

makes implementation of the plan difficult. At this stage, the planning team determines the life of the plan so that the stakeholders and subsequent managers of the area know when to evaluate and review the effectiveness of the plan.

- Endorse the plan
  The plan should be forwarded to relevant agencies for endorsement. The agencies which should endorse the plan will generally be made known during the plan production process.

- Implement the plan
  A series of action plans (e.g. capital works programme), with tangible benefits, are formulated so that implementation proceeds in an effective and efficient way. These action plans can be short- or long-term depending on the level of planning. At this stage, monitoring plans should be underway or initiated to enable managers to evaluate whether the plan is meeting its goals and objectives.

- Review and monitor
  The life of a plan is usually between one and five years. New information, changes in government policy and direction, and changes in community values and attitudes will all influence the relevance of implemented plans.
  It is important that monitoring be considered as an ongoing activity throughout the life of the plan. The variables and criteria used in a monitoring programme should be identified during the setting of goals and objectives.

The complexity of estuarine systems, coupled with often intense and competing demands, has been one of the most difficult aspects of coastal management planning (van Westen and Scheele, 1996). Indeed, adaptive consensual styles of management plans for large and complex estuarine systems appears to be an emerging norm, especially in the developed world (e.g. Government of Victoria, 1990; Imperial *et al.*, 1992; Imperial and Hennessey, 1996; Inder, 1997).

(a) Administrative process

Planning does not happen spontaneously; there must be an organizational structure or force responsible for the plan's ultimate production. An administrative programme is often used to accomplish this. Administrative programmes are especially necessary for higher-order plans and serve a number of functions, such as to:

Box 5.27

Thames Estuary Management Plan production process (Kennedy, 1996b)

The Thames Estuary Management Plan (Box 2.1) was produced using a clearly defined consensual process. This method was adopted in order to encourage understanding between user groups and to promote a sense of ownership among stakeholders. The process used to produce the plan is summarized in the table.

Thames Estuary Management Plan – production process

Phases	Advantages / Considerations / Issues
SCOPING (Product: Issues Report)	– Identification of 'real issues' – Widespread agreement that there are things that can be resolved via coastal management planning – Rejection of non-issues – Establish partnerships – Building trust – Allay fears of unwanted compromise – Never present a 'fait accompli'
PLANNING (Product: Business Plan)	– The process and its design – Administrative arrangements – Estimate cost and time – Allow contingency for problem areas – Differences in organizational culture – Range of political, ethical and aesthetic perspectives Define project limitations (what can be resolved via a mechanism that is already in place) – Risk assessment – Relationship with other plans
MANAGEMENT STRUCTURES	– Allocate time for this realistically – Make room for minority views – Ensure effective lines of communication – up, down & across
INFORMATION GATHERING (Product: Topic Papers)	– Try to get agreed technical information – Information and draft policies provided by the practitioners themselves – Joint fact finding – Agreeing what is technically viable & then deciding what to do within the context of the wider economic, environmental & social framework – Listening to the range of views

- establish committees to assist in the planning process;
- coordinate and integrate existing plans and studies;
- involve relevant government agencies, industries, user groups and the general community;
- ensure that the legislative requirements for planning are met;
- provide a mechanism for ensuring that the decision-making process is made by, and supported by, representatives from the community (this extends the decision-making process beyond the legislative or government process);
- provide funding and other resources for the planning process; and
- provide secretarial and administrative support.

The administrative process is generally initiated when the decision to formulate a plan is made. The next decision concerns the administrative structure to be used to guide the process. A commonly used structure of a local steering committee and/or working groups, and their role in decision making on the plan's production, is shown in Box 5.28.

Membership of local plan steering committees is usually kept under 10 if possible, simply for administrative efficiency; however, efficiency needs to be balanced against demands to be as inclusive as possible. In some cases there may be a need for large steering committees. These can reduce administrative time and effort in the long term by being able to reach agreement among the major stakeholders relatively quickly through steering committee meetings. Often steering committees are made up of elected government representatives. These representatives assist in making sure that the drafted plan is politically acceptable, does not conflict with government policies, and where possible complements government initiatives.

Steering committee members evaluate their need for working groups to support them in undertaking the study. Working groups are usually formed to investigate specific issues or aspects of the planning process and to provide advice to the steering committee, and commonly consist of interest groups and technical people from government. Attendance at working group meetings may change according to the issues discussed.

(b) Public participation

A public participation programme ensures that the local community and user groups have the opportunity to participate fully in the plan production process (Figure 5.15). Ideally they should be a part of the entire planning process, but this is difficult to attain for a number of reasons – including the funding and resources needed for such a high level of involvement. Where involvement is possible, the public should be given several opportunities to be involved in all aspects of the plan's

---

Box 5.28

Typical membership and decision-making roles of local integrated coastal plan steering committees

Steering committee membership is typically made up of:

- chairperson – someone who has standing within the community, should be aware of coastal management issues and be dedicated to the task;
- local elected representatives and/or elders;
- members of key community groups – e.g. coastal ratepayers, progress associations;
- members of key community groups – e.g. recreation fishers, surfers, retirement clubs;
- representatives of government agencies; and
- other members as required.

Local plan steering committee usually makes decisions on:

- terms and definitions associated with the plan; these should be specified with mutual agreement reached on the intentions of terms;
- the spatial boundaries, scope (degree of local/regional content and nature of the plan need to be specified early in the process);
- what planning techniques will be used – could include various coastal management and planning techniques (Chapter 4);
- what reporting procedures are required – clarification on the powers of the steering committee and those who report to it; and
- the resources, funds and staff needed to support the committee's operation.

---

formulation. The initial stages of the public participation process should ensure that the community is aware of and understands the processes. A number of useful principles in guiding public participation processes for large infrastructure developments in Western Australia provide a useful introduction to this section (Box 5.29).

Public participation in integrated coastal planning comes from many perspectives. People living in the study area will be the most affected by planning (Figure 5.16). They should have a say in the region's future and feel confident that decision makers are aware of their views, and have considered them, before plans are finalised. The public can also often identify values which need to be managed and priority issues which need to be addressed in planning. Public involvement is essential if a plan is to be supported and easily implemented.

As mentioned throughout this chapter, planning varies with scale. This

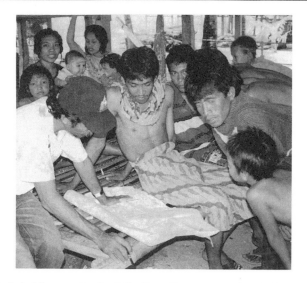

Figure 5.15 Stakeholder meeting for Take Bone Rate marine plan, Indonesia.

concept also applies in public participation. Public participation at the national level is very different from that at the site level. At the national level it is both logistically and practically impossible to directly involve all individuals with interests in planning. To ensure that individuals do have the opportunity to influence the planning process, representatives are usually included on the planning team, and where possible public meetings and written submissions are used. At the site level it is possible to have individuals directly involved in planning either through membership on the planning team or through extensive liaison and ongoing workshops within the area.

Public participation is a continuing process. The community should be involved from the time the decision to formulate a plan is taken through to its implementation and review. Ways of achieving this can include combinations of:

- initial advice of the intent to undertake a planning process with periodic briefings on the progress of the project;
- representation on steering and/or work committees;
- community workshops using facilitators who help participants to identify issues and values and management options;
- media campaigns which use press releases, newspaper, radio and television to encourage community involvement;
- surveys;
- meetings with specific interest groups;
- public submissions throughout the planning process.

Box 5.29

Guiding principles for public participation for large infrastructure projects in Western Australia (Department of Resources Development, 1994)

The following principles are essential in the design and implementation of a public participation programme.

- Public participation is an integral part of, and complementary to, planning and decision-making processes.
- Public participation programmes should occur throughout the life of a proposal.
- A public participation programme should recognize the diversity of values and opinions that exist within and between communities.
- A public participation programme must be designed to deal with controversy.
- Specialized public participation techniques are required for contentious or complex issues.
- The timing of the public participation programme is crucial to its success.
- The information content of the public participation programme must be comprehensive, balanced and accurate.
- A public participation programme must be custom designed.
- Public participation should always be a two-way process between the proponent and community.

A public participation programme requires adequate amounts of time, money and skilled staff.

The techniques to apply will depend on funding, staffing, social and political acceptability, and the complexity of the issues and the community.

Stakeholders within the study area should ideally have the opportunity to participate in the planning process, and where possible other agencies and interested parties from outside the study area should be consulted. The community should be consulted on all aspects of the planning process and invited to assist in the collection and collation of information. Community members can also be involved in setting goals and objectives, selecting preferred options, and determining implementation actions for the plan. The level of participation will depend on the issues being addressed, the planning approach and the resources available.

The consensual planning process aims to produce a coastal management plan which has broad community ownership, especially of its recommended actions. Local needs and the issues covered by the plan will determine how this aim will be achieved. The factor which perhaps more

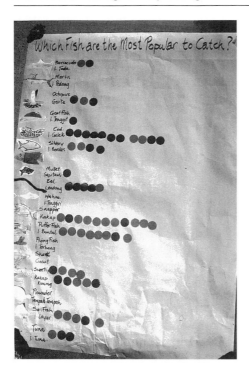

Figure 5.16 Involving children in coastal planning, Cocos Island.

than any other dictates the style and extent of participation is the type and intensity of use of the coast under study. If a section of coast is used by millions of people each year there is little point trying to involve everyone in the production of the plan, but rather to focus on representatives of key user groups. In contrast, a plan for a small section of a more isolated or less intensively used part of the coast could involve the majority of its users. Indeed, depending on the local perceptions of the impact of planning decisions, public participation can be quite extensive and active at the local level compared with national programmes (City of Mandurah, Peel Development Commission and Western Australian Planning Commission, 1993).

Central to a participatory style of coastal management plans is a series of community workshops or meetings (Table 5.6). A simple coastal management plan may have only one phase of public workshops in order to establish coastal resident and user perceptions of the coast under study, and the relationship of this region to the broader district. These workshops can then be used as input into the structure as well as the content of the plan. Issues raised at the workshops may change the opinions of the steering committee about the priority issues in the area.

It is important to recognize that there are basically two types of audiences involved in public participation: individuals and groups; and

Table 5.6 Matrix of community participation techniques in the planning process (Department Planning and Urban Development, 1993)

Technique	Regional scale plan	Local scale management plan	Specific development proposal	Discussion/ background paper	Policy statement	Technical paper
GROUP TECHNIQUES						
Community Consultative Committee	R	O	O	NA	O	NA
Technical Consultative Committee	R	O	O	O	O	O
Workshop	R	R	NA	NA	O	NA
Seminar	O	O	NA	NA	O	NA
Search Conferences	R	O	NA	NA	O	NA
Public Meetings	R	R	O	NA	O	NA
Public Forum	O	NA	O	NA	NA	NA
Small Group Meetings	O	O	O	NA	NA	NA
Design-In	O	NA	O	NA	NA	NA
INDIVIDUAL TECHNIQUES						
Public Submissions	O	O	R	NA	O	NA
Individual Discussion	O	O	R	NA	O	NA
Contact with affected Residents	R	R	R	NA	O	NA
Project Team Contact	R	R	R	O	O	O
Site Office	O	O	O	NA	NA	NA
Participant Observation	NA	NA	O	O	NA	NA
Surveys and Questionnaires	O	O	O	O	O	O
Opinion Polls	NA	NA	R	NA	NA	NA
Telephone Hotline	O	O	O	NA	NA	NA

Key: R = recommended; O = optional; NA = considered not applicable that the same techniques are not necessarily applicable to both groups. Table 5.6 outlines what techniques are the most appropriate for these audiences at various planning scales.

For example, at the site plan level of coastal management planning common methods for public participation include:

- notification of proposed works;
- displays;
- call for formal submissions or comments;
- series of formal and informal meetings and workshops with key individuals, the community or groups with a particular interest in the study; and
- advertisements in the media.

With more elaborate plans, or those which address issues of conflict and/or controversy, there may be the need for further public workshops. These can be held at key times in the development of the plan to:

- provide an opportunity to review outcomes to date together with possible actions to address issues raised; and
- review and evaluate the plan's recommendations and to seek input to the implementation of key recommendations.

Efforts should be made to ensure the widest representation at workshops of coastal residents and users. Advertisements or features in the local press are the usual means of achieving this. In addition, invitations should be sent to key bodies such as ratepayers' organizations, progress associations, local non-government organizations and user groups. During workshops the names and addresses of attendees should be recorded in order to ensure that they receive invitations to subsequent workshops and are sent copies of draft and final plans. Depending on the social and cultural context of the area, other innovative techniques such as compensating fishers who participate instead of fishing, or involving religious leaders to facilitate participation, need to be considered.

Effective community workshops need careful planning. Thought should be given to the structure and content of the workshop, as well as the best location and time. A range of techniques can be used to ensure that the maximum benefit is gained from the workshops, including new analytical techniques such as saliency testing.

If required, a number of smaller forums or one-on-one interviews may be held with key users and non-users. These can include high school students, commercial/business groups, resident associations and coastal-based sporting groups. Consultations should also be held with key government agencies if they are not represented on the steering committee. Industry association representatives or representatives from specific companies may also be interviewed; this is essential if they will be affected by the plan.

Findings of the consultations, public workshops and other meetings should be summarized in a written report and be made available to the public. The report should highlight the issues raised and the management options available to address the issue.

(c) Producing the plan

The responsibility for producing the plan – the written document or map – is generally undertaken by the steering committe with the assistance of a technical and professional staff. At lower planning levels the steering committee may be actively involved, but at higher levels such as the national level they rarely do the groundwork of data collections, mapping, etc. This is done by technical and professional staff.

Irrespective of who produces the plan, the production of the actual document generally follows Figure 5.14, with the technical and professional staff or working groups undertaking a range of activities such as revising goals and objectives, data analysis, mapping, producing various reports and drafting the plan text. The team assists in setting and revising goals and objectives based on advice and direction from the steering committee and analysis of information. Regard should be given to including measures for evaluating the success of the plan in meeting these objectives. Tangible objectives such as maintaining ecosystems or reducing impacts to dune systems are easy to measure, but intangible objectives such as increasing stakeholders' enjoyment of a particular site are difficult to measure and monitor. As more information is obtained, goals and objectives need to be revised accordingly.

In producing the plan information is collected, collated and analysed. This includes all types of information – biophysical, social and economic – in a range of formats: statistics, digital, maps and Geographic Information Systems, and information obtained through public participation programmes. There may be areas where information is lacking. If funding or staffing is available, developers or managers can initiate those studies to obtain the required information. If the issues are significant or the study area is diverse, background papers on particular issues or subjects can be prepared so that stakeholders or participants are better placed to make informed decisions.

Once the information is analysed, including the community's views, the planning team prepares a report of the various options for managing the issues and meeting objectives. This report is often released for public comment after endorsement at a steering committee meeting. Comment may be sought through additional workshops, requests for written comments or both. The report should summarize the major issues raised, and present the steering committee's opinion on management actions required. There are generally two styles for such reports, each with their advantages and disadvantages:

- an Issues and Options Paper which is followed by a Draft Coastal Management Plan; or
- a Draft Coastal Management Plan.

Draft plans, which summarize findings and make a series of specific recommendations for action, have traditionally been the preferred choice. Comment is sought on each recommendation. An issues and options paper, as the name implies, gives a range of options, and comment is sought on the preferred options.

The main advantage of releasing a draft plan is that the opinion of a steering committee is clearly stated. Another advantage is that if public comment on the draft plan is favourable, there can simply be final endorsement of the draft plan, which then becomes the final plan.

The main disadvantage of releasing a draft plan, especially if it is professionally printed and designed, is that can look too final and people can feel that, whatever the merit of their comments, it will not be changed. The danger is that a sense of ownership or commitment to the implementation of the plan is not engendered in the plan's stakeholders.

Whatever choice is made about the form of the report, effort is generally made to ensure that:

- the response period is long enough for all to adequately comment (this is usually 2 or 3 months);
- distribution is wide – to relevant state government departments, politicians, libraries of educational institutions, study workshop participants, local public libraries, local community groups and clubs;
- copies are made available free of cost, in order to encourage the widest opportunity for comment; and
- local newspapers are used to advise the public that the document is available for comment.

Once the responses from stakeholders and interested parties are collated and analysed, and any additional information is analysed, the planning team selects a final option. The option selected will depend on a number of factors such as available resources, community attitudes, existing plans and policies, and feasibility of implementation. Ease of implementation of the selected option is critical to the long-term success of the planning project. Relevant to this is ensuring that recommended management actions and guidelines are consistent, integrated and coordinated with other strategies and management plans (e.g. regional transport strategy or local tourism strategy). Finally, the plan can be forwarded to relevant agencies for endorsement if required.

5.6 The implementation of coastal management plans

> In most nations it is generally much more difficult to secure commitment to management than for the creation or establishment of management plans.
>
> (Kenchington, 1990)

> Two conclusions are constantly reinforced whenever public programme implementation is subjected to scrutiny:
>
> 1. No one is clearly in charge of implementation; and
> 2. Domestic programmes virtually never achieve all that is expected of them.
>
> (Ripley and Franklin, 1986, p. 2)

It is worth concluding this chapter with a brief section on the implementation of coastal management plans. The ever-present danger is that the production of the plan is viewed as the end of the planning process, instead of the beginning of the real actions for its implementation. After all, most coastal management plans were started in the first place to help solve problems, not to just sit on the shelf.

There are many reasons why more plans grace bookshelves than become the dog-eared guide for the people on the ground. The implementation of plans can be difficult to achieve. Perhaps the most important of these is the emotions felt by those who are actually undertaking the planning exercise. Finishing and publishing a plan can be an exciting and satisfying experience, one giving a real sense of achievement. There can be, quite naturally, a lull in the motivation of these individuals after the plan is finished. Picking themselves up and implementing the plan can be tough going.

A related phenomenon is the division of responsibilities which can occur when a specialist planning group that has produced the plan is distinct from those who are charged with its implementation. The result of planners not involving the staff who will implement the plan is that implementation staff lack the background knowledge and rationale, and that sense of ownership that ensures effective implementation. In turn, staff charged with implementing the plan may be unsure about the plan's objectives. This is especially true if the implementation staff were peripheral to the planning process. These apprehensions and mistrusts can be managed by involving implementing agencies and staff throughout the planning process, and through a well designed implementation programme which makes a conscious effort to ensure that all involved understand why the plan was developed and what it is aiming to achieve.

Implementing plans formulated at strategic and operational levels involves translating objectives into actions. At the strategic level, implementation can involve several diverse activities: the drafting of relevant legislation; the formulation of policy for a wide range of issues; the establishment and resourcing of programmes to identify and declare potential protected areas; and the development of other programmes. An important failure of implementation at this level is the setting of unrealistic expectations for what the plan can actually assist in achieving (OECD, 1993).

At the operational level, the translation of management objectives and guidelines into on-the-ground and on-the-water management is known as operational or day-to-day management. It is these activities, such as enforcement, surveillance and EIA, which provide most of the active management of current and future uses of an area as well as meeting the wider community's perception of coastal management. Many stakeholders perceive this form of management as the 'doing' part of planning. These practical actions, together with communication and education, research, monitoring and site planning, are all considered by managers to be necessary for sound implementation.

The problem of allocating responsibilities for plan implementation increases as the complexity and/or geographic coverage of the plan increases. A wide-ranging plan will inevitably require the involvement of a wider field of stakeholders.

Implementation can be considered as having three major components:

- managing the resource and resource users;
- ensuring that stakeholder expectations are met; and
- meeting statutory requirements in a cost effective manner.

Figure 5.17 is a simple representation of the interactions of the three major components in the implementation of management plans.

The three components of users, resources and statutory requirements shown in Figure 5.17 interact, the areas of intersection representing management activities. Effective implementation is achieved by developing a balance between the three components. The area of intersection of all three components represents optimal implementation, where effective and efficient management is achieved. Here the resources are effectively managed, while users' expectations are satisfied and statutory requirements met. Implementation therefore tries to minimize regulation and resource costs while maximizing community support and participation.

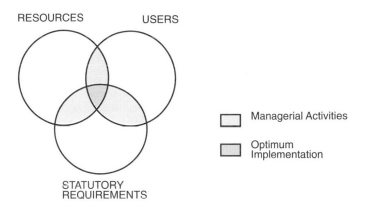

RESOURCES   USERS

Managerial Activities

Optimum
Implementation

STATUTORY
REQUIREMENTS

Figure 5.17 Interaction of the major components in implementing management plans.

Managers have at their disposal a variety of tools for plan implementation, including planning and programming, staff training, education, stakeholder involvement, surveillance and enforcement, environmental impact assessment, research, and project monitoring and evaluation. Many of these tools are described in Chapter 4 and illustrated in a general model for implementing plans in Box 5.30.

An innovative approach to the problem of bridging the often large chasm between the finalization of a plan and its implementation is shown by the Thames Estuary Management Plan (Box 5.31).

Two important tools in assessing the effectiveness of a plan or programme are programme monitoring and evaluation, which are discussed in section 5.7.

The political aspect of coastal management implementation should also be borne in mind. If a coastal management planning exercise has attained a high political profile then the political interest in the plan usually peaks at the time the plan is released. Politicians are often given the opportunity (and they sometimes demand) to release the plan publicly. This inevitably gives a politician the chance to show he or she is doing something positive, innovative and forward-looking for the coast. Political interest is then likely to drop off until such time as tangible results of the plan's implementation can be seen. This first stage of the plan's implementation is often the most fragile, especially if funds are required. Tangible outcomes should be included early in the plan's implementation to maintain the interest and support of politicians and stakeholders. More politically astute coastal managers have been known to deliberately stage the implementation of plans to give some early 'wins' to ensure the ongoing interest of politicians and other senior stakeholders.

Implementation of plans and strategies may also be distributed across

Box 5.30

A general model of plan implementation

Many of the tools described in this chapter can be used to implement a plan. A general model for implementing plans is illustrated in the figure. This model is based on a generalized regional or local-level integrated plan, and assumes that reasonable resources are available for implementation.

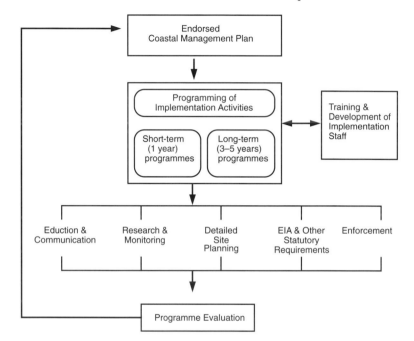

Generalized flow diagram of plan implementation.

Once a plan is endorsed, several ongoing activities take place to implement the plan. Some take place simultaneously while others take place in a sequence. Implementation usually commences with the programming of implementation requirements. This stage focuses on the establishment and management of day-to-day activities and projects. The requirements of this stage depend on the focus and scale of the plan – site-level plans will usually have focused on the direction of tangible management activities, and hence implementation programming will be minimal.

Staff who will be involved in implementation activities should have ideally been involved in producing the plan. This ensures that

continued . . .

all staff are aware of the reasons behind the various planning outcomes and subsequent management activities used in their day-to-day responsibilities. However, different staff have varying levels of involvement in the range of implementation activities. They should be given training and job preparation. This assists in team building, and ensures that all staff understand the reason for the programmes and the basis of their personal work plan.

Training and staff development are important components of the implementation process and they serve two main functions. First, they create and build an effective management team. Usually, no one person has all the skills and expertise required in the operational management of an area. By building a team and focusing on a common goal of plan implementation, one person's deficiencies may be compensated by another's strengths. Team building ensures that everyone knows their role in management of an area. Second, it ensures that staff have the skills and expertise to perform their expected tasks and to contribute to the overall management of the area.

An implementation programme will use a number of activities to meet management objectives, including education, research, surveillance and enforcement. Education should be an ongoing programme of activities to raise awareness of issues, alter user behaviour and facilitate involvement in management throughout the life of the management plan. The focus of education will change as the level of awareness of stakeholders improves, and issues are managed. Each programme will need to be developed for the area's needs and the available resources.

Research programmes, within a plan implementation framework, should be designed to fill in the information gaps identified in the planning process. Research activities outside of the implementation programme (e.g. university programmes) should be coordinated to ensure that research is relevant to management, and that staff and financial resources dedicated to plan implementation are maximized.

Surveillance and enforcement requirements for implementing a plan are established in relation to the management objectives and priorities, the needs of various monitoring programmes, and available resources. Surveillance is a multi-functional monitoring tool in day-to-day management. It not only detects and deters infringers, but it can gather information for monitoring and research purposes. The objectives of an enforcement programme usually include improving user compliance of rules and regulations – to help 'do the right thing'. Similarly, education and communication programmes seek the same objective, and therefore when enforcement programmes are formulated managers should ensure that they are complementary to education programmes to maximize their effectiveness.

Box 5.31

Moving from planning to implementation for the Thames Estuary Management Plan (Kennedy, 1996b)

One way of ensuring successful plan implementation is to 'Reality Test' policies as they are devised. The following reality-testing questions were asked in the implementation of the Thames Estuary Management Plan (EMP).

- Who will support this policy?
- Who is the lead agency?
- Where will the funding come from?
- Do we need a trial?
- Is this a priority?
- How can we monitor its success/failure?
- When will the policy need to be reviewed?
- HOW CAN WE MAKE SURE IT WORKS?

A Strategy for implementation of the Thames Estuary Management Plan (English Nature, 1996) was launched in July 1996 in parallel to public consultation on the estuary management plan itself. The objectives of the Implementation Strategy were to:

- define the key elements and priorities for implementation;
- outline the initial work programme and time-scale for implementation;
- suggest appropriate management and administrative structures; and
- identify the financial and other key resources required to support implementation.

The Implementation Strategy was intended as a 'prospectus' on how the Estuary Management Plan would work. The Plan consists of a series of aims, principles and supporting recommendations framed as action plans for the various issues on the estuary (see Boxes 2.1 and 3.10, and Figure 3.8). The strategy provides a road map for initiating, coordinating and monitoring the various action plans described in the plan.

Four strands of implementation activities were identified as needing to be delivered to ensure successful implementation of the plan:

- rationalizing Action Plan recommendations into manageable elements, capable of being monitored and which may attract funding (see figure);
- a more formalized system of consensus building and conflict resolution;
- increasing the EMP profile to the public, political organizations and senior management in the public and private sectors; and
- a system for monitoring success and failure of the EMP and its continuous improvement.

continued . . .

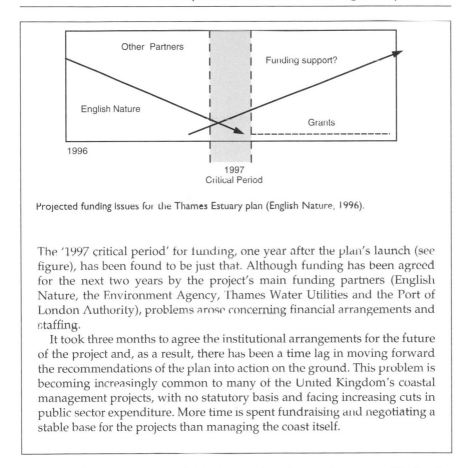

Projected funding issues for the Thames Estuary plan (English Nature, 1996).

The '1997 critical period' for funding, one year after the plan's launch (see figure), has been found to be just that. Although funding has been agreed for the next two years by the project's main funding partners (English Nature, the Environment Agency, Thames Water Utilities and the Port of London Authority), problems arose concerning financial arrangements and staffing.

It took three months to agree the institutional arrangements for the future of the project and, as a result, there has been a time lag in moving forward the recommendations of the plan into action on the ground. This problem is becoming increasingly common to many of the United Kingdom's coastal management projects, with no statutory basis and facing increasing cuts in public sector expenditure. More time is spent fundraising and negotiating a stable base for the projects than managing the coast itself.

a range of programmes or initiatives rather than using a specific implementation programme. This approach requires fewer resources but it does not ensure implementation will be coordinated or complete. It is an approach that is used to implement components of Indonesia's coastal management plans (Box 5.32).

Finally, the costs of plan implementation can often be realized only after the plan has been completed. Clearly, this can lead to major implementation problems. Explicitly including costings for implementation within the planning process is emerging as one mechanism to counter the 'this costs too much to implement' argument. In the case of the Thames Estuary (Box 5.31), a plan implementation 'prospectus' was developed in order to clarify the requirements and costs for implementation. In the case of the Hikkaduwa Special Area Management Plan in Sri Lanka, the plan's authors used environmental economics techniques to assess the benefits and costs of implementing the plan, as opposed to leaving the area to degrade without mangement interventions (White *et al.*, 1997). The

Box 5.32

Implementing coastal management planning in Indonesia

The policies, zoning plans and subject plans developed as part of Indonesia's national Marine Resources Evaluation Programme (Box 5.9) will be implemented through a number of government initiatives.

One of these initiatives is COREMAP, a national programme focused on coral reef rehabilitation and management. The COREMAP programme has four major components:

- public awareness and participation;
- locally based management of priority coral reef sites;
- institutional strengthening, planning and policy, human resources capacity building; and
- establishment of a Coral Reef Information Network.

These components of COREMAP are focused on implementing management objectives at varying government levels, as well as within industry and the community. Although this programme is focused on coral reefs, other coastal environments such as mangroves and seagrasses are considered since they play an important role in maintaining coral reefs.

Whereas MREP is primarily a resource information management and planning programme, COREMAP is a programme which can assist in implementing MREP outcomes. Sulawesi Selatan province is one of the first provinces to participate in COREMAP (Box 5.13). Province-wide COREMAP initiatives such as multi-agency enforcement patrols to address the problem of blast fishing and cyanide fishing will contribute towards the provincial policy of sustainable resource use. In addition, a number of initiatives at the community level are also proposed. Awareness programmes for cyanide fishers and developing alternative income-generating activities for fishers are two examples which will assist in implementing a number of provincial MREP policies such as 'raising public awareness of the value of resources and processes so as to encourage responsible resource use' and 'all coral reefs in Sulawesi Selatan waters will be protected from unsustainable exploitation and damage due to human activity'.

The COREMAP programme also proposes the formation of community management groups to provide a bottom-up approach to management. These groups can identify issues and problems which may be widespread and better managed at the regional or provincial level as part of MREP. The MREP and COREMAP programmes are a good example of how top-down and bottom-up approaches to coastal management can be undertaken simultaneously and merged at the provincial planning level.

Box 5.33

The costs on plan implementation of the Hikkaduwa Special Area Management Plan, Sri Lanka (White et al., 1997)

Many coastal plans are not implemented because the cost of doing so is not fully considered during the plan's production. Governments and coastal users are increasingly unwilling to commit to the costs of implementation unless they can perceive a real benefit in doing so.

An interesting method for clarifying the costs and benefits of plan implementation was used in the Hikkaduwa (Sri Lanka) Special Area Management Plan, described in Box 5.16. The authors of the plan used direct and indirect economic analysis techniques (Chapter 4) to calculate the net present benefits and costs of plan implementation together with appropriate sources of funds for cost recapture (see table).

Absolute and percentage share of Hikkaduwa Special Area Management (SAM) plan beneficiaries/costs burdens

Project beneficiary/ burden holder	Net present value of benefits/costs (US$ million)	Percentage share
Local tourism industry	18.86	72.5
International community	5.64	21.7
Local community	0.52	2.0
National economic welfare	0.98	3.8
SAM plan costs	3.47	72.6
Coral miner costs	1.31	27.4

The results of the economic analysis outlined in the table show that the costs of implementing the plan are significantly less than implementation benefits and the main benefits are to local tourist operators and the greatest loss is to coral miners (through restriction of coral mining). Both of these issues highlighted the need to create alternative employment opportunities to coral miners, and demonstrated to local tourism operators the benefits of implementing the management plan. The analysis is currently forming the basis of discussion in the local plan steering committee (Box 5.16) regarding the potential to raise and manage revenue locally through targeted taxes and fees and some additional donor and national government support.

economic analysis clearly demonstrated the economic efficiency of plan implementation (Box 5.33).

5.7 Monitoring and evaluation

This last section of this chapter, on monitoring and evaluation, completes the management–planning cycle shown in Figure 5.13. However, as discussed in Chapter 3, this does not mean the cycle ends, but rather a new planning and management cycle begins. Figure 5.18 shows two interpretations of how planning initiatives cycle. The top row shows three separate planning cycles. In this example, Plan 2 occurs a number of years after Plan 1 is completed and so on. This separated plan cycling is demonstrated in the Shark Bay regional planning example shown in Box 5.23. In contrast, the second row in Figure 5.18 shows that Plan 2 is initiated soon after the completion of Plan 1, effectively linking them. This close linkage between plans may be required where issues are particularly complex, or where evaluation requirements are stringent.

Monitoring and evaluation are processes which assist in answering the question, 'Is the plan working?' and, if it is not working, what future actions are needed to make it work. If a coastal plan has included

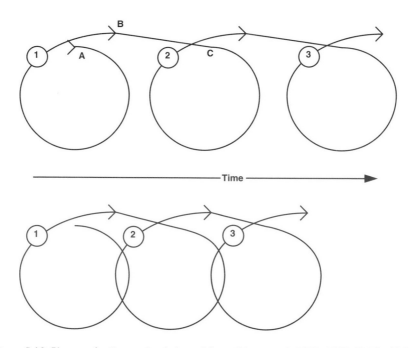

Figure 5.18 Plan production cycles (adapted from Olsen *et al.*, 1996, 1997; Zeitlin-Hale, 1996).

measurable objectives and criteria to evaluate the plan's ability to meet those objectives, ongoing monitoring may provide some of the information required to evaluate the effectiveness of the plan. Monitoring and evaluation also contribute to decision making; as noted by the National Research Council (1995b):

> monitoring can narrow the uncertainty associated with decision making, but it can not eliminate it, and monitoring contributes to understanding change and ascribing causes to these changes. Monitoring results are also useful in weighting the societal benefits of management alternatives.

The need to define measurable objectives and criteria early in the planning process and the design of monitoring programmes was discussed in section 3.5. This section outlines how a programme or plan can be evaluated in terms of meeting measurable objectives, especially when it is linked to monitoring activities.

The terms monitoring and programme evaluation need to be defined in a management planning framework for this section. Monitoring is a process where repetitive measurements in time and space are recorded to indicate natural variability, and changes in environmental, social and economic parameters. Measuring these changes contributes to the information base needed by managers to evaluate the plan's effectiveness. Evaluation is analysing information, some of it gained through monitoring, then comparing the results of the analysis against predetermined criteria. Optimal programme evaluation therefore depends on using quality information, and much of this information is a result of monitoring. Hence, a well designed, ongoing monitoring programme is fundamental to programme evaluation.

5.7.1 Monitoring

Monitoring allows managers to evaluate the effectiveness of plans at the national, regional and local level. It can be used for a range of plans and strategies such as zoning plans, subject strategies and EIA. The design of an effective monitoring programme depends on the plan's objectives, resources (funding and staff) and available technology. When designing a monitoring programme, there are a number of factors to consider (Box 5.34).

In designing a monitoring programme, the variables to measure and desired levels of information must be balanced against costs. As with any programme, resources are limited and managers need to ensure optimal use is made of funds and staff. The sampling strategies used for environmental parameters are quite different from strategies used to sample

---

Box 5.34

General factors in the design of a coastal monitoring programme

- A monitoring programme should be designed to reflect management objectives and to detect changes which facilitate or constrain the meeting of these objectives. If an objective is to maintain natural processes, then monitoring should detect as soon as possible signs of stress and the source of the stress.
- Includes baseline or control sites to eliminate the normal range of variation encountered if changes had not been made.
- Measuring variables which will indicate stress or change. Here indicator species may be useful in measuring changes; when these species are showing signs of stress, the whole ecosystem should be considered at risk.
- Consider sampling frequency, sample sizes, sampling precision and temporal changes.
- The information recorded, collation methods and reporting need to be determined and to be consistent between agencies and levels of government.
- The criteria to use for evaluating the programme. These can be limits or thresholds so that managers can signal the possibility of significant changes or stress in the system.
- The mechanisms which can be used to modify the plan or change management practices if significant changes are detected.

---

communities. The level of information collected needs to be selected so that changes can be detected: if there is a need to collect information at the species level, using the manta tow technique which collects broad-scale habitat information may not detect the changes.

Who monitors what again depends on costs, the variable to measure and levels of information. Monitoring can be done by other components of the community as well as the managing agencies. Consultants, research and community organizations can also participate in ongoing monitoring. There are advantages and disadvantages in using these various groups. Management agencies may have some of the resources needed to undertake monitoring and may be able to reduce the costs, but they may lack the expertise or time; researchers and consultants have the expertise and facilities to undertake intensive and detailed studies, but they generally cost more than other organizations. Community groups do not always have the expertise to conduct detailed investigations but with training they can be involved in collecting information on an ongoing basis while keeping costs to a minimum. Often their involvement can also lead to community support for management. A good example of this is the Citizens' Monitoring Program for Chesapeake Bay. Members of the group who live along the bay and its tributaries measure selected water quality variables and report their results to the Chesapeake data centres. The citizens are able to track the success of clean-up efforts, and as a consequence they become strong advocates of monitoring activities. In addition they are able to bridge the gap between the community and researchers (National Research Council, 1995b).

5.7.2 Programme evaluation

Programme evaluation is well developed in the field of business and organizational management (e.g. Mukhi *et al.*, 1988), but in the area of coastal planning evaluating the effectiveness of plans is a recent development. The evaluation of coastal plans and programmes is a result of progress in several fields of coastal management, increased community participation and government accountability. Advances in other fundamental aspects of management such as administrative arrangements, institutional development, planning approaches and monitoring methods, have provided managers with the scope to address the question of programme evaluation at the coast. As community involvement in decision making has increased, so has their questioning of the consequences of decision making, including planning, increased. Accountability in government has also increased with the general trend of evaluating government programmes and strategies now being the policy of many agencies. These factors have all contributed to the increased interest in programme evaluation.

Designing an evaluation programme for coastal initiatives is so recent that managers are still experimenting on how to approach the problem. Many of their questions centre on what criteria to use and what variables should be measured. In evaluating environmental objectives such as maintaining water quality, international and national standards can provide some guidance, but for less tangible objectives, such as maintaining biodiversity, deciding what to measure and the criteria to use is more difficult; when social and economic parameters are included, the problem becomes even more difficult. A recent study of the Perth South Coast Metropolitan Waters (Box 3.14) gives some insight on how the problem of programme evaluation is being addressed at a local level. The problems and issues at regional and whole-of-jurisdiction levels are also discussed in Chapter 3.

A potential tool for assisting in programme evaluation is adapting 'state of the environment reporting' (SER) processes to coastal programmes. State of the environment reports are becoming increasingly common at international (e.g. ESCAP, 1990), national (e.g. Bird and Rapport, 1986;

Zann, 1995; CSIRO, 1996) and sometimes sub-national levels (e.g. Environmental Protection Authority, 1992). SER documents aim to report on environmental information in a similar way to that in which economic or demographic statistics are reported. An early classification of SER frameworks (UN, 1982) identified four basic approaches (Sheerin, 1991):

- media – describes the state of the environment by collecting statistics on its basic environmental components (media) such as air, water, land/soil, human settlement, etc.;
- stress response – attempts to statistically measure environmental change (response) brought about by human intervention or natural event (stress on the environment);
- ecological – closely linked to stress response but describes the state of the environment by reference to ecoregions or ecosystems; and
- accounting – assesses stocks and flows of natural resources.

While each of the above frameworks has applications in coastal planning and management, it is the pressure response model (now commonly referred to as the pressure-state-response model) which has begun to be adopted as a useful basis for coastal programme evaluation (Olsen and Tobey, 1997; Olsen *et al.*, in press).

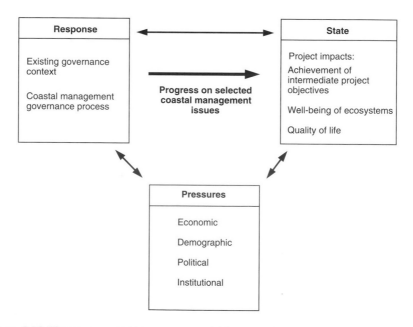

Figure 5.19 The pressure-state-response model for coastal programme evaluation and learning (Olsen *et al.*, 1997).

This simple model (Figure 5.13) is indeed appealing. However, the application of this model remains largely untested, especially at more local levels. In addition, the integration of the results of pressure-state-response evaluation into day-to-day decision making will require further work if it is to become common practice in coastal management.

Monitoring is a well established component of many environmental management programmes. Its application in planning has been well recognized, but much less so in coastal management. Programme evaluation is a very recent component of coastal planning and managers are still developing approaches and techniques. Both components are closely linked and contribute to answering the question, 'Is the plan working?' The question needs well defined objectives (preferably operational ones) to facilitate the identification of criteria to measure plan performance and to determine what variables to include in a monitoring programme. Just as planning takes place at various spatial scales, so does monitoring. The variables to measure are determined by the plan's objectives and the issues addressed. Similarly those most affected should decide what is a balance between the level of precision needed, costs and community interest within a monitoring or evaluation programme.

5.8 Chapter summary

This chapter focused on taking the theory and concepts of coastal management and planning discussed in Chapter 3 and the tools for management described in Chapter 4, and translating them into a structured planning framework. We have shown the power of planning as a management tool. A hierarchical geographic planning framework was selected as the basis for structuring this chapter for a number of reasons, most notably its wide-ranging application and generic nature. This approach to planning is not necessarily the answer for all circumstances. Consequently, we discussed alternative frameworks – strategic, operational, subject, integrated, statutory and non-statutory.

Much of this chapter has focused on how to develop integrated coastal plans from the whole-of-government level to site planning, how to integrate these plans vertically or horizontally and the advantages and disadvantages of each.

Finally, and perhaps most importantly to ensure that plans leave the shelf and become part of the ongoing management of coastal areas, we discussed their implementation and evaluation.

Chapter 6

Conclusions and future directions

In this book we have endeavoured to provide a fresh foundation for coastal planning and management by mixing theory with examples of best-practice from around the world. What has this approach told us about the current status of coastal planning and management; and what pointers has it given us to possible directions for the future?

The main theme of the book has been that the coast, with its intensity of land and water use, is a place where the issues of economic development and environmental management, and their interactions with social and cultural values, are brought into sharp relief. If there are problems with any of these issues, in any area of a coastal nation, the symptoms are likely to show up at the coast first.

Several other themes and principles emerge from the book. They are necessarily broad in scope, given the enormity of the issues and challenges facing coastal managers, but we summarize them in order to stimulate further discussion and research.

THE CENTRAL ROLE OF SUSTAINABLE DEVELOPMENT PRINCIPLES AND APPROACHES

Coastal programmes are now generally based on principles of sustainable development, the precautionary principle, and inter-generational equity. The challenge for coastal planners and managers is to transfer sustainable development principles into tangible management outcomes. We hope that the tools and techniques described in this book go some way towards meeting this challenge.

THE INSEPARABLE NATURE OF COASTAL PLANNING AND MANAGEMENT

Coastal planning and management activities are generally so strongly linked that in successful coastal programmes they are almost indistinguishable. The interweaving of planning and management to create a single coastal programme can help to break down institutional boundaries or possible professional rivalries between planners and managers, and is to be encouraged.

THE INCREASING EMPHASIS ON CONSENSUAL STYLES OF COASTAL PLANNING AND MANAGEMENT

Consensual plan production has become the most widely used approach for integrated coastal plans at the regional and local levels. Increased community empowerment and the problems caused by more directive planning styles have led to different community-based, collaborative and co-management methods of coastal management planning. Innovative consensus-building tools have to be used to ensure that consensus does not equate to 'lowest common denominator', resulting in bland outcomes. This is especially so where conflict arises, often in the case of siting hazardous and/or polluting industries. Early indications of the use of consensual styles of planning in these cases suggest that they can be successful when adequate resources are allocated to them, although they are yet to be used in extreme cases of conflict.

COASTAL MANAGEMENT IS A SHARED CONCERN

Responsibility for sustainably managing the coast is shared by all levels of government, from international to local, along with coastal users, residents, private companies and advocacy groups. Governments are increasingly realizing the long-term benefits of engaging all stakeholders on the coast in coastal programme development. This partnership approach is rapidly evolving from just a 'good idea' into a cornerstone of many coastal initiatives around the world.

GOOD COASTAL MANAGEMENT IS FOUNDED ON AN APPRECIATION OF LOCAL CULTURAL FACTORS

Western approaches to coastal planning and management, while successful in many countries, especially those with European land-tenure systems, may require modification if they are to be successfully integrated into local cultural settings. Traditional knowledge about coastal resources and their management can be invaluable in formulating management prescriptions. The bringing together of western and traditional management tools and techniques is showing increasing signs of success in many developing countries. Indeed, there are increasing signs of a genuine two-way flow of knowledge and experience in coastal management between developed and developing countries.

THE CROSSING OVER OF PLANNING AND MANAGEMENT TECHNIQUES

Coastal programmes have become a melting pot for various planning and management techniques which have crossed over from other disciplines. Land-use planning techniques, such as separating conflicting uses through zoning, blend with economic analysis and risk management, co-management and a host of other approaches to help address coastal issues. Coastal planners and managers are increasingly being encouraged to add to – and occasionally stir – this melting pot to find innovative ways of addressing coastal problems and opportunities.

DESIGNING A MIXED COASTAL PLANNING SYSTEM CAN BE SUCCESSFUL

Issues requiring coastal management and planning cut across jurisdictions, occur at widely different scales, and involve a diversity of stakeholders. No single plan can be expected to cope with all coastal issues, but management practices and plans can be substantially improved by mixing integrated coastal plans at different scales, orientations and statutory bases. Cascading planning systems designed to link broad strategic plans to detailed local planning initiatives are an example of such integration.

THE IMPORTANCE OF MEASURING SUCCESS

A plethora of coastal plans exists around the world, addressing vastly different issues, often in very different ways. But a common feature of most of these plans is the absence of quantitative evidence of their effectiveness – this despite the often considerable resources used in their formulation and implementation. Politicians, government departments and the public are increasingly expecting coastal programmes to provide clear demonstrations of success. Performance measures, evaluation criteria and success indicators have become part of the coastal manager's lexicon. Yet measuring how successful coastal programmes are is only just beginning in earnest. Coastal programme managers are increasingly required to include monitoring and evaluation measures in programme design at the outset – a difficult task without a set of commonly accepted coastal management performance measures.

Facing the future

Chapter 2 could invoke contrasting responses in the reader: pessimism at its its rather depressing list of often chronic problems, painting a not too bright future for the coast; or excitement and optimism about the challenges that these problems present. A realistic coastal planner/ manager is one who would absorb a little of both and plan to tackle pragmatically the major challenges facing the world's coast, while being creative and flexible in dealing with the inherent limitations of the workings of government and private sector bureaucracies. To this planner/manager we offer our Six Virtues of Coastal Planning: to seek, to understand, to develop, to link, to bring into mainstream, to sustain. And the challenges that go with them.

- To seek
  - the money and willingness to implement plans;
  - true economic values of coastal resources and implementing management responses which reflect those values;
  - an appropriate balance between traditional management practices and knowledge, and typical western approaches to coastal management;
  - the mechanisms which allow developing countries to sustainably exploit coastal resources and avoid many of the mistakes of developed countries;
  - optimal solutions to resource sharing on the coast, especially emerging industries and uses, such as recreational and tourism demands;
  - workable strategies for ensuring equitable access to coastal resources for all sections of society;
  - locally sustainable and tailored economic and social growth for the coastlines of developing countries.

- To understand
  - the values and expectations of all stakeholders in coastal management;
  - the role of traditional and user knowledge;
  - and deal with uncertainties in decision making;
  - the social and biophysical interrelationships between catchments and coasts and oceans;
  - the capacities required for coastal management, including training, monitoring and scientific studies.

- To develop
  - and maintain appropriate stewardship of coastal resources tailored to social and cultural settings of coastal nations;
  - meaningful indicators for the evaluation of coastal initiatives.

- To link
  - coastal initiatives at all scales – from international to local;
  - integrated and subject plans.

- To bring into mainstream
  - monitoring and evaluation at all stages of coastal management;
  - sustainable economic and social development.
- To sustain
  - community expectations after plans are completed;
  - the momentum going from Agenda 21, and related international initiatives;
  - the energy of local coastal managers.

Afterword

The enormous problems facing the world's coastlines are unlikely to diminish with time. Population increase, technological change, economic growth and ever more waste generation make it likely that the problems will become even more acute. The key question is thus not if, or when, these pressures will occur, but whether the coast can be managed to sustainably absorb them. And in this of course lies the fascination of being involved with the management of the coast – the huge challenge it presents to forge creative and innovative solutions to apparently intractable management problems.

We do not pretend with this book to have offered all the solutions, but rather to have provided a wide selection of methods and models to guide the search for environmentally, culturally and economically appropriate planning and management outcomes. We will judge our success by the extent to which we have stimulated the searchers and helped them to meet their challenges with optimism.

Appendix A

Some definitions of the coastal zone for planning and management

The following are definitions of the coastal zone used to define areas within which coastal management policies apply. Chapter 1 describes the advantages and disadvantages of each type of definition.

Fixed distance definitions

Country	Inland boundary	Ocean boundary
Australia (State of New South Wales)	1 km from LWM	3 n mile from coastal baseline
Brazil	2 km from MHW	12 km from MHW
Costa Rica	200 m from MHW	MLW
China	10 km from MHW	15 m isobath (depth)
Spain	500 m from highest storm or tide line	12 n mile (limit of territorial sea)
Sri Lanka	300 m from MHW	2 km from MLW

Example Fixed Definition Boundaries of the Coastal Zone (Coastal Committee of New South Wales, 1990; Sorensen and McCreary, 1990)

Sri Lanka Coast Conservation Act (1990)

The area lying within a limit of three hundred metres landward of the Mean High Water Line and a limit of two kilometres seaward of the Mean Low Water Line and in the case of rivers, streams, lagoons, or any other body of water connected to the sea either permanently or periodically, the landward boundary shall extend to a limit of two kilometres measured perpendicular to the straight line base line drawn between the natural entrance points (defined by the Mean Low Water Line) thereof and shall include waters of such rivers, streams and lagoons or any other body of water so connected to the sea.

Variable distance definitions

The South Pacific Regional Environment Programme (1993)

The coastal zone is a region of indeterminate and variable width. It extends from and includes, the wholly marine (i.e. the seabed, the overlying waters and their resources) to the wholly terrestrial (i.e. beyond the limits of marine incursion and the reach of salt spray). Linking these two environments is the tidal area which forms a transition between land and the sea.

Definition according to use

United States Federal Coastal Zone Management Act (1990) Section 304

(Note that each coastal State must interpret the Federal definition through the production of maps and charts):

The term 'coastal zone' means the coastal waters (including the lands therein and thereunder) and the adjacent shorelands (including the lands therein and thereunder), strongly influenced by each other and proximity to the shorelines of the several coastal states, and includes islands, transitional and intertidal areas, salt marshes, wetlands and beaches. The zone extends, in Great Lakes waters, to the international boundary between the United States and Canada and, in other areas, seaward to the outer limit of State title and ownership . . . [continues with list of Acts] . . . The zone extends inland from the shorelines only to the extent necessary to control shorelands, the uses of which have a direct and significant impact on the coastal waters.

Australian Commonwealth Coastal Policy (1995)

For the purpose and actions of the Commonwealth, the boundaries of the coastal zone are considered to extend as far inland and as far seaward as necessary to achieve the Coastal Policy objectives, with a primary focus on the land–sea interface.

United Kingdom Government Environment Committee
Report on Coastal Zone Protection and Planning (1992)

We conclude that definitions of the coastal zone may vary from area to area and from issue to issue, and that a pragmatic approach must therefore be taken at the appropriate national, regional or local level.

World Bank Environment Department (1993)

For practical planning purposes, the coastal zone is the **special area** [original bolding], endowed with special characteristics, of which the boundaries are often determined by the special problems to be tackled.

OECD Environment Directorate (1991, 1993)

What constitutes the coastal zone depends on the purpose at hand. From both the administrative and scientific viewpoints, the extent of the zone will vary depending on the nature of the problem. Accordingly, the boundaries of the coastal zone should extend as far inland and as far seaward as necessary to achieve the objectives of management.

Australian Commonwealth House of Representatives
Inquiry (1980)

Any definition of the coastal zone should be flexible, and should depend on the issue being confronted.

New South Wales Government Draft Revised Coastal
Policy (1994) – Option 5

an issues based definition where the boundaries of the coastal zone extend as far inland and as far seaward as necessary to achieve the policy's objectives, with a focus on the land sea interface . . .

Appendix B

Examples of texts, conference and workshop proceedings that outline coastal problems around the world

(Only the most recent published references of conference series are shown.)

- Asia-Pacific: various workshops and conferences (Chua and Pauly, 1989; McLean and Mimura, 1993; Hotta and Dutton, 1994).
- Africa (e.g. Sowman, 1993; Linden, 1994; World Bank, 1994; Kimani, 1995).
- Australasia: proceedings of the Coast to Coast (Australia) conference series (Kriwoken and McAdam, 1994; Harvey, 1996), the Institution of Engineers (Australia and New Zealand) (Australian Institute of Civil Engineers, 1993a,b).
- Europe (including Eastern Europe and Scandinavia): proceedings of the Eurocoast conferences (Taussik and Mitchell, 1997) and the European Union for Coastal Conservation (Healy and Doody, 1995; Jones et al., 1996).
- North America: proceedings of the Coastal Zone (e.g. Magoon et al., 1993) and Coastal Zone Canada (e.g. Coastal Zone Canada '94, 1994), conference series book of Beatley et al. (1993) and many individual articles in the journals *Coastal Management* and *Ocean and Coastal Management*.

In addition, there are conferences on various coastal management problems on a sector-by-sector or subject-by-subject basis.

References

Agardy, M. (1990) Integrating tourism in multiple use planning for coastal and marine protected areas. In *Proceedings of 1990 Congress on Coastal and Marine Tourism: A Symposium and Workshop on Balancing Conservation and Economic Development, Honolulu, Hawaii, 21–31 May*, (eds M. Miller and J. Auyong), pp. 204–210, National Coastal Resources Research and Development Institute, Newport, OR.

Alcala, A. and Russ, G. (1990) A direct test of the effects of protective management on abundance and yield of tropical marine resources. *Journal du Conseil International pour l'Exploration de la Mer*, **46**, 40–47.

Alcock, D. (1991) Education and extension: management's best strategy. *Australian Parks & Recreation*, **27**(1), 15–17.

Alder, J. (1993) Permits, an evolving tool in the day-to-day management of the Cairns section of the Great Barrier Reef Marine Park. *Coastal Management*, **21**(1), 25–36.

Alder, J. (1994) Have six years of public education changed community awareness of and attitudes towards marine park management? In *Proceedings of the 7th International Coral Reef Symposium*, pp. 1043–1051.

Alder, J. (1996) Education or enforcement – the better deal?. In *Great Barrier Reef Science, Use and Management*, Vol. II, p. 221, James Cook University, Townsville.

Alder, J., Sloan, N.A. and Uktolseya, H. (1994) A comparison of management planning and implementation in three Indonesian Marine Protected Areas. *Ocean and Coastal Management*, **24**, 179–198.

Alder, J., Raymakers, C., Laudi, S. and Purbasari, I. (1995a) Taka Bone Rate National Park – progressing towards community based management. *Coastal Management in Tropical Asia*, 8–11.

Alder, J., Sloan, N. and Uktolseya, H. (1995b) Advances in Marine Protected Area management in Indonesia: 1988–1993. *Ocean and Coastal Management*, **25**(1), 63–75.

Alder, J.A., Kay, R.C., Clegg, S. *et al.* (in preparation) *Coastal Planning Manual for Western Australia*, West Australian Ministry for Planning, Perth, Australia.

Alexander, E.R. (1986) *Approaches to Planning: Introducing Current Planning Theories, Concepts and Issues*, Gordon and Breach, New York.

Alix, J.C. (1989) Community-based resources management: the experience of the Central Visayas Regional Project-I. In *Coastal Area Management in Southeast Asia:*

Policies, Management Strategies and Case Studies (eds T.-E. Chua and D. Pauly), pp. 185–190, ICLARM, Manila, Philippines.

Ambraseys, N.N. (1983) Evaluation of seismic risk. In *Seismicity and Seismic Risk in the Offshore North Sea Area* (eds A.R. Ritsema and A. Gurpinar), pp. 317–345, Reidel, Dordrecht, The Netherlands.

AMCORD (1995) *A National Resource Document for Residential Development*, Australian Government Printing Service, Canberra.

Anderson, J.E., Brady, D.W., Bullock, C.S. and Stewart, J.S. (1984) *Public Policy and Politics in America*, Brooks Cole, Monterey, California.

Anon. (1997) Gill-netting banned to protect endangered dugong. *Courier Mail*, 31 March 1997, Brisbane.

Anthias (1994) *Anthias: The Ras Mohammed National Park Newsletter*, Egyptian Environmental Affairs Agency, Cairo.

Anutha, K. and Johnson, D. (1996) Aquaculture planning and coastal management in Tasmania. *Ocean and Coastal Management*, **33**(1–3), 167–192

Anutha, K. and O'Sullivan, D. (1994) *Aquaculture and Coastal Zone Management in Australia and New Zealand*, Turtle Press, Hobart, Tasmania.

Armstrong, M. (1986) *A Handbook of Management Techniques*, Kogan Page, London.

Arnstein, S. (1969) A ladder of citizen participation. *American Institute of Planners*, **35**, 216–224.

Asian Development Bank (1987) *Indonesia Environmental and Natural Resources Briefing Profile*, Environment Unit, Asian Development Bank, Manila, Philippines.

Asian Development Bank (1991a) *Environmental Evaluation of Coastal Zone Projects: Methods and Approaches*, Asian Development Bank, Manila, Philippines.

Asian Development Bank (1991b) *Remote Sensing and Geographic Information System for Natural Resource Management*, Asian Development Bank, Manila, Philippines.

Australian Institute of Civil Engineers (ed.) (1993) *12th Australasian Conference on Coastal and Ocean Engineering*, Australian Institute of Civil Engineers, Melbourne.

Baines, G. (1985) Coastal area conservation in Tropical Islands. *IUCN Bulletin*, **16**(7–9), 79–80.

Bandara, C.M.M. (1995) The Network for Environmental Training at Tertiary-Level in Asia and the Pacific (NETTLAP): an experience in environmental networking. In *Educating Coastal Managers* (eds B.R. Crawford, J.S. Cobb and C.L. Ming), pp. 119–122, University of Rhode Island, Coastal Resources Centre.

Bangda (1996) *Technical Manual for Preparation of a Provincial Coastal and Marine Management Strategy* (English version), Marine Resource Evaluation and Planning Project (MREP) Report, Government of Indonesia, Jakarta, 35 pp.

Bason, D. (1996) *Bom!*, The Nature Conservancy, Jakarta.

Bateman, I. (1995) Environmental and economic appraisal. In *Environmental Science for Environmental Management* (ed. T. O'Riordan), pp. 45–65, Longman, Harlow.

Beatley, T., Brower, D.J. and Schwab, A.K. (1993) *An Introduction to Coastal Zone Management*, Island Press, Washington DC.

Bergin, T. (1993) Marine and Estuarine Protected Areas (MEPAs): where did

Australia get it wrong? In *Proceedings of the Fourth Fenner Conference on the Environment 9–11 October 1991, Sydney, Australia*, pp. 148–152, IUCN.

Berkes, F. (ed.) (1989) *Common Property Resources: Ecology and Community-Based Sustainable Development*, Belhaven Press, London.

Beukenkamp, P., Gunther, P., Klein, R. *et al.* (eds) (1993) *Proceedings World Coast '93*, National Institute for Coastal and Marine Management, Coastal Zone Management Centre, Noordwijk, The Netherlands.

Bird, P. and Rapport, D. (1986) *State of the Environment Report for Canada*, Statistics Canada, Ottawa.

Black, P.F. (1981) *Environmental Impact Analysis*, Praeger, New York.

Blair, S. and Truscott, M. (1989) Cultural landscapes – their scope and their recognition. *Historic Environment*, **7**(2), 3–8.

Blowers, A. (ed.) (1993) *Planning for a Sustainable Environment*, Earthscan–Town and Country Planning Association, London.

Boelaert-Suominen, S. and Cullinan, C. (1994) *Legal and Institutional Aspects of Integrated Coastal Area Management in National Legislation*, Development Law Service Legal Office, Food and Agriculture Organization of the United Nations, Rome.

Born, S.M. and Miller, A.H. (1988) Assessing networked coastal zone management programs. *Coastal Management*, **16**, 229–243.

Bower, B.T. (1992) *Producing Information for Integrated Coastal Management Decisions: An Annotated Seminar Outline*, NOAA, Washington, DC.

Brandon, K. (1996) *Ecotourism and Conservation: A Review of Key Issues*, The World Bank, Washington, DC.

Bregman, J.I. and Mackenthun, K.M. (1992) *Environmental Impact Statements*, Lewis Publishers, Chelsea, MI.

British Medical Association (1988) *Living with Risk*, Wiley Interscience, London.

Brown, A.L. and McDonald, G.T. (1995) From environmental impact assessment to environmental design and planning. *Australian Journal of Environmental Management*, **2**, 65–77.

Brown, V.A. (1995) *Turning the Tide: Integrated Local Area Management for Australia's Coastal Zone*, Commonwealth Department of the Environment, Sports and Territories, Canberra.

Brush, R.O. (1976) Perceived quality of scenic and recreational environments: some methodological issues. In *Perceiving Environmental Quality: Research Applications* (eds K.H. Craik and E.H. Zube), pp. 47–58, Plenum Press, New York.

Bryner, G.C. (1987) *Bureaucratic Discretion: Law and Policy in Federal Regulatory Agencies*, Pergamon, New York.

Buckingham-Hatfield, S. and Evans, B. (1996a) Achieving sustainability through environmental planning. In *Environmental Planning and Sustainability* (eds S. Buckingham-Hatfield and B. Evans), pp. 1–18, Wiley, Chichester, UK.

Buckingham-Hatfield, S. and Evans, B. (ed.) (1996b) *Environmental Planning and Sustainability*, Wiley, Chichester, UK.

Buhat, D. (1994) Community-based coral reef fisheries management, Philippines. Unpublished manuscript.

Burby, R.J., Cigler, B.A., French, S.F. *et al.* (1991) *Sharing Environmental Risks: How to Control Governments' Losses from Natural Disasters*, Westview Press, Boulder.

Burton, I., Kates, R.W. and White, R.F. (1978) *The Environment as Hazard*, Oxford University Press, New York.

Butler, R.W. (1980) The concept of a tourist area cycle of evolution: implications for management of resources. *Canadian Geographer*, **24**(1), 5–12.

Callan, S. and Thomas, J.M. (1996) *Environmental Economics and Management: Theory, Policy, and Applications*, Irwin, Chicago.

CALM (1997) *Reading the Remote*, West Australian Government Department of Conservation and Land Management, Perth, Australia.

Cameron, J. (1991) The precautionary principle: a fundamental principle of law and policy for the protection of the global environment. *Boston College of International and Comparative Law Review*, **14**(1), 1–27.

Campbell, S. and Fainstein, S. (eds) (1996) *Readings in Planning Theory*, Blackwell, Cambridge, Massachusetts.

Castledine, G. and Herrick, R. (1995) Subdivision: assessment of landscaped value and public interest. *Australian Environmental Law News*, **2**, 40–44.

Cesar, H. (1996) *Economic Analysis of Indonesian Coral Reefs*, Environment Department, World Bank, Washington, DC.

Chou, L.M. (ed.) (1994) *UNEP/NETTLAP Training and Resources Development Workshop in Coastal Zone Management, Kandy, Sri Lanka*, United Nations Regional Office for Asia and the Pacific, Bangkok, Thailand.

Christensen, C.R. (1982) *Business Policies and Case Studies*, 5th edn, R.D. Irwin, Homewood, Ill.

Christensin, H.H. (1987) Vandalism and depreciative behavior in Americans outdoors: legacy, the challenge, with case studies. In *The Report of the President's Commission on Americans Outdoors (US)*, Island Press, Washington DC.

Christie, P. and White, A. (1994) Community-based coral reef management on San Salvador Island, the Philippines. *Society and Natural Resources*, **7**, 103–117.

Chua, T.E. (ed.) (1991) *Coastal Area Management Education in the ASEAN Region*, International Center for Living Aquatic Resources Management, Manila, Philippines.

Chua, T.E. and Pauly, D. (ed.) (1989) *Coastal Area Management in Southeast Asia: Policies, Management Strategies and Case Studies: International Center for Living Aquatic Resources Management Conference Proceedings*, ICLARM, Manila.

Cicin-Sain, B. (1993) Sustainable development and integrated coastal zone management. *Ocean and Coastal Management*, **21**, 11–44.

City of Mandurah, Peel Development Commission and Western Australian Planning Commission (1993) *The Mandurah Coastal Strategy Technical Appendix, June 1993*, Western Australian Planning Commission, Perth, Australia.

Clark, J.R. (1977) *Coastal Ecosystem Management: A Technical Manual for the Conservation of Coastal Zone Resources*, John Wiley, New York.

Clark, J.R. (1996) *Coastal Zone Management Handbook*, CRC Press, Boca Raton.

Clark, R.N. and Stankey, G.H. (1979) *The Recreation Opportunity Spectrum: A Framework for Planning, Management and Research*, USDA Forest Service, Seattle, Washington.

Cleary, J. (1997) An 'all of government' approach to landscape management in Western Australia. Unpublished report, West Australian Government Department of Conservation and Land Management, Perth, Australia.

Coast Conservation Department (1996) *Revised Coastal Management Plan, Sri Lanka 1996–2000*, Coast Conservation Department, Colombo, Sri Lanka.

Coastal Committee of New South Wales (1990) *The NSW Coast: Have Your Say in its Future – NSW Coast Government Policy 1990*, Government of New South Wales, Sydney.

Coastal Committee of New South Wales (1994) *Draft Revised Coastal Policy for New South Wales*, Government of New South Wales, Sydney.

Coastal Zone Canada '94 (1994) Conference Statement and Call for Action. In *Coastal Zone Canada '94, Halifax, Nova Scotia, September 21–23*.

Coccossis, H. and Nijkamp, P. (ed.) (1995) *Sustainable Tourism Development*, Avebury, Aldershot.

Colebatch, H.K. (1993) Policy-making and volatility: what is the problem? In *Policy-making in Volatile Times* (eds A. Hede and S. Prasser), pp. 29–46, Hale & Iremonger, Sydney.

Coles, T.F. and Tarling, J.P. (1991) *Environmental Assessment: Experience to Date*, Institute of Environmental Assessment, Lincolnshire, UK.

Commonwealth House of Representatives Standing Committee on Environment and Conservation (1980) *Australian Coastal Zone Management*, Australian Government Publishing Service, Canberra.

Commonwealth of Australia (1992) *National Strategy for Ecologically Sustainable Development*, Australian Government Publishing Service, Canberra.

Commonwealth of Australia (1995) *Living on the Coast: The Commonwealth Coastal Policy*, Australian Government Printing Service, Canberra.

Considine, M. (1994) *Public Policy: A Critical Approach*, Macmillan Education Australia, South Melbourne.

Cornforth, R. (1992) Bridging the gap: the distance between customary and parliamentary law making in the management of natural resources in Western Samoa. In *SPREP/UNEP Workshop on Strengthening Environmental Legislation in the South Pacific, 23–27 November 1992, Apia, Western Samoa*.

Costannza, R.E. (1992) *Ecological Economics*, Columbia University Press, New York.

Couper, A. (ed.) (1983) *The Times Atlas of the Oceans*, Times Books, London.

Court, J., Wright, C. and Guthrie, A. (1994) *Assessment of Cumulative Impacts and Strategic Assessment in Environmental Impact Assessment*, Commonwealth Environmental Protection Agency, Canberra, Australia.

Crawford, B., Cobb, J.S. and Friedman, A. (1993) Building capacity for integrated coastal management in developing countries. *Ocean and Coastal Management*, **21**, 311–337.

Crocombe, R. (ed.) (1995) *Customary Land Tenure and Sustainable Development: Complementarity or Conflict?*, South Pacific Commission and University of the South Pacific, Noumea and Suva.

Cronbach, L.J. (1992) *Designing Evaluations of Educational and Social Programs*, Jossey Bass, San Francisco.

CSIRO (1996) *Australia: State of the Environment 1996*, Commonwealth Scientific and Industrial Research Organisation, Melbourne.

Cullingworth, J.B. (1993) *The Political Culture of Planning: American Land Use Planning in Comparative Perspective*, Routledge, New York.

Daniel, T.C. (1976) Criteria for development and application of perceived environmental indices. In *Perceiving Environmental Quality: Research Applications* (eds K.H. Craik and E.H. Zube), pp. 27–46, Plenum Press, New York.

Davis, G., Wanna, J., Warhurst, J. and Weller, P. (1993) *Public Policy in Australia*, 2nd edn, Allen & Unwin, Sydney.

Dean, R.F. (1988) Realistic economic benefits from beach nourishment. *Coastal Engineering*, ASCE. In Proceedings 21st International Conference on Coastal Engineering, 1158–1172.

deGorges, A.P. (1990) *An Environmental Appraisal of Impacts from Dredge Fill and Reclamation on Coastal Marine Waters and Coral Reefs of Mahe, Seychelles*, USAID/REDSO/ESA, Washington, DC.

deHaven-Smith, L. and Wodraska, J.R. (1996) Consensus-building for integrated resource management. *Public Administration Review*, **56**(4), 367–371.

Department of Information (1992) *Indonesia 1992: An Official Handbook*, Republic of Indonesia, Jakarta, Indonesia.

Department of Planning and Urban Development (1993) *Public Participation in the Planning Process*, Western Australian Government Department of Planning and Urban Development, Perth, Australia.

Department of Resources Development (1994) *Working With Communities: A Guide for Proponents*, Western Australian Government Department of Resources Development, Perth, Australia.

Ditton, R.B., Seymour, J.L. and Swanson, G.C. (1978) *Coastal Resources Management: Beyond Bureaucracy and the Market*, Lexington Books, Lexington, MA.

Dixon, J.A. and Sherman, P.B. (1990) *Economics of Protected Areas*, Earthscan Publications, London.

Dolan, R. and Goodell, H.G. (1986) Sinking cities. *American Scientist*, **74**, 38–47.

Donaldson, B., Eliot, I. and Kay, R. C. (1994) *Review of Coastal Management in Western Australia: Issues and Options Paper*, Coastal Management Review Committee, Perth.

Donaldson, B., Eliot, I. and Kay, R.C. (1995) *Final Report of the Review of Coastal Management in Western Australia: A Report to the Minister for Planning*, Coastal Management Review Committee, Perth.

Dowling, R. and Alder, J. (1996) Shark Bay, Western Australia: managing a Coastal World Heritage Area. *Coastal Management in Tropical Asia*, March, 17–21.

Doxey, G.V. (1975) A causation theory visitor–resident irritations: methodology and research inferences. In *Proceedings of the Travel Research Association Sixth Annual Conference, San Diego, California*.

Drake, S.F. (1996) The International Coral Reef Initiative: a strategy for the sustainable management of coral reefs and related ecosystems. *Coastal Management*, **24**(4), 279–300.

Drijver, C. and Sajise, P. (1993) Community-based resource management and environmental action research. In *Proceedings of the Experts' Workshop on Community Based Resource Management: Perspectives, Experiences and Policy Issues, Los Banos, Philippines (UPLB)*, Environmental and Resource Management Project & UPLB.

Dutton, I.M. and Hotta, K. (1994) Introduction. In *Coastal Management in the Asia-Pacific Region: Issues and Approaches* (eds K. Hotta and I. M. Dutton), pp. 3–18, Japan International Marine Science and Technology Federation, Tokyo.

Edgren, G. (1993) Expected economic and demographic developments in coastal zones world wide. In *World Coast '93* (eds P. Beukenkamp, P. Gunther, R. Klein *et al.*), pp. 367–370, National Institute for Coastal and Marine Management, Coastal Zone Management Centre, Noordwijk, The Netherlands.

Edwards, S.F. (1987) *An Introduction to Coastal Zone Economics: Concepts, Methods and Case Studies*, Taylor & Francis, New York.

Ehler, C. (1995) Integrated coastal ocean space management: challenges for the next decade. In *Coastal Ocean Space Utilization III*, (eds N.D. Croce, S. Connell and R. Abel), pp. 175–188, E. & F.N. Spon, London.

Ehler, C. and Basta, D. (1993) Integrated management of coastal areas and marine sanctuaries. *Oceanus*, **36**(3), 6–14.

Elder, D. (1993) International developments in marine conservation and the World Conservation Union Marine Agenda. In *Proceedings of the Fourth Fenner Conference on the Environment, Sydney, Australia*, pp. 30–35, IUCN.

English Nature (1996a) *Thames Estuary Management Plan: Draft for Consultation*, English Nature, London.

English Nature (1996b) *Thames Estuary Mangement Plan: Strategy for Implementation*, English Nature, London.

Environment Committee (1992) *Coastal Zone Protection and Planning. House of Commons Environment Select Committee: Volume I, Report, Together with the Proceedings of the Committee Relating to the Report*, Report 17-I, HMSO, London.

Environmental Protection Authority (1992) *State of the Environment Report*, Government of Western Australia Environmental Protection Authority, Perth.

Erickson, N.J. (1986) *Creating Flood Disasters? New Zealand's Need for a New Approach to Urban Flood Hazard*, National Water and Soil Conservation Authority, Wellington.

ESCAP (1990) *State of the Environment in Asia and the Pacific*, United Nations Economic and Social Commission for Asia and the Pacific, Bangkok, Thailand.

Evans, M.D. and Burgen, B. (1992) *The Economic Value of Adelaide Beaches*, Centre for South Australian Economic Studies (The University of Adelaide and Flinders University), Adelaide.

Fabbri, P. (ed.) (1990) *Recreational Uses of Coastal Areas: A Research Project of the Commission on the Coastal Environment, International Geographical Union*, Kluwer Academic, Dordrecht.

Fabos, J.G. and McGregor, A. (1979) *Assessment of Visual/Aesthetic Landscape Qualities*, Centre For Environmental Studies, University of Melbourne, Melbourne.

Faludi, A. (ed.) (1973) *A Reader in Planning Theory*, Urban and Regional Planning Series, Vol. 5, Pergamon, Oxford.

Feeny, D., Berkes, F., McKay, B. and Acheson, J. (1990) The tragedy of the commons: twenty-two years later. *Human Ecology*, **18**(1), 1–19.

Feldman, M. (1989) *Order without Design: Information Production and Policy Making*, Stanford University Press, Stanford, California.

Ferrer, E.M. (1992) *Learning and Working Together: Towards a Community-Based Coastal Resources Management*, College of Social Work and Community Development, University of Philippines, Quezon City, Philippines.

Field, B.C. (1994) *Environmental Economics*, McGraw-Hill, New York.

Fisk, G.W. (1996a) Integrated coastal management in developed countries: the case of Australia. Unpublished Masters thesis, School of Marine Policy, University of Delaware, Newark, Delaware.

Fisk, G.W. (1996b) [Unpublished material supplied by the author to this book.]

Folmer, H., Gabel, H.L. and Opschoor, H. (ed.) (1995) *Principles of Environmental and Resource Economics*, New Horizons in Environmental Economics, Edward Elgar, Cheltenham, England.

Frassetto, R. (ed.) (1989) *Impact of Sea-Level Rise on Cities and Regions. Proceedings of the First International Meeting 'Cities on Water' Venice, December 11–13 1989*, p. 238, Centro Internazionale Citta d'Acqua, Venice, Italy.

Gascoyne Development Commission (1996) *Gascoyne Aquaculture Development Plan*, Western Australian Government Department of Fisheries and Gascoyne Development Commission, Perth, Australia.

Gavaghan, H. (1990) The dangers faced by ships in port. *New Scientist*, September, 19.

Gerrard, S. (1994) Managing risks in the coastal zone: assessment, perception and communication. In *Coast to Coast '94*, pp. 82–89, Government of South Tasmania, Hobart, Tasmania.

Gerrard, S. (1995) Environmental risk management. In *Environmental Science for Environmental Management* (ed. T. O'Riordan), pp. 296–316, Longman, Harlow.

Gerrard, S. (1997) [Unpublished material supplied by the author to this book.]

GESAMP (1996) *The Contributions of Science to Integrated Coastal Management*, FAO, Rome.

Gilpin, A. (1995) *Environmental Impact Assessment: Cutting Edge for the Twenty-First Century*, Cambridge University Press, Cambridge.

Glasson, J., Therivel, R. and Andrew, C. (1994) *Introduction to Environmental Impact Assessment*, UCL Press, London.

Global Vision (1996) *Preparation Document for the Awareness and Participation Component of COREMAP: Draft No. 2*, Global Vision Inc., Silver Springs, MD.

Godschalk, D.R. (1992) Implementing coastal zone management: 1972–1990. *Coastal Management*, **20**, 93–116.

Godschalk, D.R., Brower, D.J. and Beatley, T. (1989) *Catastrophic Coastal Storms: Hazard Mitigation and Development Management*, Duke University Press, Durham, North Carolina.

Goldberg, E.B. (1994) *Coastal Zone Space – Prelude to Conflict?*, UNESCO, Paris.

Goldin, I. and Winters, L.A. (eds) (1995) *The Economics of Sustainable Development*, OECD Centre for Economic Policy Research, Cambridge University Press, Cambridge, UK.

Goodhead, T. and Johnson, D. (eds) (1996) *Coastal Recreation Management: The Sustainable Development of Maritime Leisure*, E. & F.N. Spon, London.

Government of Australia (1975) *Great Barrier Reef Marine Park Act 1975*, Government of Australia, Canberra, Australia.

Government of Victoria (1990) *Making the Most of the Bay: A Plan for the Protection and Development of Port Phillip and Corio Bays*, Government of Victoria, Melbourne.

Gray, W.M. (1968) Global review of the origin of tropical disturbances and storms. *Monthly Weather Review*, **96**, 669–700.

Grigalunas, T.A. and Congar, R. (1995) *Environmental Economics for Integrated Coastal Area Management: Valuation Methods and Policy Instruments*, United Nations Environment Programme, Nairobi, Kenya.

Grimble, A. (1972) *Migrations, Myths and Magic from the Gilbert Islands*, Routledge and Kegan Paul, London.

Gubbay, S. (1989) *Coastal and Sea Use Management: A Review of Approaches and Techniques*, Marine Conservation Society, Ross-on-Wye, UK.

Gubbay, S. (1994) Local authorities and integrated coastal zone management plans. *Marine Update: Newsletter of the World Wide Fund for Nature*, (June), 1–4.

Gubbay, S. (ed.) (1995) *Marine Protected Areas*, Conservation Biology Series, Chapman & Hall, London.

Haar, C.M. (1977) *Land-Use Planning: A Casebook on the Use, Misuse and Re-use of Urban Land*, 3rd edn, Little, Brown, Boston.

Haeruman Js, H. (1988) Conservation in Indonesia. *Ambio*, **17**(2), 218–222.

Hall, D., Hebbert, M. and Lusser, H. (1993) The planning background. In *Planning for a Sustainable Environment* (ed. A. Blowers), pp. 19–35, Earthscan–Town and Country Planning Association, London.

Harger, J.R.E. (1986) Community structure as a response to natural and man-made environmental variable in the Pulau Seribu Island chain. In *Proceedings of MAB-COMAR Regional Workshop on Coral Reef Ecosystems: Their Management Practices and Research/Training Needs* (ed. Soemodihardjo), pp. 34–85, UNESCO: MAB-COMAR and LIPI, Jakarta.

Harvey, N.C. (ed.) (1996) *Proceedings of the Coast to Coast '96 Conference*, University of Adelaide, Adelaide, 16–19 April.

Haub, C. (1996) Future global population growth. In *Population Growth and Environmental Issues* (eds S. Ramphal and S.W. Sinding), pp. 53–62, Praeger, Westport, Connecticut.

Hawkins, J.P. and Roberts, C.M. (1992) Can Egypt's coral reefs support ambitious plans for diving tourism? In *7th International Coral Reef Symposium*, pp. 1007–1013, University of Guam Marine Laboratory, Guam, USA.

Hawkins, J.P. and Roberts, C.M. (1993) The growth of coastal tourism in the Red Sea: present and possible future effects on coral reefs. In *Proceedings of the Colloquium on Global Aspects of Coral Reefs: Health, Hazards and History*, University of Miami, Miami, Florida.

Hay, J. and Ming, C.L. (ed.) (1993) *Contributions to Training in Coastal Zone Management in the Asia-Pacific Region and Report of the First NETTLAP Resources Development Workshop for Education and Training at Tertiary Level in Coastal Zone Management*, United Nations Regional Office for Asia and the Pacific, Bangkok, Thailand.

Hay, J.E. (1994) Education, training and networking in coastal zone management. In *Coastal Management in the Asia-Pacific Region: Issues and Approaches* (eds K. Hotta and I.M. Dutton), pp. 133–152, Japan International Marine Science and Technology Federation, Tokyo.

Hay, J.E. and Kay, R.C. (1993) Possible future directions for integrated coastal zone management: a discussion paper. In *Intergovernmental Panel on Climate Change, Coastal Zone Management Sub-Group, Eastern Hemisphere Workshop, Tsukuba, Japan, 3–6 August* (eds R.F. McLean and N. Mimura), pp. 351–366, Secretariat of the Eastern Hemisphere Workshop on Sea-level Rise and Coastal Zone Management.

Hay, J.E., Chou, L.M., Sharp, B. and Thom, N.G. (ed.) (1994) *Environmental and*

Related Issues in the Asia-Pacific Region: Implications for Tertiary-Level Environmental Training, United Nations Environment Programme (UNEP), Regional Office for Asia and the Pacific (ROAP), Network for Environmental Training at Tertiary Level in Asia and the Pacific (NETTLAP), Bangkok, Thailand.

Heady, F. (1996) *Public Administration: A Comparative Perpective*, 5th edn, Marcel Dekker, New York.

Healy, I. and Doody, P. (eds) (1995) *Directions in European Coastal Management: Proceedings of the European Union for Coastal Conservation, 3–7 July 1995*, Samara Publishing, Swansea, UK.

Hedgcock, D. and Yiftachel, O. (ed.) (1992) *Urban and Regional Planning in Western Australia*, Paridigm Press, Perth, Western Australia.

Herman, J.L., Morris, L.L. and Fitz-Gibbon, C.T. (1987) *Evaluator's Handbook*, Sage, Newbury Park, CA.

Hershey, T. and Wilson, J. (1991) *Dynamite Fishing*, Melanesian Environment Foundation and Papua New Guinea Integrated Human Development Trust, Port Moresby, Papua New Guinea.

Hikkaduwa Special Area Management and Marine Sanctuary Coordination Committee (1996) *Special Area Management Plan for Hikkaduwa Marine Sanctuary and Surrounding Area*, Coastal Resources Management Project, Coast Conservation Department, National Aquatic Resources Agency, Colombo.

Hildebrand, L.P. and Norrena, E.J. (1992) Approaches and progress toward effective integrated coastal zone management. *Marine Pollution Bulletin*, **25**(1–4), 94–97.

Hinrichsen, D. (1994) Coasts under pressure. *People and the Planet*, **3**(1), 6–9.

Hinrichsen, D. (1996) *Coasts Under Pressure*, http://www.enews.com/magazines/issues/current/960601- 001.html

HMSO (1988) *The Tolerability of Risk at Nuclear Power Stations*, Health and Safety Executive, Her Majesty's Stationery Office, London.

Holling, C.S. (1978) *Adaptive Environmental Assessment and Management*, John Wiley and Sons, New York.

Hossain, H., Dodge, C.P. and Abed, F.H. (eds) (1992) *From Crisis to Development: Coping with Disasters in Bangladesh*, University Press, Dhaka, Bangladesh.

Hotta, K. and Dutton, I.M. (eds) (1994) *Coastal Management in the Asia-Pacific Region: Issues and Approaches*, Japan International Marine Science and Technology Federation, Tokyo.

Houghton, J.T., Meirs Filho, L.G., Callander, B.A. *et al.* (eds) (1996) *Climate Change 1995: The Science of Climate Change*, Contribution of Working Group I to the Second Assessment Report of the Intergovernmental Panel on Climate Change, Cambridge University Press, Cambridge.

House, P.W. and Shull, R.D. (1988) *Rush to Policy: Using Analytic Techniques in Public Sector Decision Making*, Transaction Books, New Brunswick, NJ.

Hussey, D.E. (1991) *Introducing Corporate Planning: Guide to Strategic Management*, 4th edn, Pergamon, Oxford.

Huxhold, W.E. and Levinsohn, A.G. (1995) *Managing Geographic Information System Projects*, Oxford University Press, New York.

Imperial, M.T. and Hennessey, T.M. (1996) An ecosystem-based approach to managing estuaries: an assessment of the National Estuary Program. *Coastal Management*, **24**(2), 115–140.

Imperial, M.T., Robadue Jr, D. and Hennessey, T.M. (1992) An evolutionary perspective on the development and assessment of the National Estuary Program. *Coastal Management*, **20**(4), 311–342.

Inder, A. (1997) Partnership in planning and management of the Solent. In *Partnership in Coastal Zone Management: Proceedings of the Eurocoast '96 Conference* (eds J. Taussik and J. Mitchell), pp. 405–412, Samara Publishing, Swansea, UK.

Inmon, W.H. and Hackathorn, R.D. (1994) *Using the Data Warehouse*, Wiley, New York.

Innes, J.E. (1996) Planning through consensus building: a new view of the comprehensive planning ideal. *Journal of the American Planning Association*, **62**(4), 460–472.

Institute of Environmental Assessment (UK) and The Landscape Institute (UK) (1995) *Guidelines for Landscape and Visual Impact*, E. & F.N. Spon, London.

Intercoast (1995) Coastal management initiatives in Kenya and Zanzibar. *Intercoast Network*, **22**, 1–2.

International Offshore Oil and Natural Gas Exploration and Production Industry (1996) *Decommissioning Offshore Oil and Gas Installations: Finding the Right Balance*, International Offshore Oil and Natural Gas Exploration and Production Industry, London.

IPCC – Intergovernmental Panel on Climate Change (1990) *Strategies for Adaption to Sea Level Rise*, Ministry for Transport and Public Works, The Hague, The Netherlands.

IPCC (1992) *Global Climate Change and the Challenge of the Rising Sea*, Ministry for Transport and Public Works, The Hague, The Netherlands.

Iqbal, M.S. (1992) *Assessment of the Implementation of the East African Action Plan and the Effectiveness of its Legal Instruments*, UNEP, Nairobi, Kenya.

IWICM (1996) Enhancing the success of integrated coastal management: good practices in the formulation, design and implementation of integrated coastal management initiatives. In *International Workshop on Integrated Coastal Management in Tropical Developing Countries: Lessons Learned from Successes and Failures, Xiamen, China, 24–28 May*, p. 21, GEF/UNDP/IMO Regional Programme for the Prevention and Management of Marine Pollution in the East Asian Seas and Coastal Management Centre, Quezon City, Philippines.

Jackson, E.L. and Burton, I. (1978) The process of human adjustment to earthquake risk. In *The Assessment and Mitigation of Earthquake Risk*, pp. 241–260, UNESCO, Paris.

Jacobs, M. (1991) *The Green Economy: Environment, Sustainable Development and the Politics of the Future*, Pluto Press, London.

Jentoft, S. (1989) Fisheries co-management: delegating government responsibility to fisherman's organizations. *Marine Policy*, **13**(2), 137–154.

Johannes, R. (ed.) (1989) *Traditional Ecological Knowledge: A Collection of Essays*, International Union for the Conservation of Nature, Gland, Switzerland and Cambridge, UK.

Johannes, R.E. (1984) Traditional conservation methods and protected areas in Oceania. In *Proceedings on the World Congress on National Parks*, pp. 344–347, Smithsonian Institution, Washington DC.

Johannes, R.E. (1995) Fishery for live reef food fish is spreading death and environmental degradation. *Coastal Management in Tropical Asia*, **5**, 8–9.

Johnson, D. (1974) *The Alps at the Crossroads*, Victorian National Parks Association, Victoria, Australia.

Johnston, C. (1989) Whose views count? Achieving community support for landscape conservation. *Historic Environment*, **7**(2), 33–37.

Jones, G., Prentice, D.R. and Stansfield, J.T. (1995) The management of safety in marine transportation. In *Recent Advances in Marine Science and Technology '94*, pp. 581–589, James Cook University, Townsville.

Jones, P.S., Healy, M.G. and Williams, A.T. (1996) Studies in European Coastal Management. In *Proceedings of Directions in European Coastal Management, 3–7 July, 1995*, Samara Publishing, Swansea, UK.

Jones, V. and Westmacott, S.E. (1993) *Management Arrangements for the Development and Implementation of Coastal Zone Management Programmes*, National Institute for Coastal and Marine Management, Coastal Zone Management Centre, The Netherlands, Noordwijk.

Jubenville, A., Twight, B.W. and Becker, R.H. (1987) *Outdoor Recreation Management: Theory and Application*, Venture Publishing, State College, PA.

Kahawita, B.S. (1993) *Coastal Zone Management in Sri Lanka*. In *World Coast '93* (eds P. Beukenkamp, P. Gunther, R. Klein *et al.*), pp. 669–675, National Institute for Coastal and Marine Management, Coastal Zone Management Centre, Noordwijk, The Netherlands.

Kaluwin, C. (1996) ICM takes different approach in Pacific Islands. *Intercoast Network*, **27**(Summer), 4,10.

Kausher, A., Kay, R.C., Asaduzzaman, M. and Paul, S. (1994) *Climate Change and Sea-Level Rise: The Case of the Bangladesh Coast*, Bangladesh Unnayan Parishad, Dhaka; Centre for Environmental and Resource Studies, New Zealand; Climatic Research Unit, England; Dhaka, Bangladesh.

Kausher, A., Kay, R.C., Asaduzzaman, M. and Paul, S. (1996) Climate change and sea-level rise: the case of the Bangladesh Coast. In *The Implications of Climate Change and Sea-Level Change for Bangladesh* (eds R.A. Warrick and Q.K. Ahmad), pp. 335–396, Kluwer Academic, Dordrecht.

Kay, R.C. and Lester, C. (1997) Benchmarking Australian coastal management. *Coastal Management*, **25**(3), 265–292.

Kay, R.C., Cole, R.G., Elisara-Laulu, F.M. and Yamada, K. (1993) *Assessment of Coastal Vulnerability and Resilience to Sea-level Rise and Climate Change. Case Study: 'Upolu Island, Western Samoa. Phase I: Concepts and Approach*, South Pacific Regional Environment Programme (SPREP), Apia, Western Samoa.

Kay, R.C., Ericksen, N.J., Foster, G.A. *et al.* (1994) *Assessment of Coastal Hazards and their Management for Selected Parts of the Coastal Zone Administered by the Tauranga District Council*, University of Waikato Centre for Environmental and Resource Studies, Waikato, Auckland, New Zealand.

Kay, R.C., Carman-Brown, A. and King, G. (1995) Western Australian experiences in preparing and implementing coastal management plans: some implications for shoreline management planning in England and Wales. In *Proceedings of the Ministry of Agriculture, Fisheries and Food Annual Conference of River and Coastal Engineers, Keele, United Kingdom, 6–7 July 1995*, pp. 8.1.1–8.1.15.

Kay, R.C., Eliot, I., Caton, B. *et al.* (1996a) A review of the Intergovernmental Panel on Climate Change's *Common Methodology for Assessing the Vulnerability of Coastal Areas to Sea-Level Rise. Coastal Management*, **24**, 165–188.

Kay, R.C., Kirkland, A. and Stewart, I. (1996b) Planning for future climate change and sea-level rise induced coastal change in Australia and New Zealand. In *Greenhouse: Coping With Climate Change* (eds W.J. Bouma, G.I. Pearman and M.M. Manning), pp. 377–398, CSIRO Publishing, Wellington, New Zealand.

Kay, R.C., Panizza, V., Eliot, I.E. and Donaldson, B. (1997) Reforming coastal management in Western Australia. *Ocean and Coastal Management*, **35**(1), 1–29.

Keeley, D. (1994) Balancing coastal resource use: in search of sustainability. In *Coast to Coast '94, Hobart, Tasmania* (eds L. Kriwoken and S. McAdam), pp. 127–133, Tasmanian Department of Environment and Land Management, Hobart, Tasmania.

Kelleher, G. (1993) Progress towards a global system of Marine Protected Areas. In *Proceedings of the Fourth Fenner Conference on the Environment, Sydney, Australia*, pp. 36–38, IUCN.

Kelly, P.M., Granich, S.L.V. and Secrett, C.M. (1994) Global warming: responding to an uncertain future. *Asia Pacific Journal on Environment and Development*, **1**(1), 28–45.

Kenchington, R.A. (1990) *Managing Marine Environments*, Taylor & Francis, New York.

Kenchington, R. (1993) Tourism in coastal and marine environments – a recreational perspective. *Ocean and Coastal Management*, **19**, 1–16.

Kenchington, R.A. and Crawford, D. (1993) On the meaning of integration of coastal zone management. *Ocean and Coastal Management*, **21**, 109–127.

Kennedy, K. (1996a) From planning to implementation: lessons from San Fransisco Bay. Unpublished report to English Nature, London.

Kennedy, K. (1996b) [Unpublished material supplied by the author to this book.]

Ketchum, B.H. (ed.) (1972) The water's edge: critical problems of the coastal zone. In *Proceedings of the Coastal Zone Workshop, Woods Hole, Massachusetts, 22 May–3 June 1972*, MIT Press, Cambridge, MA.

Kimani, E.N. (1995) Coral reef resources of East Africa: Kenya, Tanzania and the Seychelles. *NAGA The ICLARM Quarterly*, **18**(4), 4–7.

King, G. (1996) [Unpublished material supplied by the author to this book.]

King, G. and Bridge, L. (1994) *Directory of Coastal Planning and Management Initiatives in England*, National Coasts and Estuaries Advisory Group, Maidstone, Kent.

King, J.A., Morris, L.L. and Fitz-Gibbon, C.T. (1987) *How to Assess Program Implementation*, Sage, Newbury Park, CA.

Kirkby, J., O'Keefe, P. and Timberlake, L. (ed.) (1991) *The Earthscan Reader in Sustainable Development*, Earthscan, London.

Knecht, R., Cicin-Sain, B. and Fisk, G.W. (1996) Perceptions on the performance of State coastal zone management programs in the United States. *Coastal Management*, **24**(2), 141–164.

Kraus, R.G. and Curtis, J.E. (1986) *Creative Management Recreation, Parks and Lesiure Services*, Times Mirror/Mosby, St Louis.

Kriwoken, L. and McAdam, S. (eds) (1994) *Proceedings of Coast to Coast '94*, Tasmanian Department of Environment and Land Management, Hobart, Tasmania.

Lange, E. (1994) Integration of computerised visual simulation and visual assessment in environmental planning. *Landscape and Urban Planning*, **30**, 99–112.

Latin, H.A. (1993) Reef conservation: disciplinary conflicts between scientists and policymakers. In *Proceedings of the Seventh International Coral Reef Symposium, Guam*, pp. 1101–1108, University of Guam Marine Laboratory, Mangilao, Guam.

Laurie, I.C. (1975) Aesthetic factors in visual evaluation. In *Landscape Assessment: Values, Perceptions and Resources* (eds E.H. Zube, R.O. Brush and J.G. Fabos), pp. 102–117, Dowden, Hutchinson and Ross, Stroudsberg, Penn.

Leung, H.L. (1989) *Land Use Planning Made Plain*, Ronald P. Fryre & Co., Kingston, Canada.

Lieber, S.R. and Fesenmaier, D.R. (eds) (1983) *Recreation Planning and Management*, E. & F.N. Spon, London.

Lim, C.P., Matsuda, Y. and Shigemi, Y. (1995) Co-management in marine fisheries: the Japanese experience. *Coastal Management*, **23**(3), 195–222.

Linden, O. (1994) Resolution on integrated coastal zone management in East Africa signed in Arusha, Tanzania. *AMBIO*, **22**(6), 408–409.

Lipton, D.W. and Wellman, K.F. (1995) *Economic Valuation of Natural Resources*, US National Oceanic and Atmospheric Administration, Coastal Ocean Program, Washington, DC.

Litton, R.B. (1979) Descriptive approaches to landscape analysis. In *Proceedings of Our National Landscape, A Conference on Applied Techniques for Analysis and Management of the Visual Resource, PSW Forest and Range Experiment Station*, USDA Forest Service, Berkeley, California.

Local Government Management Board (1995) *Action on the Coast*, Local Government Management Board, Luton, England.

Lowenthal, D. (1978) *Finding Valued Landscapes*, Institute of Environmental Studies, University of Toronto, Toronto.

Lowry, K. and Sadacharan, D. (1993) Coastal management in Sri Lanka. *Coastal Management in Tropical Asia*, **1**(September), 1–7.

Lowry, K. and Wickramaratne, H.J.M. (1987) Coastal area management in Sri Lanka. In *Ocean Yearbook 7*, pp. 263–293, University of Chicago, Chicago.

Mace, P.M. (1996) Developing and sustaining world fisheries resources: the state of science and management. In *Proceedings of the 2nd World Fisheries Congress, Brisbane* (eds D.A. Hancock, D.C. Smith, A. Grant and J.P. Beumer), pp. 1–20, CSIRO Publishing, Collingwood, Victoria.

MacEwen, A., and MacEwen, R. (1982) *National Parks: Conservation or Cosmetics?*, George Allen & Unwin, Sydney, Australia.

Magoon, O., Wilson, W.S., Converse, H. and Tobin, L.T. (eds) (1993) *Coastal Zone '93 – Proceedings*, pp. 615–623, American Society of Civil Engineers, New York.

Malafant, K. and Radke, S. (1995) The terabyte problem in environmental databases. In *Proceedings of the PACON Conference, Townsville, Australia* (eds O. Bellwood, H. Choat and N. Saxena), James Cook University.

Marris, C., Langford, I. and O'Riordan, T. (1997) *Integrating Psychological Approaches to Public Perceptions of Environmental Risks: Detailed Results from a Questionnaire Survey*, Centre for Social and Economic Research on the Global Environment, Norwich, UK.

Mathieson, A. and Wall, G. (1982) *Tourism: Economic and Social Impacts*, Longman, New York.

McCay, B. and Acheson, J. (1987) *The Question of the Commons: The Culture and Ecology of Communal Resources*, The University of Arizona Press, Tucson, Arizona.

McCloskey, M. (1979) Litigation and landscape aesthetics. In *Our National Landscape, A Conference on Applied Techniques for Analysis and Management of the Visual Resource, PSW Forest and Range Experiment Station*, USDA Forest Service, Berkeley, California.

McCool, S.F., Stankey, G.H. and Clark, R.N. (1985) *Proceedings – Symposium on Recreational Choice Behavior*, General Technical Report INT-184, US Forest Service, Missoula, Montana.

McHarg, I. (1969) *Design With Nature*, Natural History Press, Garden City, New York.

McLain, R.J. and Lee, R.G. (1996) Adaptive management: promises and pitfalls. *Environmental Management*, **20**(4), 437–448.

McLean, R. and Mimura, N. (1993) Vulnerability assessment to sea-level rise and coastal zone management. In *Proceedings of the IPCC Eastern Hemisphere Workshop*, p. 427, Tsukuba, Japan.

Meinig, D.W. (ed.) (1979) *The Interpretation of Ordinary Landscapes, Geographical Essays*, Oxford University Press, New York.

Mercer, D. (1995) *A Question of Balance: Natural Resources Conflicts and Issues in Australia*, The Federation Press, Sydney.

Miles, E.L. (1989) Concepts, approaches and applications in sea use planning and management. *Ocean Development and International Law*, **20**(3), 213–238.

Miller, M.L. (1993) The rise of coastal and marine tourism. *Coastal and Ocean Management*, **20**, 181–199.

Miller, M.L. and Auyomg, J. (1991) Tourism in the coastal zone: portents, problems and possibilities. In *Congress on Coastal and Marine Tourism, 21 31 May, Honolulu*, pp. 1–8, National Coastal Resources and Development Institute.

Ministry for Population and Environment (1992) *Strategy on Coral Reef Ecosystem Conservation and Management*, Government of Indonesia, Jakarta, Indonesia.

Ministry for the Environment (1988) Impact assessment in resource management. In *Working Paper 20*, p. 27, New Zealand Ministry for the Environment, Wellington.

Ministry of Home Affairs (1996) *Technical Manual for Preparation of a Provincial Coastal and Marine Management Strategy (English Version)*, Government of Indonesia, Jakarta, Indonesia.

Miossec, J.M. (1976) Eléments pour une Theorie de l'Espace Touristique, Les Cahiers due Tourisme C-36 CHET, Aix-en-Provence.

Mitchell, J.K. (1982) Coastal zone management: a comparative analysis of national programs. In *Ocean Yearbook 3* (eds E.M. Borgese and N. Ginsburg), pp. 258–319, University of Chicago Press, Chicago.

Moll, H. and Suharsono (1986) Distribution, diversity and abundance of coral reefs in Jakarta Bay and Kepulauan Seribu. In *Human Induced Damage to Coral Reefs: Results of a Regional UNESCO (COMAR) Workshop with Advanced Training*, (ed. B.E. Brown), UNESCO Reports in Marine Science 40, pp. 112–125.

Mukhi, S., Hampton, D. and Barnwell, N. (1988) *Australian Management*, McGraw-Hill, Sydney, Australia.

National Research Council (1990) *Managing Troubled Waters: the Role of Marine Environmental Monitoring*, National Academy Press, Washington DC.

National Research Council (1993) *Toward a Coordinated Spatial Data Infrastructure for the Nation*, National Academy Press, Washington, DC.

National Research Council (1995a) *Promoting the National Spatial Data Infrastructure Through Partnerships*, National Academy Press, Washington, DC.

National Research Council (1995b) *Science, Policy and the Coast: Improving Decision Making*, National Academy Press, Washington DC.

National Research Council (1997) *The Future of Spatial Data and Society*, National Academy Press, Washington, DC.

New South Wales Government Department of Public Works (1990) *Coastal Management Manual, 102*, NSW Public Works Department, Sydney.

NOAA (1994a) *River to Reef Newsletter*, NOAA, Washington, DC.

NOAA (1994b) *1992–1993 Biennial Report to Congress on the Administration of the Coastal Zone Management Act*, Vol. II, National Oceanic and Atmospheric Administration, Washington DC.

O'Riordan, T. (1981) *Environmentalism*, 2nd edn, Psion, London.

O'Riordan, T. (ed) (1995) *Environmental Science for Environmental Management*, Longman, Harlow, UK.

O'Riordan, T. and Cameron, J. (ed.) (1994) *Interpreting the Precautionary Principle*, Earthscan, London.

O'Riordan, T. and Sewell, W.R.D. (1981) *Project Appraisal and Policy Review*, Wiley, Chichester, UK.

O'Riordan, T. and Vellinga, P. (1993) Integrated coastal zone management: the next steps. In *World Coast '93* (eds P. Beukenkamp, P. Gunther, R. Klein *et al.*), pp. 409–413, National Institute for Coastal and Marine Management, Coastal Zone Management Centre, Noordwijk, The Netherlands.

O'Riordan, T., Kemp, R. and Perdue, M. (1987) *Sizewell B: The Anatomy of An Enquiry*, Macmillan, London.

OECD (1989a) *Economic Instruments for Environmental Protection*, Organisation for Economic Co-operation and Development, Paris.

OECD (1989b) *Environmental Policy: How to Apply Economic Instruments*, Organisation for Economic Co-operation and Development, Paris.

OECD (1992) *Recommendation of the Council on Integrated Coastal Zone Management*, Organisation for Economic Co-operation and Development, Paris.

OECD (1993) *Coastal Zone Management: Integrated Policies*, Organisation for Economic Co-operation and Development, Paris.

Olsen, S. (1995) The skills, knowledge, and attitudes of an ideal coastal manager. In *Educating Coastal Managers* (eds B.R. Crawford, J.S. Cobb and C.L. Ming), p. 170, University of Rhode Island, Coastal Resources Center.

Olsen, S. and Tobey, J. (1997) *Final Evaluation of the Global Environment Facility Patagonian Coastal Zone Management Plan*, University of Rhode Island, Rhode Island.

Olsen, S., Sadacharan, D., Samarakoon, J.I. *et al.* (1992) *Coastal 2000: Recommendations for a Resource Management Strategy for Sri Lanka's Coastal Regions*, Vols I and II, Coast Conservation Department, Coastal Resources Management Project, Sri Lanka and Coastal Resources Center, University of Rhode Island, Colombo, Sri Lanka.

Olsen, S., Tobey, J., Robadue, D. and Ochoa, E. (1996) *Coastal Management in Latin America and the Caribbean: Lessons Learned and Opportunities for the Inter-American Development Bank*, University of Rhode Island, Narragansett, USA.

Olsen, S., Tobey, J. and Kerr, M. (1997) A common framework for learning from ICM experience. *Ocean and Coastal Management*, **37**(2), 155–174.

Oma, V.P.M., Clayton, D.M., Broun, J.B. and Keating, C.D.M. (1992) *Coastal Rehabilitation Manual*, Western Australian Government Department of Agriculture, Perth.

Owens, D.W. (1992) National goals, state flexibility, and accountability in coastal zone management. *Coastal Management*, **20**, 143–165.

Owens, J.M. (1993) *Program Evaluation: Forms and Approaches*, Allen & Unwin, Melbourne.

Paris, C. (ed.) (1982) *Critical Readings in Planning Theory*, Urban and Regional Planning Series, Volume 27, Pergamon, Oxford.

Patmore, J.A. (1983) *Recreation and Resources*, Blackwell, London.

Pearce, D., Markandya, A. and Barbier, E. (eds) (1989) *Blueprint for a Green Economy*, Earthscan, London.

Pearce, D.G. (1987) *Tourist Today: A Geographical Analysis*, Longman, New York.

Pearce, D.G. (1989) *Tourist Development*, 2nd edn, Longman, New York.

Pearson, M. (1995) Sustainable financing of protected areas in Southern Sinai. In *Sustainable Financing Mechanisms for Coral Reef Conservation* (ed. J. Hooten and M.E. Hatziolos), pp. 72–77, World Bank, Washington DC.

Pernetta, J.C. and Elder, J.L. (1993) *Cross-Sectoral, Integrated Coastal Area Planning (CICAP): Guidelines and Principles for Coastal Area Development*, IUCN, Gland, Switzerland.

Platt, R.H. (1991) *Land Use Control: Geography, Law and Public Policy*, Prentice Hall, Englwood Cliffs, New Jersey.

Platt, R.H., Miller, H.C., Beatley, T. *et al.* (1992) *Coastal Erosion: Has Retreat Sounded?*, Institute of Behavioural Science, University of Colorado, Boulder.

Pomeroy, R.S., Pollnac, R.B., Predo, C.D. and Katon, B.M. (1997) Impact evaluation of community-based coastal resource management projects in the Philippines. *NAGA The ICLARM Quarterly*, **19**(4), 9–12.

Porcher, M. and Millon, L. (1991) *Compte Redu de Fin de Ghantier Relatif a l'Utilisation de Silt-Screens pour la Protection du Milieu Recifal*, CETE Mediterranée/Société Mater/Jan de Nul, Mahe, Seychelles.

Post, J.C. and Lundin, C.G. (eds) (1996) *Guidelines for Integrated Coastal Zone Management*, Environmentally Sustainable Development Series and Monographs Series No. 9, World Bank Environment Department, Land Water and Natural Habitats Division, Washington DC.

Potapchuk, W. (1995) Resolving disputes the kinder, gentler way. *Planning*, (May), 16–20.

Prosser, G. (1986) An introduction to a framework for natural area planning. *Australian Parks and Recreation*, **22**(2), 3–10.

Quilty Environmental Consulting (1991) *Warnbro Dunes Foreshore Stabilisation and Management Plan*, Quilty Environmental Consulting for Australian Housing and Land, Perth, Western Australia.

Redclift, M. (1987) *Sustainable Development: Exploring the Contradictions*, Routledge, London.

Reid, D. (ed.) (1995) *Sustainable Development: An Introductory Guide*, Earthscan, London.

Ren, M.-E. (1992) Human impact on coastal landform and sedimentation – the Yellow River example. *GeoJournal*, **28**(4), 443–448.

Rennie, H.G. (1993) The coastal environment. In *Environmental Planning in New Zealand* (eds P.A. Memom and H.C. Perkins), pp. 150–168, Dunmore Press, Palmerston North.

Republic of Indonesia (1990) *Conservation of Living Resources and Their Ecosystems. Act No.5 of 1990*, Ministry of Forestry, Jakarta.

Ribe, R.G. (1989) The aesthetics of forestry: what has empirical preference research taught us? *Environmental Management*, **13**(1), 55–74.

Ripley, R.B. and Franklin, G.A. (1986) *Policy Implementation and Bureaucracy*, 2nd edn, Dorsey Press, Chicago.

Robadue, D. (ed.) (1995) *Eight Years in Equador: The Road to Integrated Coastal Management*, University of Rhode Island and US Agency for International Development, Narragansett.

Rogers, D.L. and Whetten, D.A. (1982) *Interorganizational Coordination: Theory, Research and Implementation*, Iowa State University Press, Ames, Iowa.

Rolden, R. and Sievert, R. (1993) *Coastal Resources Management: A Manual for Government Officials and Community Organizers*, Department of Agriculture, Manila, Philippines.

Rossi, P.H. and Freeman, H.E. (1989) *Evaluation: A Systematic Approach*, 4th edn, Sage, Nebury Park, CA.

Royal Society of London (1983) *Risk Assessment: Report of a Royal Society Study Group*, Royal Society of London, London.

Rubenstein, H.M. (1987) *A Guide to Site and Environmental Planning*, 3rd edn, Wiley, New York.

Ruddle, K. and Johannes, R.E. (1983) *The Traditional Knowledge and Management of Coastal Systems in Asia and the Pacific*, UNESCO, Jakarta, Indonesia.

Ruitenbeek, H.J. (1991) *Mangrove Management: An Economic Analysis of Management Options with a Focus on Bintuni Bay, Irian Jaya*, Report of Dalhousie University to the Government of Indonesia, Jakarta, Indonesia.

Ruitenbeek, H.J. (1994) Modelling economy–ecological linkages in mangroves: economic evidence for promoting conservation in Bintuni Bay, Indonesia. *Ecological Economics*, **10**, 233–247.

Russ, G. and Alcala, A. (1994) Sumilon Island Marine Reserve: 20 years of hopes and frustrations. *NAGA The ICLARM Quarterly*, **17**(3), 8–12.

Rutledge, A.J. (1971) *Anatomy of a Park : The Essentials of Recreation Area Planning and Design*, McGraw-Hill, New York.

Salam, M., Irawan, D. and Tomboelu, N. (1996) *Integrated Management Planning of Kapoposang Marine Tourism Park*, MREP, Jakarta.

Savina, G.C. and White, A.T. (1986) A tale of two islands: some lessons for marine resource management. *Environmental Conservation*, **13**(2), 107–113.

Schmidt, W. (ed.) (1996) *Advanced Recreation Planning and Management Course: Course Notes*, Department of Conservation and Land Managmement, Perth, Western Australia.

Schoen, R.-J., and Djohani, R. (1992) *A Communication Strategy to Support Marine*

Conservation Policies, Programmes and Projects in Indonesia for the Years 1992–1995, World Wide Fund for Nature Indonesia, Jakarta, Indonesia.

Schröder, P.C. (1993) OCA/PAC Activities related to climate change and sea-level rise. In *World Coast '93* (eds P. Beukenkamp, P. Gunther, R. Klein *et al.*), pp. 347–350, National Institute for Coastal and Marine Management, Coastal Zone Management Centre, Noordwijk, The Netherlands.

Scura (1993) Review of Recent Experiences in Integrated Coastal Management. Unpublished report to the International Center for Living Coastal Aquatic Resources Management and University of Rhode Island, FI:DP/INT/91/007 Field Document 2.

Scura (1994) *Typological Framework and Strategy Elements for Integrated Coastal Fisheries Management*, Food and Agriculture Organization of the United Nations, Rome.

Shah, N. (1995) Coastal zone management in the Seychelles. In *Integrated Coastal Zone Management in the Seychelles* (eds C.G. Lundin and O. Linden), pp. 14–125, Government of the Seychelles Ministry of Foreign Affairs in cooperation with SIDA and the World Bank, Mahe, Seychelles.

Sheerin, J. (1991) Frameworks for state of the environment reporting. In *Reporting State of the Environment Information*, pp. 65–85, New Zealand Department of Statistics and Ministry for the Environment, Wellington, New Zealand.

Sherwood, A. and Howarth, R. (1996) *Coasts of Pacific Islands*, South Pacific Geoscience Commission (SOPAC) Miscellaneous Report 222, Suva, Fiji.

Sloan, N.A. and Sugandhy, A. (1994) An overview of Indonesian coastal environmental management. *Coastal Management*, **22**(3), 215–234.

Smith, A.H. and Homer, F. (1994) Collaborative coral reef monitoring in the Caribbean Region. In *Proceedings of the Seventh International Coral Reef Symposium, Guam*, University of Guam Marine Laboratory.

Smith, L.G. (1993) *Impact Assessment and Sustainable Resource Management*, Longman, Harlow, UK.

Smith, L.G., Nell, C.Y. and Prystupa, M.V. (1997) The converging dynamics of interest representation in resources management. *Environment Management*, **21**(2), 139–146.

Smith, R. (1992) Beach resort evolution. *Annals of Tourism Research*, **19**, 304–322.

Smith, V.K. (1996) *Estimating Economic Values for Nature: Methods for Non Market Valuation*, Edward Elgar, Cheltenham, UK.

Smyth, D. (1991) *Aboriginal Maritime Culture in the Far Northern Section of the Great Barrier Reef Marine Park. 73*, Great Barrier Reef Marine Park Authority, Townsville, Australia.

Smyth, D. (1993) *A Voice in All Places: Aboriginal and Torres Strait Islander Interests in Australia's Coastal Zone*, Commonwealth of Australia, Canberra.

Soby, B.A., Simpson, A.C.D. and Ives, D.P. (1993) *Integrating Public and Scientific Judgements into a Tool Kit for Managing Food-related Risks, Stage I: Literature Review and Feasibility Study*, University of East Anglia, Norwich, UK.

Soegiarto, A. (1981) The development of a marine park system in Indonesia. In *The Reef and Man: Proceedings of the Fourth International Coral Reef Symposium, Quezon City, Philippines*, Marine Sciences Centre, University of the Philippines.

Sorensen, J. (1993) The international proliferation of integrated coastal zone management efforts. *Ocean and Coastal Management*, **21**(1–3), 45–80.

Sorensen, J. (1997) National and international efforts at integrated coastal zone management: definitions, achievements, and lessons. *Coastal Management*, **25**(1), 3–41.

Sorensen, J.C., and McCreary, S.T. (1990) *Institutional Arrangements for Managing Coastal Resources and Environments*, 2nd edn, University of Rhode Island, Narragansett.

Sorensen, J.C. and West, N. (1992) *A Guide to Impact Assessment in Coastal Environments*, Coastal Resources Center, University of Rhode Island, Narragansett.

Sorensen, J.C., McCreary, S.T. and Hershman, M.J. (1984) *Institutional Arrangements for Coastal Resources Management*, prepared by Research Planning Institute for National Park Service, US Dept of Interior, Columbia, SC.

Sowman, M. (1993) The status of coastal zone management in South Africa. *Coastal Management*, **21**, 163–184.

SPREP (1992) *Environmental Impact Assessment Training in the South Pacific Region: Meeting Report 31 August – 4 September 1992*, South Pacific Regional Environment Programme, Apia, Western Samoa.

SPREP (1993) *Project Proposal: SPREP Integrated Coastal Zone Management in the Pacific Islands Region (Draft)*, South Pacific Regional Environment Programme, Apia, Western Samoa.

Standards Australia and Standards New Zealand (1995) *Australian/New Zealand Standard: Risk Management*, Standards Australia and Standards New Zealand, Sydney and Wellington.

Stankey, G.H. and Wood, J. (1982) The recreation opportunity spectrum: an introduction. *Australian Parks and Recreation*, February, 6–14.

Steers, R.M., Ungson, G.R. and Mowday, R.T. (1985) *Managing Effective Organisations: An Introduction*, Kent, Boston, Massachusetts.

Stone, R. (1988) Conservation and development in St Lucia. *World Wildlife Fund Letter*, **3**, 1–8.

Stronge, W.B. (1994) Beaches, Tourism and Economic Development. *Jounal of the American Shore and Beach Preservation Association*, **62**(2), 6–8.

Talukder, J. and Ahmad, M. (eds) (1992) *The April Disaster: Study on Cylcone Affected Region in Bangladesh*, Community Development Library, Dhaka.

Talukder, J., Roy, G.D. and Ahmad, M. (ed.) (1992) *Living With Cylcone: Study on Storm Surge Prediction and Disaster Preparedness*, Community Development Library, Dhaka.

Tasque Consultants (1994) *Break O'Day Regional Marine and Coastal Strategy*, Dorset, Break O'Day, Glamorgan and Spring Bay Councils, Hobart, Tasmania.

Taussik, J. and Gubbay, S. (1997) Networking in integrated coastal zone management. In *Partnership in Coastal Zone Management: Proceedings of the Eurocoast '96 Conference* (eds J. Taussik and J. Mitchell), pp. 57–63, Samara Publishing, Swansea, UK.

Taussik, J. and Mitchell, J. (eds) (1997) *Partnership in Coastal Zone Management: Proceedings of the Eurocoast '96 Conference*, Samara Publishing, Swansea, UK.

Thomas, I. (1996) *Environmental Impact Assessment in Australia*, Federation Press, Sydney.

Thompson, G.F. and Steiner, F.R. (eds) (1997) *Ecological Design and Planning*, Wiley, New York.

Thorman, R. (1995) *Suggested Framework for Preparing Regional Environmental Strategies (Draft Guidelines)*, Australian Local Government Association, Canberra, Australia.

Tomasick, T., Suharsono and Mah, A. J. (1993) Global histories: a historical perspective of the natural and anthropogenic impacts in the Indonesian Archipelego with a focus on Kepulauan Seibu, Java Sea. In *Colloquium on Global Aspects of Coral Reefs: Health, Hazards and History, University of Miami*.

Torkildsen, G. (1992) *Leisure and Recreation Management*, 3rd edn, E. & F.N. Spon, London.

Tri, N.H., Adger, N., Kelly, M. *et al.* (1996) *The Role of Natural Resource Management in Mitigating Climate Impacts: Mangrove Restoration in Vietnam*, Centre for Social and Economic Research on the Global Environment, Norwich, UK.

Turner, R.K. (1991) Environment, economics and ethics. In *Blueprint 2: Greening the World Economy* (ed. D. Pearce), pp. 209–224, Earthscan, London.

Turner, R.K. (ed.) (1993) *Sustainable Environmental Economics and Management*, Belhaven, London.

Turner, R.K. (1995) Environmental economics and management. In *Environmental Science for Environmental Management* (ed. T. O'Riordan), pp. 30–44, Longman, Harlow, UK.

Turner, R.K., Pearce, D.W. and Bateman, I.J. (1993) *Environmental Economics: An Elementary Introduction*, Harvester Wheatsheaf, Hemel Hempstead, UK.

Turner, R.K., Subak, S. and Adger, N. (1995) *Pressures, Trends and Impacts in the Coastal Zones. Interactions Between Socio-Economic and Natural Systems*, Centre for Social and Economic Research on the Global Environment, Norwich, UK.

UN (1982) *Survey of Environmental Statistics: Frameworks, Approaches and Statistical Publications*, United Nations, New York.

UN Department of International Economic and Social Affairs (1982) *Coastal Area Management and Development*, Pergamon, Oxford, UK.

UNCED (1992) *Agenda 21 – United Nations Conference on Environment and Development: Outcomes of the Conference*, Rio de Janeiro, Brazil 3–14 June.

Underdahl, A. (1980) Integrated marine policy: what? why? how? *Marine Policy*, July, 159–169.

UNEP (1982) *Environmental Problems of the East African Region*, UNEP, Nairobi, Kenya.

UNEP (1995) *Guidelines for Integrated Management of Coastal and Marine Areas – with Special Reference to the Mediterranean Basin*, United Nations Environment Programme, Nairobi, Kenya.

UNEP OCA/PAC (1982) *Marine and Coastal Area Development in the East African Region*, UNEP, Nairobi, Kenya.

UNEP OCA/PAC (1986) *Action Plan for the Conservation of the Marine Environment and Coastal Areas of the Red Sea and Gulf of Aden*, UNEP, Nairobi, Kenya.

UNEP OCA/PAC (1996) *Regional Seas Programme: Document Ref. No. 150120 OCA/PAC*, United Nations Environment Programme Oceans and Coastal Areas Programme Activity Centre, Nairobi, Kenya.

United States Agency for International Development (1996) *Learning from Experience: Progress in Integrated Coastal Zone Management*, United States Agency for International Development, Washington, DC.

University of North Carolina Center for Urban and Regional Studies (1991) *Evaluation of the National Coastal Zone Management Program*, National Coastal Resources Research and Development Institute, Washington DC.

USDA Forest Service (1973) *National Forest Landscape Management Manual*, USDA Forest Service, Washington DC.

van Lier, H.N., Jaarsma, C.F., Jurgens, C.F. and de Buck, A.J. (ed.) (1994) *Sustainable Land Use Planning*, Developments in Landscape Management and Urban Planning, Elsevier, Amsterdam.

van Westen, C.-J. and Scheele, R. (1996) *Planning Estuaries*, Plenum Press, New York.

Veal, A.J. (1992) Definitions of leisure and recreation. *Australian Journal of Leisure and Recreation*, **2**(4), 44–52.

Vestal, B., Rieser, A., Ludwig, M. *et al.* (1995) *Methodologies and Mechanisms for Management of Cumulative Coastal Environmental Impacts*, US National Oceanic and Atmospheric Administration, Coastal Ocean Program, Washington, DC.

Wager, R.G. (1964) *The Carrying Capacity of Wild Lands for Recreation*, American Society of Foresters, Washington, DC.

Watson, R.T., Zinyowera, M.C., Moss, R.H., and Dokken, D.J. (1996) *Climate Change 1995: Impacts, Adaptations and Mitigation of Climate Change: Scientific–Technical Analyses*, Cambridge University Press, Cambridge, UK.

Wegener, M. and Masser, I. (1996) Brave new GIS worlds. In *GIS Diffusion: The Adoption and Use of Geographic Systems in Local Government in Europe* (eds I. Masser, H. Campbell and M. Craglia), Taylor & Francis, London.

Welch, D. (1991) Information needs for resource management in Australian Marine Protected Areas, 1989–90. In *Proceedings of International Conference on Science and Management of Protected Areas, Halifax, Nova Scotia, May 14–19, 1991*.

Wells, M. and Brandon, K. (1992) *People and Parks: Linking Protected Area Management with Local Communities*, World Bank, Washington, DC.

Western Australian Department of Conservation and Land Management (1991) *South Coast Region: Regional Management Plan: 1992–2002*, West Australian Department of Conservation and Land Management, Perth, Australia.

Western Australian Government (1983) *Coastal Planning and Management in Western Australia: a position paper*, Western Australian Government, Perth, Australia.

Western Australian Planning Commission (1995a) *The Economy*, Western Australian Government, Western Australian Planning Commission, Perth, Australia.

Western Australian Planning Commission (1995b) *Population*, Western Australian Government, Western Australian Planning Commission, Perth, Australia.

Western Australian Planning Commission (1996a) *Central Coast Regional Strategy: A Strategy to Guide Land Use in the Next Decade*, Government of West Australia, Perth, Australia.

Western Australian Planning Commission (1996b) *Shark Bay Regional Strategy: A Review of the 1988 Shark Bay Region Plan*, Western Australian Government Western Australian Planning Commission, Perth, Australia.

Western Australian Planning Commission (1996c) *State Planning Strategy*, Western Australian Government Western Australian Planning Commission, Perth, Australia.

Westman, W.E. (1985) *Ecology, Impact Assessment and Environmental Planning*, Wiley, New York.

White, A., Zeitlin-Hale, L., Renard, Y. and Cortesi, L. (1994) *Collaborative and Community-Based Management of Coral Reefs: Lessons from Experience*, Kumarian Press, West Hartford, CT.

White, A.T. (1986) Philippine Marine Park pilot site benefits and management conflicts. *Environmental Conservation*, 13(4), 355–359.

White, A.T. (1988a) The effect of community-managed marine reserves in the Philippines on their associated coral reef fish populations. *Asian Fisheries Science*, 2, 27–41.

White, A.T. (1988b) *Marine Parks and Reserves: Management for Coastal Environments in Southeast Asia*, ICLARM Education Series 2, International Center for Living Coastal Aquatic Resources Management, Manila, Philippines.

White, A.T. (1989) Two community-based marine reserves: lessons for coastal management. In *Coastal Area Management in Southeast Asia: Policies, Management Strategies and Case Studies*. *ICLARM Conference Proceedings 19, Manila, Philippines* (eds E. Chua and D. Pauly), ICLARM,

White, A.T. (1995) Comments on coastal zone management and ecosystem protection. In *SEAPOL Singapore Conference on Sustainable Development of Coastal and Ocean Areas in Southeast Asia: Post-Rio Perpectives* (eds K.L. Koh, R.C. Beckman and C.L. Sien), pp. 207–220, National University of Singapore, Singapore.

White, A.T. (1996) Philippines: community management of coral reef resources. In *Coastal Zone Management Handbook* (ed. J.R. Clark), pp. 561–567, CRC Press, Boca Raton.

White, A.T. and Samarakoon, J.I. (1994) Special Area management for coastal resources: a first for Sri Lanka. *Coastal Management in Tropical Asia*, March, 20–24.

White, A.T. and Savina, G. (1987) Community based marine reserves, a Philippine first. In *Proceedings of Coastal Zone '87* (ed. O. Magoon), American Society of Civil Engineers.

White, A.T., Barker, V. and Tantrigama, G. (1997) Using integrated coastal management and economics to conserve tourism resources in Sri Lanka. *Ambio*, 26(6), 335–344.

White, G.F. (1945) *Human Adjustments to Floods: A Geographical Approach to the Flood Problem in the United States*, University of Chicago, Department of Geography, Chicago.

White, G.F. (ed.) (1974) *Natural Hazards: Local, National and Global*, Oxford University Press, London.

Williams, L. (1995) Resolving planning conflicts. *Town and Country Planning*, (October), 263–265.

Williamson, D.N. (1978) *Landscape Perception Research: Defining a Direction*, Forests Commission, Victoria, Melbourne, Australia.

Williamson, D.N. and Calder, S.W. (1979) Visual Resource Management of Victoria's Forests: A New Concept for Australia. *Landscape Planning*, 6, 313–341.

Wong, P. P. (ed.) (1993) *Tourism vs Environment: The Case for Coastal Areas*, The GeoJournal Library. Kluwer Academic, Dordrecht.

Wood, C., and Dejeddour, M. (1992) Strategic environmental assessment of policies, plans and programmes. *Impact Assessment Bulletin*, 10, 3–22.

World Bank (1993) *Noordwijk Guidelines for Integrated Coastal Zone Management*, World Bank Environment Department, Land Water and Natural Habitats Division, Washington, DC.

World Bank (1994) *Africa: A Framework for Integrated Coastal Zone Management*, World Bank, Washington DC.

World Bank (1996) *World Development Report*, World Bank, Washington, DC.

World Commission on Environment and Development (ed.) (1987) *Our Common Future*, Oxford University Press, Oxford.

World Resources Institute (1992) *World Resources 1996–97*, Oxford University Press, New York.

World Travel and Tourism Council (1995) *Travel and Tourism's Economic Perspective*. Special Report of the World Travel and Tourism Council, Brussels, Belgium.

World Wide Fund for Nature (1996) *Fisheries crisis overshadows World Oceans Day*, World Wide Fund for Nature News Release.

Yodmani, S. (1995) Capacity building for regional cooperation in Asia-Pacific. In *Report of the Asia and the Pacific Meeting of National Councils for Sustainable Development, Manila, Philippines – 18–19 June*, United Nations Environment Program.

Young, M.D. (1992) *Sustainable Investment and Resource Use: Equity, Environmental Integrity and Economic Efficiency*, UNESCO, Paris.

Zann, L.P. (1984) Traditional management and conservation of fisheries in Kiribati and Tuvalu Atolls. In *The Traditional Knowledge and Management of Coastal Systems in Asia and the Pacific, Jakarta, Indonesia* (eds. K. Ruddle and R.E. Johannes), UNESCO, Paris.

Zann, L. P. (1995) *Our Sea, Our Future: Major Findings of the State of the Marine Environment Report for Australia*, Ocean Rescue 2000 Program: Department of the Arts, Sports and Territories, Canberra.

Zeitlin-Hale, L. (1996) Involving communities in coastal management. In *Proceedings of the Coast to Coast '96 Conference, Adelaide, 16–19 April* (ed. N.C. Harvey), University of Adelaide.

Zigterman, R. and De Campo, J. (1993) *Green Island and Reef Management Plan*, Queensland Department of Environment and Heritage, Great Barrier Reef Marine Park Authority, Cairns City Council, Cairns Port Authority and Department of Lands, Cairns, Australia.

Zube, E.H., Pitt, D.W. and Anderson, T.W. (1974) *Perception and Measurement of Scenic Resource Values in the Southern Connecticut River Valley*, University of Massachusetts, Massachusetts.

Index

The sole focus of this book is on the planning and management of the coast. To limit the need for repetition and cross-referencing of entries, the index includes the leading terms 'coast' and 'coastal' only where necessary for clarity. This allows for greater discrimination of subjects and concepts. Unless otherwise stated, entries such as 'resources' or 'pollution' refer to matters relevant to the planning and management of the coastal area.

Index entries with an asterisk (*) refer to examples or case studies in boxed text.